半导体与集成电路关键技术丛书

异构集成技术
Heterogeneous Integrations

[美] 刘汉诚（John H. Lau）著

吴向东 雷 剑 冯 慧 王 琛 李林森 周国云 李力一 刘 杰
周万丰 朱 喆 邢朝洋 单伟伟 徐春林 刘俊夫 李玖娟 译
杨银堂 张 崎 王传声 审校

机械工业出版社

本书主要内容涉及异构集成技术的基本构成、技术体系、工艺细节及其应用，涵盖有机基板上的异构集成、硅基板（TSV转接板、桥）上的异构集成、扇出型晶圆级/板级封装、扇出型RDL基板上的异构集成、PoP异构集成、内存堆叠的异构集成、芯片到芯片堆叠的异构集成、CIS、LED、MEMS和VCSEL异构集成等方面的基础知识，最后介绍了异构集成的发展趋势。本书图文并茂，既有工艺流程详解，又有电子信息行业和头部公司介绍，插图均为彩色图片，一目了然，便于阅读、理解。

本书内容对于异构集成的成功至关重要，将为我国电子信息的教学研究和产业界的研发制造提供参考，具有较强的指导价值，并将进一步推动我国先进封装技术的不断进步。

First published in English under the title
Heterogeneous Integrations
by John H Lau, edition: 1
Copyright © Springer Nature Singapore Pte Ltd., 2019
This edition has been translated and published under licence from
Springer Nature Singapore Pte Ltd..
Springer Nature Singapore Pte Ltd. takes no responsibility and shall not be made liable for the accuracy of the translation.

此版本仅限在中国大陆地区（不包括香港、澳门特别行政区及台湾地区）销售。未经出版者书面许可，不得以任何方式抄袭、复制或节录本书中的任何部分。
北京市版权局著作权合同登记　图字：01-2022-5399号。

图书在版编目（CIP）数据

异构集成技术 /（美）刘汉诚著；吴向东等译 . —北京：机械工业出版社，2023.7（2024.10 重印）
（半导体与集成电路关键技术丛书）
书名原文：Heterogeneous Integrations
ISBN 978-7-111-73273-0

Ⅰ.①异…　Ⅱ.①刘…②吴…　Ⅲ.①异构网络—研究　Ⅳ.① TP393.02

中国国家版本馆 CIP 数据核字（2023）第 098253 号

机械工业出版社（北京市百万庄大街22号　邮政编码100037）
策划编辑：吕　潇　　　　责任编辑：吕　潇　杨晓花
责任校对：张亚楠　翟天睿　责任印制：单爱军
北京虎彩文化传播有限公司印刷
2024年10月第1版第2次印刷
169mm×239mm · 20.5 印张 · 398 千字
标准书号：ISBN 978-7-111-73273-0
定价：168.00 元

电话服务　　　　　　　　网络服务
客服电话：010-88361066　机　工　官　网：www.cmpbook.com
　　　　　010-88379833　机　工　官　博：weibo.com/cmp1952
　　　　　010-68326294　金　书　网：www.golden-book.com
封底无防伪标均为盗版　　机工教育服务网：www.cmpedu.com

序

过去的半个多世纪，半导体行业一直遵循摩尔定律的轨迹高速发展，如今正面临着包括物理原理极限、技术手段极限和经济成本极限等一系列极限挑战，单纯靠提升工艺来提升芯片性能的方法已经无法充分满足时代的需求，异构集成正在成为后摩尔时代延续半导体技术的主流发展方向。异构集成采用系统设计理念，应用先进技术比如IP和小芯片（Chiplet），具有2.5D或3D高密度结构，可以融合不同半导体材料、工艺、结构和元器件，因而使得异构集成芯片可以实现强大且复杂的功能，突破单一半导体工艺的性能极限；同时具有灵活性强、可靠性高、研发周期短、可实现小型化轻质化等特点。微电子行业的领导者，如英特尔、AMD、台积电已将其视为推动半导体下一个30年发展的重要技术。

在芯片设计和制造成本越来越高的情况下，异构集成作为先进封装技术和被视为后摩尔时代新路径越来越广受关注。通过这一技术，工程师可以像搭积木一样，在芯片库里将不同工艺的小芯片组装在一起。美国在异构集成技术方面处于领先地位，诺格公司在2017年已形成代工能力，其制造的异构芯片已在诸如AESA（有源相控阵雷达）中开始应用，在以砷化镓、氮化镓、异构集成为代表的射频元器件代际发展趋势中，以异构集成技术为代表的集成微系统将有着广阔的市场应用前景。

该书主要内容涉及异构集成基本构成、技术体系、工艺细节及其应用，涵盖了有机基板上的异构集成、硅基板（TSV转接板、桥）上的异构集成、扇出型晶圆级/板级封装、扇出型RDL基板的异构集成、PoP异构集成、内存堆叠的异构集成、芯片到芯片堆叠的异构集成、CIS、LED、MEMS和VCSEL异构集成等方面的基础知识，最后介绍了异构集成的发展趋势。这些对于异构集成的成功都至关重要，将为我国电子信息的教学研究和产业界的研发制造提供参考，具有较强的指导价值。相信该书的译介将进一步推动我国先进封装技术的不断进步。

中国科学院院士

中国科学院大学党委书记、校长

译者序

在当前复杂芯片设计和制造成本越来越高的情况下，异构集成作为先进封装技术被视为后摩尔时代的新解决路径，越来越广受关注。世界半导体行业的领导者，如英特尔、AMD、台积电已将其视为推动半导体下一个三十年发展的重要技术，国内的一些主要厂商也积极投入精力研发异构集成相关工艺和产品。可以预见，异构集成技术在未来军用和民用领域有着广阔的应用前景。

本书主要涉及异构集成基本构成、技术体系、工艺细节及其应用，涵盖有机基板上的异构集成、硅基板（TSV转接板、桥）上的异构集成、扇出型晶圆级/板级封装、扇出型RDL基板上的异构集成、PoP异构集成、内存堆叠的异构集成、芯片到芯片堆叠的异构集成、CIS、LED、MEMS和VCSEL异构集成等方面的基础知识，最后介绍了异构集成的发展趋势。全书图文并茂，数据翔实，衷心希望中译文能为国内相关从业人员的研究开发提供一些借鉴和参考。

本书得以翻译出版离不开机械工业出版社电工电子分社付承桂社长和吕潇编辑的精心策划与指导。本书翻译由中国电子科技集团公司第四十三研究所吴向东所长、微系统安徽省重点实验室相关专家、清华大学材料学院王琛教授、东南大学单伟伟教授和李力一教授、电子科技大学周国云教授和李久娟博士后、航天九院13所邢朝洋研究员等人参与。为尽可能保证质量，本书采用了多人多次审校的方法。西安电子科技大学单光宝教授和贾护军教授、武汉大学郭宇锋教授参与了部分章节审校，译者对他们利用业余时间做出的无偿奉献表示衷心感谢！

异构集成技术涉及知识面较宽，新兴材料和新工艺概念术语较多，因译者翻译和学术水平有限，其中的译文表达不当之处在所难免，恳请广大读者批评指正。

译者

原书前言

在摩尔定律的驱动下，再加上诸如智能手机、平板计算机和可穿戴设备移动产品的需求，SoC（片上系统）过去10多年已经非常普及。SoC可以将不同功能的集成电路（IC）集成到单芯片上形成一个系统或子系统的单芯片。不幸的是，摩尔定律即将终结，并且制造SoC所需工艺的特征尺寸（进行缩放）的减小变得更加困难和昂贵。异构集成则与SoC形成了对照：异构集成使用封装技术，将来自不同的芯片设计公司、代工厂、不同晶圆尺寸、特征尺寸和公司的不同材料和功能的芯片、光子器件或不同材料和功能的组件（平铺、堆叠或两者兼有）集成到一个系统或子系统中。系统级封装（SiP）与异构集成非常相似，只是异构集成适用于更细的节距、更多的输入/输出（I/O）、更高的密度和更高的性能。

总体上，异构集成可以归类为基于有机基板的异构集成、基于硅基板（带TSV转接板）的异构集成、基于硅基板（无TSV，如桥）的异构集成、基于扇出RDL（再布线层）基板的异构集成和基于陶瓷基板的异构集成。未来几年里，这些不同类型基板间的异构集成将会陆续出现，无论在性能、外形尺寸、功耗、信号完整性还是在成本等因素方面都将具有更高水平。然而，大多数从业工程师和管理人员，以及科学家和研究人员，对于如何建立积层有机封装基板、积层基板上的顶部薄膜层、有机转接板、TSV转接板、TSV和RDL制造、无TSV转接板、（硅）桥、扇出RDL基板、堆叠封装、存储器堆叠、芯片到芯片堆叠、晶圆上凸点形成、热压键合、低温键合、芯片到晶圆键合以及晶圆到晶圆的键合还没有很好的理解。因此，工业界和学术界迫切需要一本对当前这些关键赋能技术知识进行全面介绍的书。本书可以为读者提供快速解决上述问题的方法和基本知识，并在做出系统级决策、进行内在权衡比较时提供指导。

本书共11章，即异构集成综述；有机基板上的异构集成；硅基板上的异构集成（TSV转接板）；硅基板（桥）上的异构集成；异构集成的扇出晶圆级/板级封装；基于扇出型RDL基板的异构集成；PoP异构集成；内存堆叠的异构集成；芯片到芯片堆叠的异构集成；CIS（CMOS图像传感器）、LED（发光二极管）、MEMS（微机电系统）和VCSEL（垂直腔面发射激光器）上的异构集成；异构集成的发展趋势。

第1章简要介绍了异构集成的定义、分类和应用领域。

第2章介绍了有机基板上的异构集成，重点介绍各种有机基板，如积层、积层顶部薄膜层、有机转接板、无芯基板、引线上凸点和用于异构集成的嵌入式迹线基板。

第 3 章详细介绍了硅基板（TSV 转接板）上的异构集成，重点介绍热性能、机械特性和在 TSV 转接板两侧进行芯片异构集成的工艺，并首先简要介绍了转接板的 TSV 和 RDL 的制造。

第 4 章介绍了硅基板（无 TSV 转接板，如桥）上的异构集成，重点介绍在桥两侧进行芯片异构集成的电性能和制造工艺。首先简要介绍了英特尔公司的 EMIB（嵌入式多芯片互连桥）和微电子研究中心（IMEC）的逻辑和内存互连桥。

第 5 章概述了扇出型晶圆级/板级封装（FOW/PLP）上的异构集成，重点介绍 FOW/PLP 的形成和 RDL 的制造，并简要介绍了 FOW/PLP 为异构集成带来的机遇。

第 6 章介绍了基于扇出型 RDL 基板的异构集成。重点介绍使用 FOW/PLP 技术制造 RDL，以消除使用 TSV 转接板进行异构集成。

第 7 章介绍了 PoP 异构集成，简要介绍了利用 PoP 制造的智能手机和智能手表的示例。

第 8 章讨论了内存堆叠的异构集成，重点介绍了在两个内存芯片堆叠在一个 ASIC 上的异构集成和内存芯片与逻辑芯片低温键合的异构集成。

第 9 章介绍了芯片到芯片和面对面异构集成的两个示例。其中一个示例是利用底部芯片中的 TSV 传输信号、电源和地，另一个示例没有 TSV，而是在更大的芯片上通过焊凸点进行传输。

第 10 章介绍了 CIS、LED、MEMS、VCSEL 的异构集成，重点介绍了每个异构集成的一些示例。

第 11 章描述了异构集成的发展趋势，重点介绍了各种异构集成的制造工艺、选择标准和应用范围（规格和引脚数）。

本书适用于对异构集成技术非常关注的以下三类专家群体：

1）积极参与或打算积极参与研究和开发异构集成的关键赋能技术的人，如积层有机封装基板、积层基板顶部薄膜层、有机转接板、TSV 转接板、TSV 和 RDL 制造、无 TSV 转接板、桥、扇出型 RDL 基板、堆叠封装（PoP）、内存堆叠、芯片与芯片堆叠、晶圆凸点、热压键合、低温键合、芯片-晶圆键合和晶圆-晶圆键合。

2）那些遇到异构集成实际问题，希望了解和学习更多解决这类问题方法的人。

3）那些必须为其产品选择高可靠、创造性、高性能、高密度、低功耗且成本高效的异构集成技术的人。

本书也可以作为有望成为异构集成技术领域未来领导者、科学家，以及电子和光电子行业工程师的大学生和研究生教材。

希望本书能成为所有面临异构集成挑战性问题的读者和对异构集成日益增长

的兴趣所带来的有价值挑战性问题的读者的宝贵参考资料，这些问题是随着对异构集成日益增长的关注而产生的。同时希望本书能有助于推动、促进对关键赋能技术的异构集成产品的研究开发，以及对异构集成产品更合理的应用。

那些已掌握半导体封装系统异构集成技术设计和制造的组织通过本书将有望在电子和光电子制造行业取得重大进展，并在性能、功能、密度、功耗、带宽、质量、尺寸和重量方面获得巨大收益。希望本书的内容有助于清除异构集成技术研发道路上的障碍、选择正确的技术路径，并加速异构集成关键赋能技术产品的设计、材料、工艺和制造开发。

John H. Lau（刘汉诚）

原书致谢

《异构集成技术》一书的筹备和出版离不开许多富有献身精神的人的工作。多位热心人士的努力促进了本书的准备和出版。在此我要感谢他们所有人，特别是 Springer Nature 科学出版服务有限公司的 Sridevi Purushothaman 女士和 Vinoth Selvamani 先生，感谢他们对我的坚定不移的支持和指导。还要特别感谢 Springer 北京代表处的 Jasmine Dou 女士，是她有效地赞助该项目并解决本书准备过程中出现的许多问题，成就了我的梦想。与他们一起工作，将我凌乱的手稿编辑成一本非常吸引人的印刷书，这是一件非常愉快、富有成效的经历。

显然，本书的素材来源广泛，包括个人、公司和组织，我试图在书中相关部分标注引用来承认我所得到的帮助。虽然我很难向在制作这本书时提供协助的每一位相关人士表示感谢，但我想要表达我应有的谢意。此外，我要感谢几个专业协会和出版商允许我在本书中引用他们的一些插图和信息，包括美国机械工程师学会（ASME）会议论文集（如国际电子封装学会会议）和杂志（如 *Journal of Electronic Packaging*），电气与电子工程师协会（IEEE）会议论文集（如电子元件与技术会议、电子封装与技术会议）和杂志（如 *IEEE Transactions on Components, Packaging and Manufactwring Technologies*），国际微电子和封装学会（IMAPS）会议论文集（如微电子国际研讨会）和杂志（如 *International Journal of Microcircuits & Electronic Packaging, Chip Scale Review*）。

我要感谢我的前雇主，中国台湾"工业技术研究院"（ITRI）、中国香港科技大学（HKUST）、微电子研究所（IME）、安捷伦公司和惠普公司为我提供的良好工作环境，并培养我成长，满足了我对就业满足感的需求，同时提高了我的职业声望。我还要对 Don Rice 博士（惠普公司）、Steve Erasmus 博士（安捷伦公司），Dim Lee Kwong（邝启明）教授（IME）、Ricky Lee（李世玮）教授（HKUST）和 Ian Yi Jen Chan（陈怡仁）博士（ITRI）在我就职期间给予的帮助和友谊表示感谢。此外，我要感谢中国香港 ASM 太平洋科技有限公司的 Lee Wai Kwong（李伟光）先生（ASM 首席执行官）和 Wong Yam Mo（黄任武）先生（ASM 首席技术官）对我工作的信任、尊重和支持。最后，我要感谢以下同事，他们对本书内容进行了热烈丰富的讨论并做出了重要贡献：N. Khan, V. Rao, D. Ho, V. Lee, X. Zhang, T. Chai, V. Kripesh, C. Lee, C. Zhan, P. Tzeng, M. Dai, H. Chien, S. Wu, R. Lo, M. Kao, L. Li, Y. Chao, R. Tain, C. Premachandran, A. Yu, C. Selvanayagam, D. Pinjala, C. Ko, M. Li, Q. Li, R. Beica, I. Xu, T. Chan, K. Tan, E. Kuah, Y. Cheung,

X. Cao、J. Ran、H. Yang、N. Lee、S. Lim、N. Fan、M. Tao、J. Lo、R. Lee、X. Qing、Z. Cheng、Y. Lei、Z. Li、Y. Chen、M. Lin、V. Sekhar、A. Kumar、P. Lim、X. Ling、T. Lim、P. Ramana、L. Lim、C. Teo、W. Liang、J. Chai、M. Zhang 和 W. Choi。当然，我要感谢我在 ASM、ITRI、HKUST、IME、安捷伦、EPS、惠普期间一起工作过的杰出同事们（不再一一列举），以及整个电子工业为本书提供的有益帮助、大力支持和启发性讨论。与他们一起工作和交流是一种荣幸和一次难忘的经历。我从他们那里学会了很多关于生活、先进半导体封装和异构集成的技术。

最后，我要感谢我的女儿 Judy 和我的妻子 Teresa，感谢她们的关爱、体贴和耐心，让我平静地完成本书撰写。她们对我在为电子工业做贡献的信任，是我完成本书的强大动力。想到女儿 Judy 嫁给了一个支持她的丈夫（Bill），有两个可爱的孩子（Allison 和 James），并一直在半导体公司做得很好，我和 Teresa 身体健康，我感恩一切拥有。

John H. Lau（刘汉诚）

目 录

序
译者序
原书前言
原书致谢

第1章 异构集成综述 ·· 1
1.1 引言 ··· 1
1.2 多芯片组件（MCM）··· 1
1.2.1 共烧陶瓷型多芯片组件（MCM-C）································· 1
1.2.2 沉积型多芯片组件（MCM-D）······································· 2
1.2.3 叠层型多芯片组件（MCM-L）······································· 2
1.3 系统级封装（SiP）··· 2
1.3.1 SiP的目的 ··· 2
1.3.2 SiP的实际应用 ··· 2
1.3.3 SiP的潜在应用 ··· 2
1.4 系统级芯片（SoC）··· 3
1.4.1 苹果应用处理器（A10）··· 3
1.4.2 苹果应用处理器（A11）··· 3
1.4.3 苹果应用处理器（A12）··· 4
1.5 异构集成 ··· 4
1.5.1 异构集成与SoC··· 4
1.5.2 异构集成的优势··· 5
1.6 有机基板上的异构集成·· 5
1.6.1 安靠科技公司的车用SiP··· 5
1.6.2 日月光半导体公司在第三代苹果手表中使用的SiP封装技术······ 6
1.6.3 思科公司基于有机基板的专用集成电路（ASIC）与高带宽内存（HBM）······································· 7
1.6.4 基于有机基板的英特尔CPU和美光科技混合立体内存······ 7
1.7 基于硅基板的异构集成（TSV转接板）·································· 8
1.7.1 莱蒂公司的SoW··· 8

  1.7.2 IME 公司的 SoW ································· 9
  1.7.3 ITRI 的异构集成 ································· 10
  1.7.4 赛灵思/台积电公司的 CoWoS ····················· 11
  1.7.5 双面带有芯片的 TSV/RDL 转接板 ················· 11
  1.7.6 双面芯片贴装的转接板 ··························· 12
  1.7.7 TSV 转接板上的 AMD 公司 GPU 和海力士 HBM ····· 13
  1.7.8 TSV 转接板上的英伟达 GPU 和三星 HBM2 ········· 13
  1.7.9 IME 基于可调谐并有硅调幅器的激光源 MEMS ······ 15
  1.7.10 美国加利福尼亚大学圣芭芭拉分校和 AMD 公司的 TSV 转接板上芯片组 ································· 15
1.8 基于硅基板（桥）的异构集成 ··························· 16
  1.8.1 英特尔公司用于异构集成的 EMIB ················· 16
  1.8.2 IMEC 用于异构集成的桥 ························· 18
  1.8.3 ITRI 用于异构集成的桥 ························· 18
1.9 用于异构集成的 FOW/PLP ······························ 19
  1.9.1 用于异构集成的 FOWLP ························· 19
  1.9.2 用于异构集成的 FOPLP ························· 20
1.10 扇出型 RDL 基板上的异构集成 ························· 22
  1.10.1 星科金朋公司的扇出型晶圆级封装 ················ 22
  1.10.2 日月光半导体公司的扇出型封装（FOCoS） ········· 22
  1.10.3 联发科公司利用扇出型晶圆级封装的 RDL 技术 ····· 23
  1.10.4 三星公司的无硅 RDL 转接板 ····················· 24
  1.10.5 台积电公司的 InFO_oS 技术 ····················· 24
1.11 封装天线（AiP）和基带芯片组的异构集成 ················ 25
  1.11.1 台积电公司利用 FOWLP 的 AiP 技术 ·············· 25
  1.11.2 AiP 和基带芯片组的异构集成 ···················· 25
1.12 PoP 的异构集成 ····································· 26
  1.12.1 安靠科技/高通/新光公司的 PoP ·················· 26
  1.12.2 苹果/台积电公司的 PoP ························· 26
  1.12.3 三星公司用于智能手表的 PoP ···················· 28
1.13 内存堆栈的异构集成 ································· 29
  1.13.1 利用引线键合的内存芯片异构集成 ················ 29
  1.13.2 利用低温键合的内存芯片异构集成 ················ 29
1.14 芯片堆叠的异构集成 ································· 30
  1.14.1 英特尔公司用于 iPhone XR 的调制解调器芯片组 ····· 30
  1.14.2 IME 基于 TSV 的芯片堆叠 ······················· 30

  1.14.3 IME 无 TSV 的芯片堆叠 31
 1.15 CMOS 图像传感器（CIS）的异构集成 32
  1.15.1 索尼公司 CIS 的异构集成 32
  1.15.2 意法半导体公司 CIS 的异构集成 32
 1.16 LED 的异构集成 33
  1.16.1 中国香港科技大学的 LED 异构集成 33
  1.16.2 江阴长电先进封装有限公司的 LED 异构封装 34
 1.17 MEMS 的异构集成 35
  1.17.1 IME 的 MEMS 异构集成 35
  1.17.2 IMEC 的 MEMS 异构集成 35
  1.17.3 德国弗劳恩霍夫 IZM 研究所的 MEMS 异构集成 37
  1.17.4 美国帝时华公司的 MEMS 异构集成 37
  1.17.5 亚德诺半导体技术有限公司的 MEMS 异构集成 37
  1.17.6 IME 在 ASIC 上的 MEMS 异构集成 38
  1.17.7 安华高科技公司的 MEMS 异构集成 39
 1.18 VCSEL 的异构集成 40
  1.18.1 IME 的 VCSEL 异构集成 40
  1.18.2 中国香港科技大学的 VCSEL 异构集成 41
 1.19 总结和建议 42
 参考文献 42

第 2 章 有机基板上的异构集成 52
 2.1 引言 52
 2.2 安靠科技公司的汽车 SiP 52
 2.3 日月光半导体公司组装的 Apple Watch Ⅲ（SiP） 53
 2.4 IBM 公司的 SLC 技术 54
 2.5 无芯基板 54
 2.6 布线上凸点（BOL） 56
 2.7 嵌入式线路基板（ETS） 57
 2.8 新光公司的具有薄膜层的积层基板 59
  2.8.1 基板结构 59
  2.8.2 制作工艺 59
  2.8.3 质量评价测试 62
  2.8.4 i-THOP 应用示例 62
 2.9 思科公司的有机转接板 63
  2.9.1 转接板层结构 63
  2.9.2 制作工艺 63

2.10 总结和建议 ··· 66
参考文献 ··· 66

第3章 硅基板上的异构集成（TSV 转接板） ································· 70
3.1 引言 ··· 70
3.2 莱蒂公司的 SoW ··· 70
3.3 台积电公司的 CoWoS 和 CoWoS-2 ································· 71
3.4 TSV 的制备 ·· 72
3.5 铜双大马士革工艺制备 RDL ··· 73
3.6 异构集成中的双面转接板 ·· 75
    3.6.1 结构 ·· 75
    3.6.2 热分析——边界条件 ·· 76
    3.6.3 热分析——TSV 等效模型 ·································· 77
    3.6.4 热分析——焊凸点/下填料等效模型 ····················· 77
    3.6.5 热分析——结果 ·· 77
    3.6.6 热机械分析——边界条件 ·································· 79
    3.6.7 热机械分析——材料特性 ·································· 80
    3.6.8 热机械分析——结果 ·· 81
    3.6.9 TSV 加工 ··· 83
    3.6.10 带有正面 RDL 的转接板加工 ···························· 86
    3.6.11 带有正面 RDL 铜填充转接板的 TSV 露头 ············ 86
    3.6.12 带有背面 RDL 的转接板加工 ···························· 88
    3.6.13 转接板的无源电学表征 ···································· 89
    3.6.14 最终组装 ·· 90
3.7 总结和建议 ··· 93
参考文献 ··· 94

第4章 硅基板（桥）上的异构集成 ··· 96
4.1 引言 ··· 96
4.2 无 TSV 转接板技术：英特尔的 EMIB 技术 ······················ 96
4.3 EMIB 的制造 ·· 98
    4.3.1 利用精细 RDL 制造硅桥 ···································· 98
    4.3.2 利用超精细 RDL 制造硅桥 ································ 99
4.4 英特尔 EMIB 有机基板的制作 ······································· 100
4.5 总体组装 ··· 101
4.6 IMEC 桥的异构集成 ··· 102
4.7 IMEC 与桥的异构集成的组装工艺步骤 ···························· 102

XIII

- 4.8 用于异构集成的低成本 TSH 转接板（桥）·················103
  - 4.8.1 结构·················103
  - 4.8.2 电学仿真·················104
  - 4.8.3 试验件·················106
  - 4.8.4 带 UBM/ 焊盘和铜柱的顶部芯片·················108
  - 4.8.5 带 UBM/ 焊盘 / 焊料的底部芯片·················110
  - 4.8.6 桥的制备·················110
  - 4.8.7 总装·················111
  - 4.8.8 可靠性评估—冲击（坠落）试验及结果·················115
  - 4.8.9 可靠性评估—热循环试验及结果·················116
- 4.9 总结和建议·················117
- 参考文献·················117

第 5 章 异构集成的扇出晶圆级 / 板级封装·················119
- 5.1 引言·················119
- 5.2 FOW/PLP 的形式·················122
- 5.3 芯片先置（芯片面朝下）·················122
  - 5.3.1 芯片先置（芯片面朝下）工艺·················122
  - 5.3.2 带有晶圆载板的芯片先置（芯片面朝下）·················124
  - 5.3.3 带有面板载体的芯片先置（芯片面朝下）·················127
  - 5.3.4 芯片先置（芯片面朝下）封装组件的热循环试验·················129
  - 5.3.5 芯片先置（芯片面朝下）FOW/PLP 的应用·················132
- 5.4 芯片先置（芯片面朝上）·················133
  - 5.4.1 芯片先置（芯片面朝上）工艺·················133
  - 5.4.2 芯片先置（芯片面朝上）封装的热循环试验·················137
  - 5.4.3 芯片先置（芯片面朝上）封装的热性能·················138
  - 5.4.4 芯片先置（芯片面朝上）FOW/PLP 的应用·················139
- 5.5 芯片后置或 RDL 先置·················140
  - 5.5.1 芯片后置或 RDL 先置的原因·················140
  - 5.5.2 芯片后置或 RDL 先置工艺·················140
  - 5.5.3 芯片后置或 RDL 先置 FOW/PLP 的应用·················142
- 5.6 RDL 制造·················143
  - 5.6.1 有机 RDL（聚合物和 ECD 铜 + 刻蚀）·················143
  - 5.6.2 无机 RDL（PECVD 和镶嵌铜 +CMP）·················144
  - 5.6.3 混合 RDL（先无机 RDL，后有机 RDL）·················146
  - 5.6.4 纯 PCB 技术的 RDL（ABF/SAP/LDI 和镀铜 + 刻蚀）·················146

5.7 翘曲 148
    5.7.1 FOW/PLP 中的各种翘曲 148
    5.7.2 允许的最大翘曲 148
    5.7.3 翘曲的测量与模拟 149

5.8 临时晶圆与面板载体 150

5.9 异构集成的 FOW/PLP 机会 151

5.10 总结和建议 155

参考文献 156

第 6 章 基于扇出型 RDL 基板的异构集成 165

6.1 引言 165

6.2 星科金朋公司的 FOFC-eWLB 技术 165

6.3 日月光半导体公司的 FOCoS 技术 166
    6.3.1 关键工艺步骤 166
    6.3.2 FOCoS 的可靠性 167

6.4 联发科公司通过 FOWLP 技术实现的 RDL 169

6.5 台积电公司的 InFO_oS 技术 171

6.6 三星公司的无硅 RDL 转接板技术 171
    6.6.1 关键工艺步骤 172
    6.6.2 无硅 RDL 转接板的可靠性 172

6.7 总结和建议 173

参考文献 174

第 7 章 PoP 异构集成 176

7.1 引言 176

7.2 引线键合 PoP 176

7.3 倒装 PoP 177

7.4 引线键合与倒装混合 PoP 封装 177

7.5 iPhone 5S 中的 PoP 177

7.6 安靠科技 / 高通 / 新科公司的 PoP 179

7.7 苹果公司的焊点倒装 PoP 封装 181

7.8 星科金朋公司的处理器 PoP 封装 181

7.9 英飞凌公司的 eWLB 上 3D eWLB 封装 182

7.10 台积电 / 苹果公司的处理器 PoP 封装 183
    7.10.1 台积电 / 苹果公司的 A10 处理器 PoP 封装 183
    7.10.2 台积电 / 苹果公司的 A11 处理器 PoP 封装 184
    7.10.3 台积电 / 苹果公司的 A12 处理器 PoP 封装 185

7.11　三星智能手表 PoP 封装 ························· 186
7.12　总结和建议 ·· 188
参考文献 ··· 188

第8章　内存堆叠的异构集成 ·························· 190
8.1　引言 ··· 190
8.2　铜低 k 芯片上堆叠裸片（存储器）的引线键合 ··· 195
　　8.2.1　测试装置 ····································· 195
　　8.2.2　铜低 k 焊盘处的应力 ···················· 196
　　8.2.3　组装和处理 ·································· 199
　　8.2.4　切割方法的测评 ··························· 199
　　8.2.5　芯片贴装工艺 ······························ 200
　　8.2.6　引线键合工艺 ······························ 202
　　8.2.7　成型工艺 ····································· 205
　　8.2.8　可靠性测试和结果 ························ 205
　　8.2.9　总结和建议 ·································· 208
8.3　存储芯片和逻辑芯片的低温键合 ················ 208
　　8.3.1　低温键合的工作过程 ····················· 208
　　8.3.2　低温 SiO_2/Ti/Au/Sn/In/Au 到 SiO_2/Ti/Au 键合 ··· 209
　　8.3.3　焊料设计 ····································· 209
　　8.3.4　试验件 ·· 209
　　8.3.5　采用 InSnAu 低温键合的 3D 集成电路芯片堆叠 ··· 211
　　8.3.6　InSnAu IMC 的 SEM、TEM、XDR 和 DSC ··· 213
　　8.3.7　InSnAu IMC 的杨氏模量和硬度 ······· 214
　　8.3.8　InSnAu IMC 的 3 次回流 ················ 214
　　8.3.9　InSnAu IMC 的剪切强度 ················ 215
　　8.3.10　InSnAu IMC 的电阻 ····················· 216
　　8.3.11　InSnAu IMC 不稳定分析 ··············· 216
　　8.3.12　总结和建议 ································· 217
参考文献 ··· 218

第9章　芯片到芯片堆叠的异构集成 ·················· 221
9.1　引言 ··· 221
9.2　带有 TSV 的芯片到芯片的异构集成 ············ 222
　　9.2.1　底层芯片的 TSV 和 UBM 焊盘设计 ··· 222
　　9.2.2　顶层芯片的焊料微凸点设计 ············· 223
　　9.2.3　TSV 制造 ···································· 223

9.2.4 底层芯片 ENIG UBM 焊盘的制造 224
9.2.5 顶层芯片铜柱和锡焊帽的制造 225
9.2.6 TSV 的 DRIE 225
9.2.7 侧壁的钝化 228
9.2.8 自下而上的电镀 229
9.2.9 ENIG 电镀结果 231
9.2.10 铜锡合金焊凸点的制造结果 232
9.2.11 组装结果 233
9.2.12 总结和建议 234
9.3 无 TSV 的芯片到芯片异构集成 235
9.3.1 试验件与制造方法 235
9.3.2 试验件的制造 237
9.3.3 芯片到晶圆的组装方法 239
9.3.4 凸点高度平面度 240
9.3.5 对齐精度 241
9.3.6 芯片到晶圆的实验设计（DoE） 242
9.3.7 可靠性试验与结果 244
9.3.8 3D IC 封装与 SnAg 互连 245
9.3.9 总结和建议 247
参考文献 247

第 10 章 CIS、LED、MEMS 和 VCSEL 的异构集成 249

10.1 引言 249
10.2 CIS 异构集成 249
10.2.1 前照式 CIS 和背照式 CIS 249
10.2.2 3D CIS 和 IC 混合集成 251
10.2.3 3D IC 和 CIS 异构集成 254
10.2.4 总结和建议 258
10.3 LED 异构集成 258
10.3.1 采用带空腔和铜填充 TSV 的硅基板 LED 封装 258
10.3.2 基于 TSV 的 LED 晶圆级封装 262
10.3.3 总结和建议 267
10.4 MEMS 异构集成 268
10.4.1 基于 TSV 的 RF MEMS 器件晶圆级封装 268
10.4.2 基于 FBAR 振荡器的晶圆级封装 272
10.4.3 基于焊料的 3D MEMS 封装低温键合 275
10.4.4 总结和建议 282

XVII

10.5　VCSEL 和 PD 的异构集成···283
　　10.5.1　嵌入式板级光互连···283
　　10.5.2　嵌入 OECB 的 3D 异构集成···293
　　10.5.3　总结和建议··301
参考文献··302

第 11 章　异构集成的发展趋势···305

11.1　引言···305
11.2　异构集成的发展趋势···305
　　11.2.1　有机基板上的异构集成···305
　　11.2.2　非有机基板上的异构集成···306
　　11.2.3　各种异构集成的应用···307
　　11.2.4　各种异构集成的应用范围···307
　　11.2.5　总结和建议··307
参考文献··309

第 1 章

异构集成综述

1.1 引言

多芯片组件(multichip module,MCM)、系统级封装(system-in-package,SiP)和异构集成均是采用封装技术将从不同 IC 设计公司、代工厂设计生产的不同晶圆尺寸、特征尺寸的不同材料、不同功能的芯片、光电器件、封装芯片集成在一个系统或不同基板的子系统中,也可以独立封装。那么 MCM、SiP 和异构集成的区别是什么呢?传统的 MCM 主要是二维集成,而 SiP 可以是三维(3D)集成,也称垂直 MCM 或 3D-MCM。异构集成与 SiP 非常类似,但采用异构集成是为了更精细的节距、更多的输入/输出、更高密度和更高性能的应用。实际上,SiP 可以被认为是异构集成的一个大的子集[1-97]。本章将逐一介绍异构集成的不同应用,在此之前,首先对 MCM 与 SiP 进行简单介绍。

1.2 多芯片组件(MCM)

MCM 在常规基板,如陶瓷基板、硅基基板、有机基板上并排(side-by-side)集成不同的芯片和分立元件,以此来组成系统或子系统,用于高端网络、电信、服务器和计算机应用。基本上,MCM 分为三个类别,分别为共烧陶瓷型多芯片组件(MCM-C)、沉积型多芯片组件(MCM-D)和叠层型多芯片组件(MCM-L)。

1.2.1 共烧陶瓷型多芯片组件(MCM–C)

MCM-C 是使用厚膜技术,如可烧结金属浆料来形成导电图形的多芯片组件,其完全由陶瓷、玻璃态陶瓷材料或其他介电常数大于 5 的材料构建而成。简而言之,MCM-C 构建在陶瓷(C)或玻璃-陶瓷基板上[98]。

1.2.2　沉积型多芯片组件（MCM-D）

MCM-D 是利用薄膜技术，通过在硅、陶瓷或金属支撑结构上沉积介电常数小于 5 的非增强介电材料及薄膜金属而构成多层信号导体而形成的多芯片组件。简而言之，MCM-D 是在多种刚性基底上使用沉积（D）金属和非增强介电材料构建而成[98]。

1.2.3　叠层型多芯片组件（MCM-L）

MCM-L 是使用层压结构和印制电路板（printed circuit board，PCB）技术形成铜导体和通孔的多芯片组件。这些结构可能有时会包含热膨胀控制金属层。简而言之，MCM-L 采用了增强有机叠层的 PCB 技术[98]。

20 世纪 90 年代，许多有关 MCM 的研究涌现而出。但由于当时陶瓷基板和硅基基板过高的成本以及线宽限制和层压基板间隔的问题，以及商业模式的问题，如获得裸片比较困难，除了在某些小的场景中应用，MCM 从未实现量产（HVM）。实际上，从那时起，MCM 已经成为一个半导体封装领域"不受欢迎"的词语。

1.3　系统级封装（SiP）

1.3.1　SiP 的目的

SiP 集成了不同的芯片和分立元件，也可以集成由封装芯片或裸片（如宽带宽内存元件和基于硅通孔的逻辑元件）并排在一个常见的（硅、陶瓷或有机）基板上形成一个系统或子系统的 3D 芯片堆叠，用于智能手机、平板计算机、高端网络、电信、服务器和计算机。SiP 技术既可以进行水平集成也可以进行纵向集成。SiP 也称垂直 MCM 或 3D MCM。

1.3.2　SiP 的实际应用

由于硅通孔（TSV）技术用于智能手机和平板计算机成本高[99, 100]，因此从未实现过。过去十年内大多数进入量产的 SiP 实际上都是针对低端应用的 MCM-L，如智能手机、平板计算机、智能手表、医疗、可穿戴电子设备、游戏系统、消费产品和物联网（IoT）相关产品[101]，如智能家居、智能能源和智能工业自动化。外包半导体组装与检测公司（OSAT）进行的大多数 SiP 实际应用是在一个常规层压基板上集成两块或更多的不同芯片、器件和一些分立元件。

1.3.3　SiP 的潜在应用

SiP 在高价格、高利润、高端产品中的应用有双镜头相机模块等。但是，目前这种 SiP 应用不能完全由 OSAT 进行，同时涉及光学设计、测试、透镜、微电机、柔性基板、系统集成等方面的能力仍有待加强。

1.4 系统级芯片（SoC）

摩尔定律[102]一直以来驱动着系统级芯片（SoC）[103-105]平台的发展。特别是在过去十年，SoC在智能手机、平板计算机方面的应用非常受欢迎。SoC在单个芯片上集成了多个功能的集成电路，形成一个系统或子系统。下面是三种典型的SoC示例。

1.4.1 苹果应用处理器（A10）

A10应用处理器（application processor，AP）是一款由苹果设计、台积电（TSMC）制造的处理器，采用16nm加工技术，包含六核图形处理单元（GPU），两个双核的中央处理单元（CPU），两块静态随机存储器（SRAM）等。芯片面积为125mm^2（11.6mm×10.8mm），如图1.1a所示。

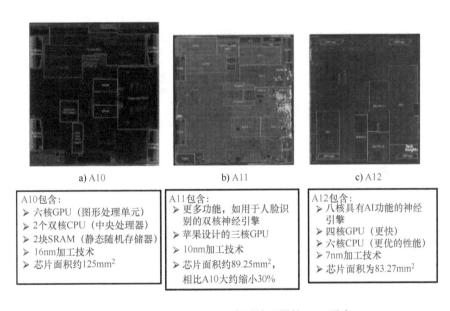

图1.1 A10、A11、A12应用处理器的SoC平台

1.4.2 苹果应用处理器（A11）

A11应用处理器也是由苹果设计、台积电制造的处理器，采用10nm加工技术，包含了更多的功能，包括苹果自主设计的三核GPU，用于人脸识别的神经引擎等。但芯片面积由于摩尔定律大约比A10缩小30%，大约为89.23mm^2，即特征尺寸由16nm缩小到了10nm，如图1.1b所示。

1.4.3 苹果应用处理器（A12）

A12应用处理器如图1.1c所示，同样由苹果设计、台积电制造，采用7nm加工技术，包含了更多功能，包含八核神经引擎（用于AI功能）、四核GPU（运行更快）、六核CPU（更优的性能）等，芯片面积为83.27mm^2，大致和A11相同。

1.5 异构集成

一些早期关于异构集成的研究由美国佐治亚理工学院提供[106-108]，其中报道了一种微分型硅基CMOS集成电路接收器（工作速率为1Gbit/s），其上也集成了大面积InGaAs/InP I-MSM（金属-半导体-金属）光电探测薄膜器件，如图1.2所示。现在，许多异构集成聚焦于更高的密度、更精细的节距及更复杂的系统。

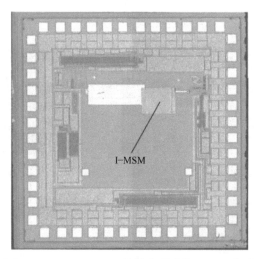

图1.2 集成有InGaAs/InP I-MSM的微分型硅基CMOS集成电路接收器

1.5.1 异构集成与SoC

研究异构集成的其中一个关键原因在于摩尔定律的终结正在快速到来，通过缩小特征尺寸（以此来进行翻倍）来制作SoC越来越困难，并且耗资巨大。异构集成与SoC相比，异构集成使用封装技术将不同芯片（无论并排还是堆叠）或者来源于不同设计公司和代工厂、不同晶圆尺寸和特征尺寸的不同材料、功能的元件集成到不同基板（如有机、硅基或RDL）或独立封装的系统、子系统中（如图1.3所示），而非将绝大部分功能集成到一个单一芯片，并不断追求更高的特征尺寸。

第 1 章　异构集成综述

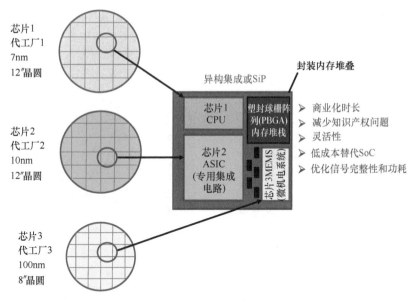

图 1.3　异构集成或 SiP

1.5.2　异构集成的优势

未来几年，无论是在商业化时长、性能、外形因素、功耗、信号完整性还是成本方面，将涌现更多高水平的异构集成。异构集成将会抢夺部分 SoC 在高端应用的市场份额，如高端智能手机、平板计算机、可穿戴设备、网络、电信和计算机。这些不同的芯片彼此之间通过再布线层（RDL）[109, 110] 建立联系。本章将讨论基于有机基板、TSV 转接硅基板、桥式硅基板和扇出基板的异构集成中的 RDL 技术，并简要介绍异构集成在堆叠封装（package-to-package，PoP）、内存堆叠、芯片对芯片堆叠、发光二极管（LED）、CMOS 图像传感器（CIS）、微机电系统（MEMS）和垂直腔面发射激光器（VCSEL）中的应用。SiP[37-97] 与异构集成[1-36] 非常类似，但 SiP 用于比较粗略的节距、更少的输入/输出、更低的密度和相对低端的性能应用。

1.6　有机基板上的异构集成

现在，最通用的异构集成应用是基于有机基板，或者称为 SiP。这种集成方法通常使用表面贴装技术（SMT），包括适合大规模回流焊的焊料凸点倒装芯片和板上引线键合芯片，一般用于中低端应用。更多关于有机基板上的异构集成内容见本书第 2 章。

1.6.1　安靠科技公司的车用 SiP

安靠科技（Amkor）的车用 SiP 聚焦于自动驾驶、信息娱乐、高级驾驶辅助

5

系统（ADAS）和车载计算机。图1.4a、b展示了一组安靠科技的车用SiP。从图1.4a可以看到42.5mm×42.5mm信息娱乐有机基板支持处理器和双数据速率（DDR）存储器。而从图1.4b可以看到55 mm×72mm有机基板支持的网络交换机、专用集成电路（ASIC）和存储器。

a) 42.5mm×42.5mm信息娱乐　　　　b) 55mm×72mm有机基板

图1.4　安靠科技（Amkor）车用SiP

1.6.2　日月光半导体公司在第三代苹果手表中使用的SiP封装技术

通过环隆电气股份有限公司（USI），日月光半导体公司（ASE）成为苹果为Apple Watch S III定制设计的S3 SiP模块的唯一后端提供商。如图1.5所示，一块有机基板上有超过40个片式元器件，这些片式元器件包括电阻器、电容器等分立无源元件以及专用集成电路、处理器、控制器、转换器、动态随机存储器（DRAM）、与非型（NAND）闪存、Wi-Fi、近距离无线通信（NFC）、全球定位系统（GPS）、传感器等。

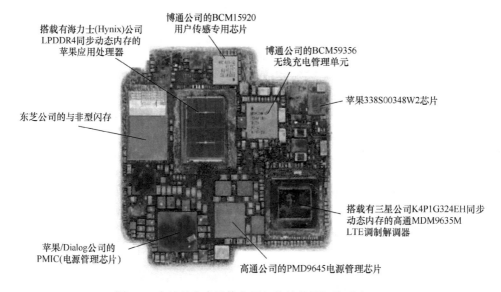

图1.5　由日月光半导体公司组装的苹果智能手表S III

1.6.3 思科公司基于有机基板的专用集成电路（ASIC）与高带宽内存（HBM）

图 1.6 为由思科（Cisco）公司设计和制造的一个具有细节距和细线互连大型有机转接板（基板）的 3D SiP[82]。有机转接板的尺寸为 38mm × 30mm × 0.4mm，其正面和背面的线宽、间距和厚度均相同，分别为 6μm、6μm 和 10μm。在有机转接板的顶部装载有 19.1mm × 24mm × 0.75mm 的高性能 ASIC 芯片和 4 个高带宽内存（HBM）DRAM 芯片堆栈。5.5mm × 7.7mm × 0.48mm 的三维 HBM 芯片堆栈包含了 1 个缓冲芯片和 4 个 DRAM 核心芯片，其通过 TSV 和带有焊锡帽的细节距微柱相互连接。这是为了高端应用而做的设计。

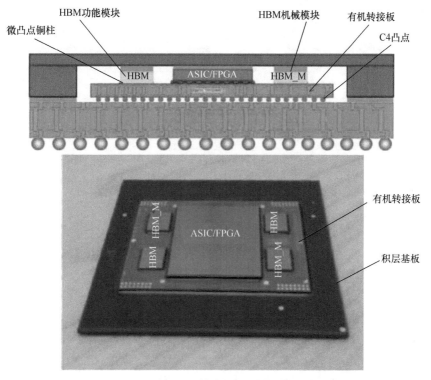

图 1.6 基于有机转接板的思科网络系统

1.6.4 基于有机基板的英特尔 CPU 和美光科技混合立体内存

图 1.7 为搭载美光科技（Micron）混合立体内存（HMC）的英特尔 Knight Landing CPU，自 2016 年下半年以来，英特尔就一直在向自己最青睐的客户发售该产品。可见，8 个基于美光科技 HMC 技术的多信道 DRAM（MCDRAM）支持 72 核处理器。每一个 HMC 包含 4 个 DRAM 和 1 个逻辑控制器（带有

异构集成技术

TSV），而且每一个 DRAM 含有超过 2000 个带有焊帽铜柱凸点的 TSV。CPU 和带有逻辑控制器堆栈的 DRAM 被装载在有机封装基板之上。这也是为了高端应用所做的设计。

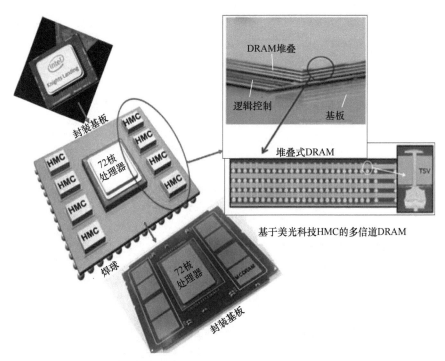

图 1.7 有机基板上的英特尔 Knights Landing 和美光科技 HMC

1.7 基于硅基板的异构集成（TSV 转接板）

一般来讲，基于硅基板的异构集成用于硅晶圆上的多芯片或系统级晶圆（SoW）集成。常用的封装方法为适合于大规模回流焊的具有 TSV 结构的晶圆级芯片（CoW）倒装，或者能实现非常细节距的热压键合。总体来说，这是为了高端应用场景而设计。想要获得更多关于基于硅基板的异构集成（TSV 转接板）信息请查阅本书第 3 章。

1.7.1 莱蒂公司的 SoW

SoW 的早期应用之一就是由莱蒂（Leti）公司[35, 36] 提供的，如图 1.8 所示，它可以看作一个在有 TSV 的硅晶片上的芯片系统，该系统可包括 ASIC 和内存，或者 PMIC 和 MEMS。在划片后，单个单元可以变成系统或者子系统，从而接在有机基板上或者单独存在。

第 1 章 异构集成综述

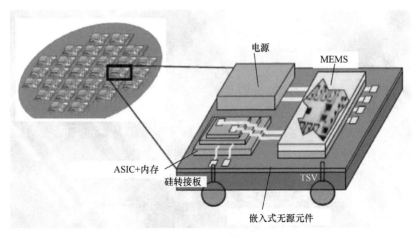

图 1.8 莱蒂公司的 SoW

1.7.2 IME 公司的 SoW

移动电子产品需要多功能模块，包括数据、射频（RF）和存储功能。TSV 技术提供了一种实现复杂、多功能的高封装密度异构集成的集成方式。图 1.9 为一个应用硅载板的 3D 模块。该封装包含了 3 个芯片的 2 个堆栈，模块尺寸为 12mm×12mm×1.3mm，硅载板尺寸为 12mm×12mm 并有 168 个外围元件通道。硅载板 1 组装有 5mm×5mm 的倒装芯片组装，硅载板 2 与 1 个 5mm×5mm 的倒装芯片和 2 个 3mm×6mm 的引线键合芯片组装，并且硅载板 2 是模塑成型以保护键合引线。硅载板制作了 2 个金属层并以 SiO_2 作为介质/钝化层，硅载板的电连接通过 TSV 形成。图 1.9 右侧展示了这种封装结构。

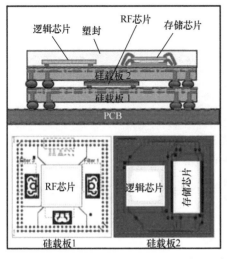
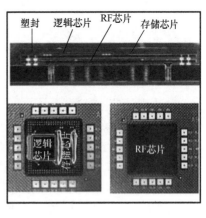

图 1.9 IME 公司 RF 芯片、逻辑芯片和存储芯片的异构集成

1.7.3 ITRI[①] 的异构集成

图 1.10 展示了一个多种芯片在 TSV 转接板上的异构集成[38, 39],它可以看作是一个支撑了 4 个存储芯片(有 10μm 的通孔)堆叠、1 个热芯片和 1 个机械芯片的转接板结构(有 15μm 的通孔)。既为拾取与放置目的,以及保护芯片不被恶劣环境所破坏,转接板采用模塑成型。在转接板顶部和底部都有再布线层,并且压力传感器也被植入到顶部,集成无源器件(IPD)也沿着转接板(12.3mm×12.2mm)厚度方向(100μm)被装配。

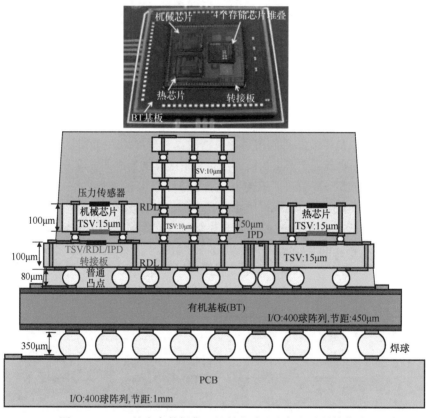

图 1.10 ITRI 的内存数据集、机械芯片和电气芯片异构集成

图 1.10 试验样品可以演化为如下情形:

1)高 I/O DRAM,没有机械/热芯片,并且转接板是 ASIC 芯片。

2)高 I/O 内存,没有存储芯片堆栈,在机械/热芯片中也没有 TSV,并且转接板为 ASIC 或者微处理器。

3)高 I/O 接口,没有存储芯片堆栈,并且没有 TSV 在热/机械芯片中。因

---

① Industrial Technology Research Institute,指中国台湾"工业技术研究院",后文同。

此，由此试验样品开发出的赋能技术有很广阔的应用。

1.7.4 赛灵思/台积电公司的 CoWoS

过去几年里，由于对高密度、高 I/O 和超细节距的需求，如切片现场可编程序门阵列（FPGA），甚至一个 12 积层（6-2-6）有机封装基板都不够支持芯片，而且还需要一个 TSV 转接板[42-54]。图 1.11 展示了赛灵思（Xilinx）/台积电（TSMC）公司的切片 FPGA 的 CoWoS（芯片-晶圆-基板，板上晶圆级封装芯片）封装[50-53]，它可以看作是 TSV（直径 10μm）转接板（深 100μm）有 4 个顶部再布线层（RDL），即 3 个铜镶嵌层和 1 个铝层。切片 FPGA 芯片间的 10000 多个横向互连主要由转接板上最小节距为 0.4μm 的再布线层连接，再布线层和钝化层的最小厚度小于 1μm。每一个 FPGA 上都有超过 50000 个微凸点（转接板上有超过 200000 个微凸点），节距 45μm，如图 1.11 所示。

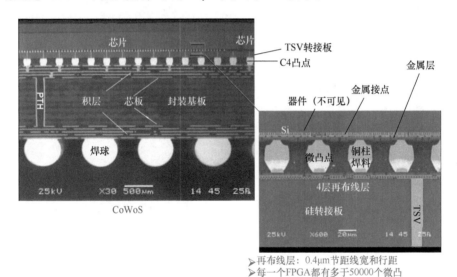

➤ 再布线层：0.4μm 节距线宽和行距
➤ 每一个 FPGA 都有多于 50000 个微凸点，节距为 45μm
➤ 转接板上可有大于 200000 个微凸点

图 1.11　赛灵思/台积电的 CoWoS

1.7.5 双面带有芯片的 TSV/RDL 转接板

图 1.12 展示了顶面载有一个 CPU 或 ASIC 并且背面载有 2 个存储芯片的转接板的 3D 集成电路异构集成[55-59]。直径 25μm、深 150μm、节距 150μm 的 TSV 被均匀地嵌入在转接板上；TSV 侧壁的钝化层（SiO_2）厚度为 0.5μm。转接板被普通凸点焊在一个有机基板上，而基板被焊球焊在一个 PCB 上。转接板尺寸为 28mm×28mm×150μm，CPU/ASIC 尺寸为 22mm×18mm×400μm，DRAM 尺寸为 10mm×10mm×100μm，封装基板尺寸为 40mm×40mm×950μm，PCB 尺

寸为 114.3mm × 101.6mm × 1600μm。

所有芯片和转接板之间的微型焊凸点的直径为 25μm，节距为 250μm。普通焊凸点直径为 150μm，节距为 250μm。焊球直径为 600μm，节距为 1000μm。

图 1.12 展示了一个 3D 集成电路异构集成封装的扫描电镜 SEM 横截面图像，可以看到在无源硅转接板顶部有一个稍大稍厚的芯片，并且在底部有一个稍小稍薄的芯片。顶部芯片与转接板间的微焊点也被放大展示在了图 1.12 中，可以看到转接板上的凸点下金属层（UBM）是铜和镍，焊料形成了 IMC（金属间化合物）Cu_6Sn_5，稍大芯片中的 UBM 是铜。同样，转接板和 TSV 及稍小芯片间的微焊点也被放大展示在了图 1.12 中。

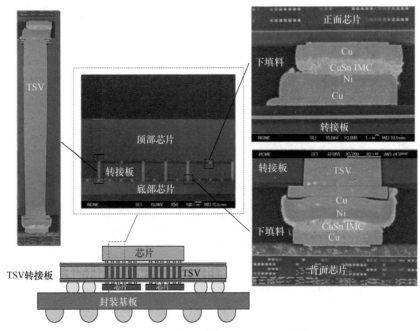

图 1.12 双面有芯片的 TSV/RDL 转接板

1.7.6 双面芯片贴装的转接板

图 1.13 的上图展示了 3D 集成电路异构集成了 2 个芯片在顶面、1 个芯片在底面的双侧转接板集成[60-62]。封装基板的尺寸为 35mm × 35 mm × 970μm。芯片和转接板以及转接板和封装基板之间使用了下填料。TSV 直径为 10μm，节距为 150μm。芯片和转接板以及转接板和封装基板之间的焊凸点直径为 90μm，节距为 125μm。封装基板和 PCB 之间的焊球直径为 600μm，节距为 1000μm。图 1.13 的下图展示了组装完成的模块横截面，可以看到 3 个芯片及其下填料完全由转接板支撑，而转接板是焊在一个 4-2-4 结构的封装基板上（有下填料）。

第 1 章　异构集成综述

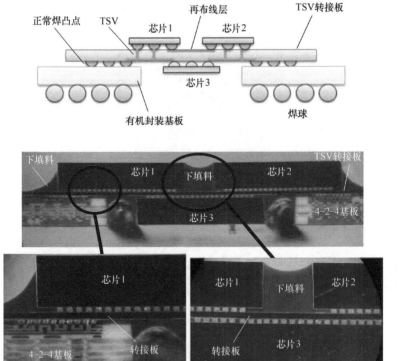

图 1.13　双面芯片贴装的硅转接板

1.7.7　TSV 转接板上的 AMD 公司 GPU 和海力士 HBM

图 1.14 展示了 AMD 公司 2015 下半年上市的 Radeon R9 Fury X 图形处理器（GPU），它基于 TSMC 的 28nm 制程技术制造，并且承载了 SK 海力士（Hynix）生产的 4 个 HBM 模块上。每一个 HBM 包含 4 个带有铜柱 + 焊凸点的 DRAM 和一个被 TSV 直接穿过的逻辑数据库，每一个 DRAM 芯片有超过 1000 个 TSV。GPU 和 HBM 模块装在一个由联华电子公司（UMC）用 64nm 制程技术生产的 TSV 转接板（28mm×35mm）的顶部。具有 C4（可控塌陷芯片连接）凸点的 TSV 安装在一个 4-2-4 有机封装基板（由日本揖斐电（Ibiden）公司生产）上，最终封装是由 ASE 完成的。

1.7.8　TSV 转接板上的英伟达 GPU 和三星 HBM2

图 1.15 展示了在 2016 年[41]下半年上市的英伟达的 Pascal 100 GPU，该 GPU 基于 TSMC 16nm 制程技术制造，且承载了三星生产的 4 个 HBM2（16GB）。每一个 HBM2 包含 4 个带有铜柱 + 焊凸点的 DRAM 和一个被 TSV 直接穿过的底层逻辑芯片，每一个 DRAM 芯片有超过 1000 个 TSV。GPU 和 HBM2 模块在

| 异构集成技术

一个由 TSMC 用 64nm 制程技术生产的 TSV 转接板（1200mm^2）的顶端。TSV 转接板被 C4 凸点连接在一个 5-2-5 有机封装基板上。

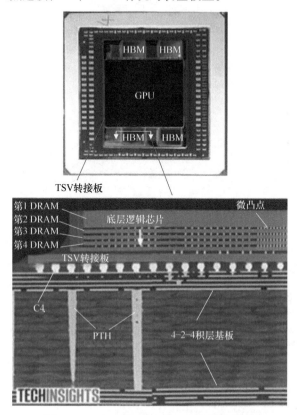

图 1.14　AMD 的 GPU 和 HBM 异构集成

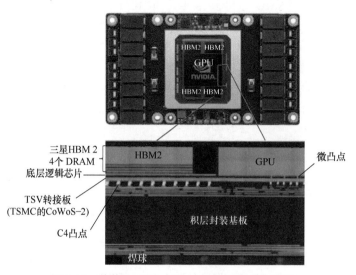

图 1.15　英伟达 GPU 与三星 HBM2 的异构集成

14

1.7.9 IME 基于可调谐并有硅调幅器的激光源 MEMS

2007 年，IME 推出了一款基于可调谐并有硅调幅器、聚合物耦合器以及由一个热电冷却器冷却的硅光具座上的Ⅲ-Ⅳ族增益芯片的激光源微机电系统，如图 1.16 所示。MEMS 驱动器的主要特征为：驱动电压为 30V 和 30μm 的电位移，谐振频率为 2.559kHz，光串扰小于 −60dB，偏振相关损耗小于 0.1dB，波长相关损耗小于 0.05dB，上升时间为 75.6μs，下降时间为 63.4μs，可靠性大于 1×10^7 次。硅调幅器的主要特征为：总长度为 3500μm，分路器为 30μm，合成器为 30μm，调幅器长大约 3440μm，横截面约为 0.4μm × 0.35μm，设计的最大频率大约 20GHz，并且设计的传输损耗小于 20dB/cm。

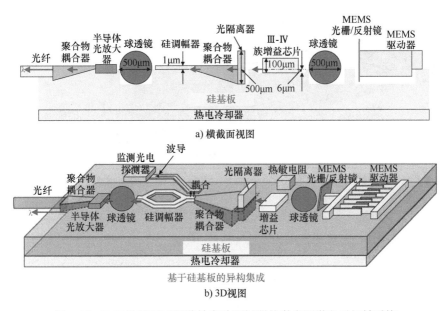

图 1.16　IME 的基于可调谐并有硅调幅器的激光源微电子机械系统

1.7.10 美国加利福尼亚大学圣芭芭拉分校和 AMD 公司的 TSV 转接板上芯片组

受美国国防部高级研究计划局（DARPA）通用异构集成和知识产权复用策略（CHIPS）项目的资助，加利福尼亚大学圣芭芭拉分校（UCSB）和 AMD 公司提出了一个未来高性能系统，如图 1.17 所示。这个系统包含 1 个 CPU 芯片组和几个 GPU 芯片组，同时也包括在一个无源 TSV 转接板和 / 或一个有再布线层的有源 TSV 转接板上的 HBM。

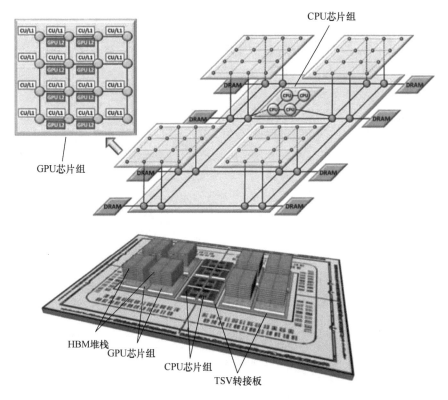

图 1.17　加利福尼亚大学圣芭芭拉分校和 AMD 公司的 GPU 芯片组和 CPU 芯片组在 TSV 转接板上的异构集成

1.8　基于硅基板（桥）的异构集成

基本上说，桥指的是一片有 RDL 和接触垫但没有 TSV 的硅片假片。通常 RDL 和接触垫会被制造在一个硅片假片晶圆上，然后再被切割成独立的桥。更多基于硅基板（桥）的异构集成信息可查阅本书第 4 章。

1.8.1　英特尔公司用于异构集成的 EMIB

英特尔公司提出了嵌入式多芯片互连桥接（EMIB）[30, 31]RDL 来取代异构集成系统中的 TSV 转接板。芯片间的横向通信将由有再布线层的硅嵌入式桥负责，并且电源/接地和一些信号将流过有机封装基板，如图 1.18 所示。

生产载有 EMIB 的有机封装基板有两个主要任务，一个任务是制造 EMIB，另一个任务是制造载有 EMIB 的有机封装基板。想要制造 EMIB 就要首先建造 RDL（包括硅晶圆上的接触垫，将在 4.3 节中讨论）。最终，将硅晶圆没有 RDL 的一侧粘接一个芯片粘贴薄膜（DAF），再把硅晶圆分离成独立的桥。为了使有机基板和 EMIB 接合，首先要将 EMIB 和 DAF 放置在有机基板腔内的铜箔上，然后在标准的有机封装基板制造所有连接到铜触板的通路（见 4.4 节）。这样载有 EMIB 的有机封装基板就可以与 GPU 和 HBM 等芯片连接了，如图 1.19 所示。

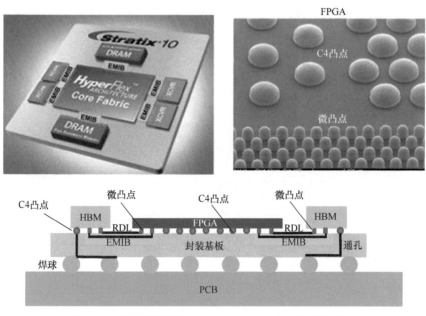

图 1.18　使用了英特尔 EMIB 和 FPGA 技术的异构集成

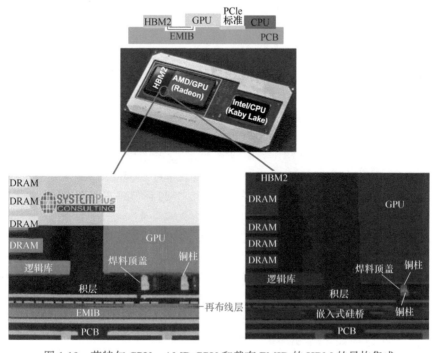

图 1.19　英特尔 CPU、AMD GPU 和载有 EMIB 的 HBM 的异构集成

1.8.2 IMEC 用于异构集成的桥

自从英特尔提出用 EMIB 作为异构集成芯片间的高密度互连,"桥"就变得非常受关注。如最近 IMEC 提出[33]使用桥 + 扇出型晶圆级封装（FOWLP）技术来互连逻辑芯片、宽 I/O DRAM 和闪存,如图 1.20 所示,他们的目的是在任何器件芯片上均不使用 TSV。

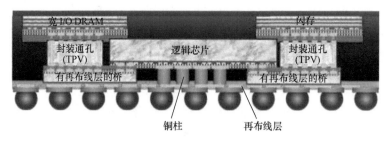

> 器件芯片上没有硅通孔
> 封装通孔是一块有硅通孔的硅片

图 1.20 IMEC 基于硅桥的异构集成

1.8.3 ITRI 用于异构集成的桥

图 1.21 所示是一个基于硅穿孔（TSH）转接板（一个桥）的异构集成,在转接板顶部和底部承载了一些芯片[34]。TSH 转接板的主要特征是孔内没有金属化,因此电介质层、阻挡层和种子层、填孔电镀、用以去除过量铜的化学机械抛光（CMP）和铜暴露都不再必要。相比 TSV 转接板,TSH 转接板仅仅需要用激光或深反应离子刻蚀（DRIE）在一个硅晶圆上打孔洞。同 TSV 转接板一样,TSH 转接板也需要 RDL。TSH 转接板可以在顶部和底部同时承载芯片,孔洞可以通过铜柱或焊料引导信号通过孔洞从底部芯片传到顶部芯片（反之亦然）。同一侧的芯片可以通过 TSH 转接板的 RDL 来实现通信,物理上顶部的芯片和底部的芯片可以通过铜柱和微焊点连接。同时,芯片的外围焊接在 TSH 转接板上,以实现结构的完整性从而抵御振动或者热干扰。另外,TSH 转接板底部的外围有连接封装基板的普通焊凸点。图 1.21 展示了一个 SiP 的电镜横截面图[34],其中包括所有的主要部件,如顶部芯片、TSH 转接板、底部芯片、封装基板、PCB、微凸点、焊凸点、焊球、TSH 和铜柱。通过 X 光和电镜图可以看出 SiP 结构的主要元件都被制造出来了。

第 1 章 异构集成综述

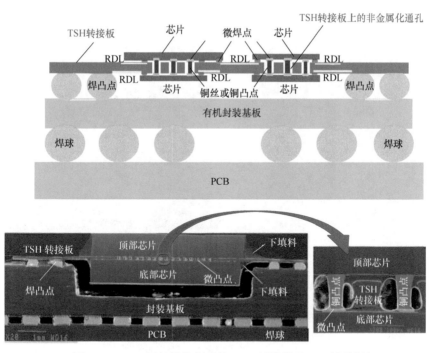

图 1.21 ITRI 用于异构集成的 ITRI 桥接器（TSH 转接板）

1.9 用于异构集成的 FOW/PLP

扇出型晶圆级 / 面板级封装（FOW/PLP）[111-131] 非常适合异构集成。本节简要介绍了两个例子，一个是 FOWLP，另一个是 FOPLP。有关异构集成的 FOW/PLP 的更多信息，可阅读本书第 5 章。

1.9.1 用于异构集成的 FOWLP

图 1.22 显示了由 629 个（10mm×10mm）封装组成的重构晶圆[7, 21]。每个封装有 4 个（1 个 5mm×5mm 和 3 个 3mm×3mm）芯片和 4 个（0402）电容器。大芯片和小芯片之间的节距为 100μm。每个封装有两个 RDL。需要强调的是，FOWLP 是一个非常高通量的工艺过程。在这种情况下，一次性可以生产 629 个 10mm×10mm 的封装。图 1.23 显示了封装的横截面。可以看出，有两个 RDL，RDL1 的金属层厚度为 3μm，RDL2 的金属层厚度为 7.5μm。RDL1 的线宽和间距为 10μm，RDL2 的线宽和间距为 15μm。DL1 和 DL2 的介电层厚度为 5μm，DL3 为 10μm。钝化层（DL3）的开口为 180μm。焊接锡球尺寸为 200μm，球节距为 0.4mm。

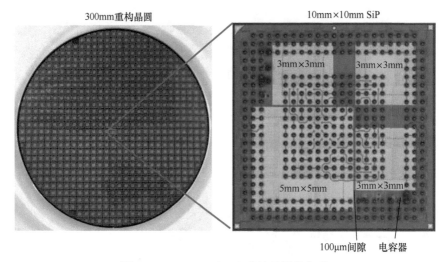

图 1.22　FOWLP 对 4 个芯片的异构集成

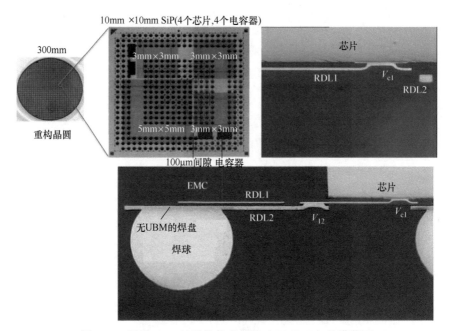

图 1.23　显示 RDL 的异构集成封装（FOWLP）的横截面

1.9.2　用于异构集成的 FOPLP

图 1.24 显示了由 1512（10mm×10mm）个封装组成的重构面板[3, 22]。同样，每个封装有 4 个（1 个 5mm×5mm 和 3 个 3mm×3mm）芯片。图 1.25 显示了 X 射线图像和封装横截面，可以看出 2 个 RDL（RDL1 和 RDL2）的金属层厚度为

10μm。其中 RDL1 的线宽和间距为 20μm，RDL2 的线宽和间距为 25μm。DL1、DL2 和 DL3 的介电层厚度为 20μm。钝化层（DL3）的开口为 180μm。锡球尺寸为 200μm，球节距为 0.4 mm。

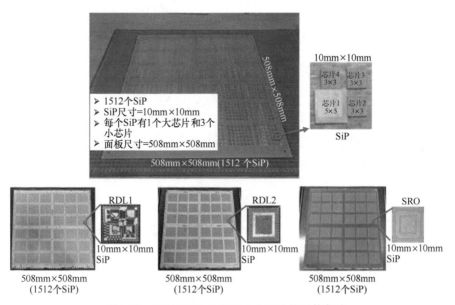

图 1.24　通过 FOPLP 实现 4 个芯片的异构集成

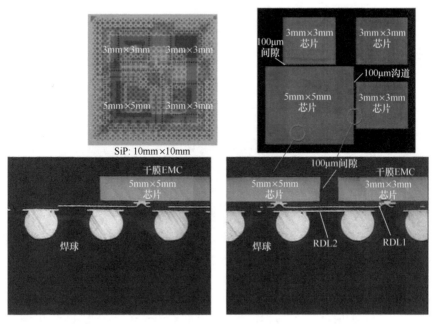

图 1.25　显示 RDL 的异构集成封装（FOPLP）横截面

1.10 扇出型 RDL 基板上的异构集成

近年来，为了降低封装剖面、提高性能和降低成本，RDL 上的异构集成得到了广泛的应用，尤其是 FOWLP 技术，一般来说，它适用于中高端的应用。有关扇出 RDL 基板上异构集成的更多信息，请阅读本书第 6 章。

1.10.1 星科金朋公司的扇出型晶圆级封装

在 ECTC 2013 会议上，长电科技子公司星科金朋（上海）有限公司提出[26,132]使用扇出型晶圆级封装（FOFC）eWLB，使芯片的 RDL 主要进行横向通信，如图 1.26 所示。可以看到，TSV 转接板、晶圆凸点、焊剂、芯片到晶圆（C2W）键合、清洁与下填料的填充、固化均已被省略。

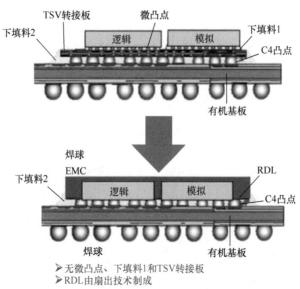

图 1.26　星科金朋公司的扇出型 RDL 基板（FOFC-eWLB）

1.10.2 日月光半导体公司的扇出型封装（FOCoS）

2016 年，日月光半导体公司[27]提出使用扇出晶圆级封装（FOWLP）技术，即先制造芯片，再在临时晶圆载体上进行模压，然后通过压缩方法二次成型，以使芯片的 RDL 主要进行横向通信，如图 1.27 所示；这项技术称为基板上的扇出晶圆级芯片。TSV 转接板、晶圆凸点、焊剂、芯片到晶圆（C2W）键合、清洁与下填料的填充、固化，均已被省略。

底部 RDL 使用凸点下金属化层（UBM）和 C4 凸点连接封装基板，如图 1.27 所示。

第 1 章 异构集成综述

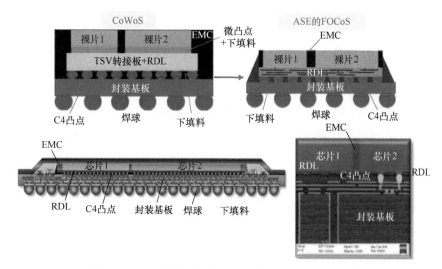

图 1.27 日月光半导体公司的扇出型封装（FOCoS）

1.10.3 联发科公司利用扇出型晶圆级封装的 RDL 技术

2016 年，联发科技股份有限公司（MediaTek Inc.）[133] 提出了类似的采用 FOWLP 技术制造的无 TSV 转接板的再布线层，如图 1.28 所示。该技术使用微凸点（铜柱＋焊帽）将底部 RDL 连接到 6-2-6 封装基板，而非 C4 凸点。

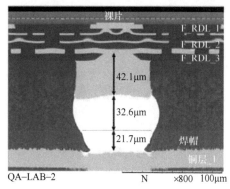

图 1.28 联发科公司制造的扇出型 RDL 基板

1.10.4 三星公司的无硅 RDL 转接板

最近，三星公司[32]提出使用扇出型晶圆级封装（FOWLP）技术，后芯片封装工艺或先再布线层封装工艺来消除 TSV 转接板，如图 1.29 所示。首先在硅或玻璃晶圆上构建 RDL，同时在逻辑和 HBM（高带宽存储器）上做出晶圆凸点；然后进行焊接、芯片到晶圆（C2W）键合、清洗与下填料填充、固化；随后是 EMC 模塑，对硅晶片和 C4 晶片凸点进行背研，并将整个模块连接在封装基板上，最后进行焊球安装与封盖。

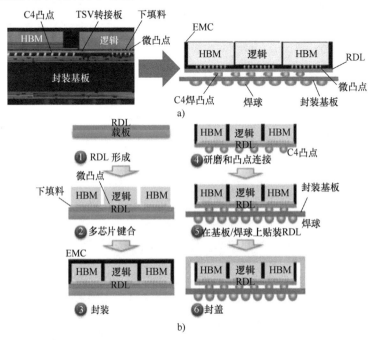

图 1.29　三星公司制造的扇出型（芯片位置）基板（无硅 RDL 转接板）

1.10.5 台积电公司的 InFO_oS 技术

图 1.30 显示了台积电的 InFO_oS（集成扇出基板）示意图。RDL 由台积电的面朝上的先芯片封装技术制造。InFO_oS 技术适用于高性能的应用，但不如使用 CoWoS 技术性能更高。

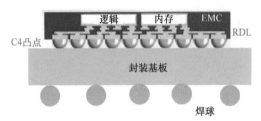

图 1.30　台积电的 InFO_oS 技术（集成扇出基板）

1.11 封装天线（AiP）和基带芯片组的异构集成

下面简要介绍 FOWLP 对 AiP 进行异构集成的两个例子。一个是台积电提出的利用 FOWLP 的 AiP 技术，另一个则为新提出的利用 FOWLP 技术对 AiP 和基带芯片组的异构集成。

1.11.1 台积电公司利用 FOWLP 的 AiP 技术

台积电公司[134]证实，用于高性能紧凑型 5G 毫米波系统集成的 InFO_AiP 技术优于图 1.31 所示的在基板上利用焊接凸点倒装焊接的 AiP 技术。可以看出，在 28GHz 频率范围内，InFO RDL 的传输损耗（0.175dB/mm）比倒装芯片基板布线上的传输损耗（0.288dB/mm）低 65%；在 38GHz 频率范围内，InFO RDL 的传输损耗（0.225dB/mm）比倒装芯片基板布线上的传输损耗（0.377dB/mm）低 53%。

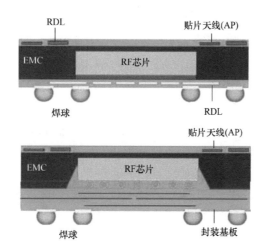

28GHz和38GHz频率范围内下再布线层和基板布线的传输损耗

频率	InFO RDL	基板布线
28GHz	0.175dB/mm	0.288dB/mm
38GHz	0.225dB/mm	0.377dB/mm

图 1.31　台积电使用 InFO 封装的 AiP 技术与基板上的倒装芯片的比较

1.11.2 AiP 和基带芯片组的异构集成

图 1.32 示意性地显示了利用 FOWLP 技术对 AiP 和基带芯片组的异构集成，可以看出，RF 芯片和基带芯片组（调制解调器 AP 和 DRAM）与 RDL 并排放置，并与贴片天线（AP）耦合。

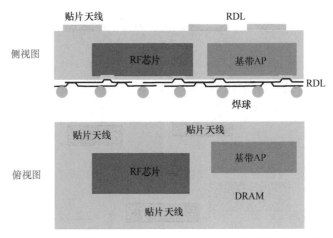

图 1.32　RF 芯片、基带 AP、DRAM 和 AiP 的异构集成方案

1.12　PoP 的异构集成

PoP（堆叠封装）是一种异构集成的方式。下面简要介绍一些 PoP 异构集成方法。了解更多 PoP 异构集成的信息，请阅读本书第 7 章。

1.12.1　安靠科技 / 高通 / 新光公司的 PoP

安靠科技（Amkor）公司首先研究了基板上带有非导电浆料（TC-NCP）[135]下填的铜柱+焊帽凸点的高结合力热压缩工艺，将其用于组装高通（Qualcomm）骁龙应用处理器，以及三星 Galaxy 智能手机的 PoP 底层封装，如图 1.33 所示。NCP 下填料可以旋涂、用针点胶、真空辅助。模制核心嵌入式封装（MCeP）基板由新光（Shinko）公司制造。

1.12.2　苹果 / 台积电公司的 PoP

图 1.34 是苹果 iPhone 6 Plus 的横截面。可以看到，A9 应用处理器以 PoP 型式封装，焊凸点倒装芯片在 2-2-2 有机封装基板上大规模回流，然后下填料。图 1.35 是苹果 iPhone 7 的横截面。

图 1.35 展示了 PoP 的示意图和扫描电子显微镜（SEM）图像。PoP 包含了苹果 A10 应用处理器和 iPhone7/7 Plus 的移动端动态随机存取存储器（DRAM），由台积电采用 InFO WLP 技术制造[94-96]。从底层封装可以看出，晶圆植球、涂刷助焊剂、清洗、下填料分配与固化以及积层封装基板（见图 1.34A9 应用处理器）已经被取消，并且被 RDL 取代（见图 1.35 A10 应用处理器）。这带来了更少的成本、更好的性能，以及更低的封装截面积。关于 A11 和 A12 应用处理器的封装，可阅读本书第 7 章。

第1章 异构集成综述

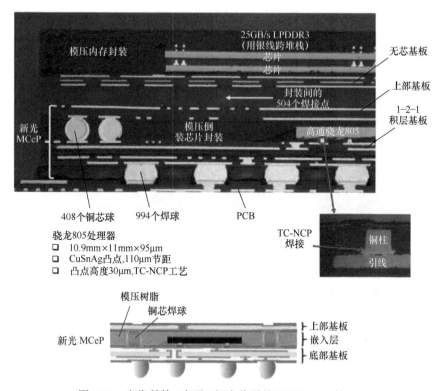

图 1.33 安靠科技 / 高通 / 新光公司基于 TC-NCP 的 PoP

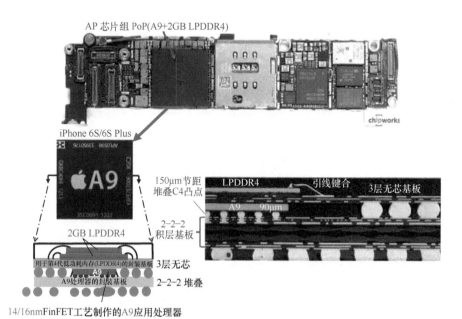

图 1.34 苹果在基板上焊凸点倒装芯片的 PoP，用于 A9 应用处理器

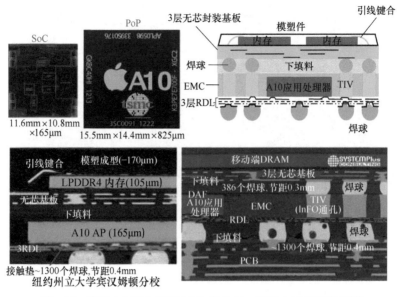

图 1.35 苹果/台积电利用 InFO 的 PoP，用于 A10 应用处理器

1.12.3 三星公司用于智能手表的 PoP

图 1.36 展示了三星公司于 2018 年 7 月推出的豪华版智能手表，它是一个 PoP。可以看到，顶部安放了内存 ePoP，由 2 个 DRAM、2 个 NAND、1 个控制器组成。底部安放了 AP 和电源管理集成芯片（PMIC），在 3 层有机基板的空腔中通过扇出面板级封装。在顶部和底部的封装之间存在下填料。AP 和 PMIC 芯片的尺寸大约是 3mm×3mm。

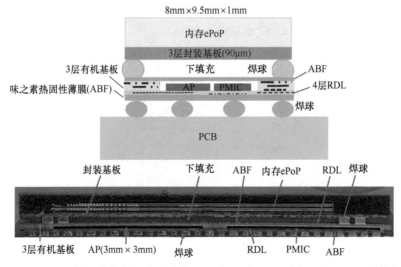

图 1.36 三星用于豪华版智能手表的 PoP（AP 和 PMIC 用三星 FOPLP 封装）

1.13 内存堆栈的异构集成

1.13.1 利用引线键合的内存芯片异构集成

图 1.37 展示了 2 个内存芯片和 1 个逻辑芯片的异构集成[136]。它们通过引线键合法堆叠。本书第 8 章将展示其可装配性和可靠性。

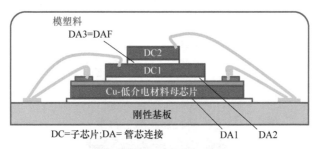

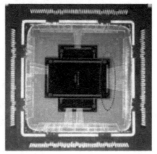

图 1.37　IME 利用引线键合法实现的 3 芯片异构集成

1.13.2 利用低温键合的内存芯片异构集成

图 1.38 展示了 2 个内存芯片和 1 个逻辑芯片的异构集成。它们通过低温键合实现堆叠[137]。其可装配性与可靠性将在第 8 章展示。

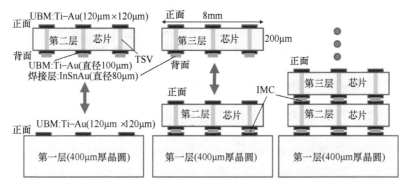

图 1.38　IME 利用低温键合法实现的若干芯片异构集成

图 1.38　IME 利用低温键合法实现的若干芯片异构集成（续）

1.14　芯片堆叠的异构集成

1.14.1　英特尔公司用于 iPhone XR 的调制解调器芯片组

图 1.39 展示了英特尔制造的调制解调器芯片组，是用于 iPhone XR 的第二重要的芯片组。可以看出，基带应用处理器是 3 层 ETS（嵌入式线路基板）上的焊凸点倒装芯片。DRAM 固定在 AP 的背面，并被导线连接到 ETS 上。这是一个芯片背对背层叠的例子。

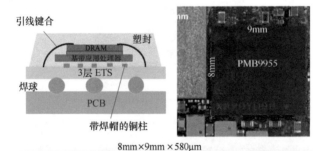

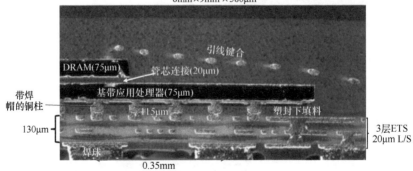

图 1.39　英特尔调制解调器芯片组的异构集成（基带 AP 与 DRAM）

1.14.2　IME 基于 TSV 的芯片堆叠

图 1.40 展示了 IME 芯片面对面堆叠的异构集成[138, 139]。顶部芯片可以是一个内存，底部芯片可以是一个带有（TSV）的逻辑芯片。其可装配性与特点可

参阅本书第9章。

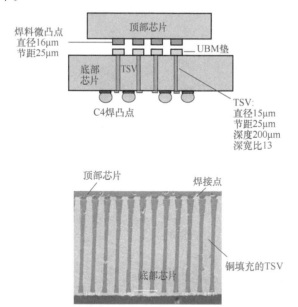

图 1.40　IME 芯片面对面堆叠（利用 TSV）的异构集成

1.14.3　IME 无 TSV 的芯片堆叠

图 1.41 展示了 IME 无 TSV 芯片面对面层叠的异构集成[140, 141]。与基底之间的连接通过更大的（母）芯片上的焊球实现。获取更多可装配性、可靠性信息可参阅本书第 9 章。

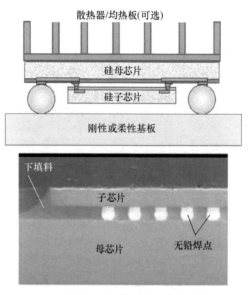

图 1.41　IME 芯片面对面堆叠的异构集成（无 TSV）

1.15 CMOS 图像传感器（CIS）的异构集成

1.15.1 索尼公司 CIS 的异构集成

索尼（Sony）公司是第一个在 HVM 中使用 Cu-Cu 直接混合键合（将金属焊盘和电介质层同时键合连接在晶圆片两侧）的公司。索尼公司生产了 IMX260 背面发光 CMOS 图像传感器（BI-CIS）用于三星 Galaxy S7。三星 Galaxy S7 于 2016 年上市。电学测试结果[142]表明，稳固的铜-铜直接混合键合实现了显著的连通性和可靠性，图像传感器的性能也很好。IMX260 BI-CIS 的一个横截面如图 1.42 所示。可以看出，与参考文献 [143] 中的索尼 ISX014 层叠相机传感器[144]不同，硅通孔消失了，并且 BI-CIS 芯片与处理器芯片的连接通过铜-铜直接键合实现。来自封装基板的信号通过键合引线传递到处理器芯片的边缘。获取更多信息请参阅本书第 10 章。

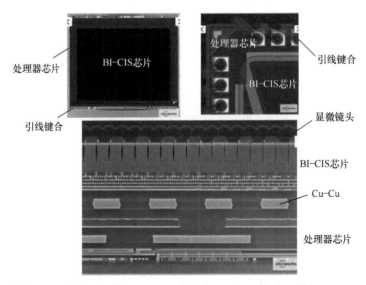

图 1.42　索尼基于铜-铜混合键合实现 CIS 异构集成的横截面图片

1.15.2 意法半导体公司 CIS 的异构集成

图 1.43 展示了意法半导体（ST）公司的 3D CIS 与 IC 集成[144]。它由 CIS、协处理器 IC、玻璃载板组成。CIS 有 80 个 I/O 位，而 IC 的 I/O 位为 164。CIS 与协处理器的尺寸并不相同。CIS 尺寸为 5mm×4.4mm，IC 尺寸为 3.4mm×3.5mm。

IC 与 CIS 通过面靠背连接，通过带有 SnAg 焊帽的铜柱实现。TSV 在 CIS 内部，通过焊凸点与 RDL 与基板相连。图 1.43 展示了装配过程中的 3D 原型（首先球化），包括安装好的 IC 协处理器与未处理的区域。获取更多信息可参阅本书第 10 章。

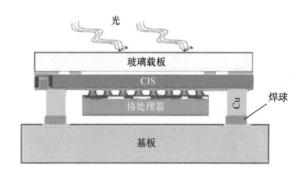

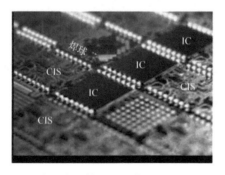

图 1.43　意法半导体（面对背）的 CIS 异构集成

1.16　LED 的异构集成

1.16.1　中国香港科技大学的 LED 异构集成

图 1.44 展示了香港科技大学（HKUST）带有用于荧光粉印制的空腔和用于互连 LED 的铜填充 TSV 的硅基板俯视图与横截面视图[145, 146]。可以看出，硅基板的厚度约为 400μm，两侧为 3μm 厚的低温氧化物。腔体尺寸为 1.3mm × 1.3mm × 0.22mm。

TSV 的直径为 100μm 并被铜填充。铜 TSV 的暴露尖端为 30μm 并用焊料电镀。图 1.44a~c 分别展示了 LED 封装的俯视图、截面图。可以清晰地看到硅基板、RDL、空腔、带有焊凸点的 TSV 尖端（在 LED 安装前）。更多信息可参见本书第 10 章。

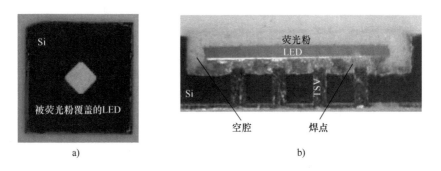

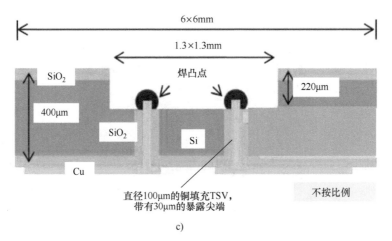

图 1.44 中国香港科技大学在 TSV 转接板上的 LED 异构集成

1.16.2 江阴长电先进封装有限公司的 LED 异构封装

图 1.45 上图示意了江阴长电先进封装有限公司（JCAP）硅基板空腔内带有 LED 器件的晶圆级封装[147, 148]。可以看到，硅基板的顶部有容纳 LED 器件的空腔，底部有连接 LED 和 RDL 的 TSV。硅基板上覆盖着一层黄色荧光粉玻璃。图 1.45a~d 为不含荧光粉的 LED 封装界面的扫描电子显微镜图像。可以看出 LED 器件连接在硅基板空腔的底部；玻璃连接在硅基板的上方；LED 板通过 TSV 连接到 RDL，接触器的大小约为 20μm；这种封装将硅基板从 RDL 中分离出来，从而防止阴极和阳极之间发生短路。更多信息请参阅本书第 10 章。

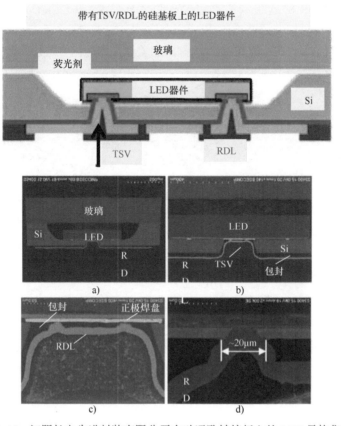

图 1.45　江阴长电先进封装有限公司在硅通孔转接板上的 LED 异构集成

1.17　MEMS 的异构集成

1.17.1　IME 的 MEMS 异构集成

图 1.46 为 IME RF-MEMS 器件晶圆级封装的横截面和俯视图[149, 150]。它由 MEMS 晶圆（RF-MEMS 器件，具有 RDL 的高电阻率硅（HR-Si）基板和用于密封的 AuSn 密封环和键合焊盘）和盖帽晶圆（带有空腔、TSV 和 RDL 的帽，以及焊凸点）构成。它是一个 2.25D MEMS 和 IC 集成。文献 [149, 150] 的主要目标是研究 RF-MEMS 晶圆在 TSV 盖帽晶圆封装过程中的插入损耗。图 1.46 两种不同类型的共面波导（CPW）结构（1mm CPW 和 2mm CPW）如被设计和制造。更多信息可阅读本书第 10 章。

1.17.2　IMEC 的 MEMS 异构集成

图 1.47 显示了 IMEC RF-MEMS 器件的零级封装[151, 152]。它由 RF-MEMS、

带有 RDL 的 HR-Si 基板和用于密封环和键合焊盘的 Cu-Sn-Cu、带有 TSV 和 RDL 的 HR-Si 帽以及焊凸点组成。同样，它是 2.25D MEMS 封装。文献 [151，152] 的主要目标之一是研究 MEMS 晶圆和盖帽晶圆的键合特性（密封环和互连凸点）。

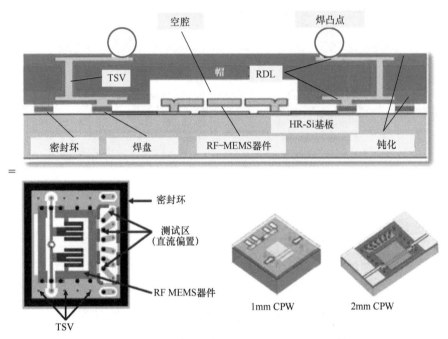

图 1.46　IME 具有横向电馈通的 3D MEMS 封装

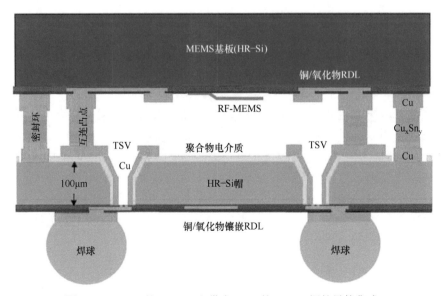

图 1.47　IMEC 的 MEMS 和带有 TSV 的 HR-Si 帽的异构集成

1.17.3 德国弗劳恩霍夫 IZM 研究所的 MEMS 异构集成

图 1.48 示意性地展示了德国弗劳恩霍夫 IZM 研究所基于转接板晶圆的 MEMS 封装[153]。可以看出，MEMS 器件连接到带有铜填充的 TSV 和 RDL 的硅转接板晶圆，并使用带腔的盖帽晶圆进行气密密封。这是一个 2.5D MEMS 和 IC 集成。最终装配的典型横截面图像如图 1.48 所示。可以看出，MEMS 封装的所有关键元素，如 MEMS 器件、硅转接板、TSV、微凸点和密封环等都在适当的位置。

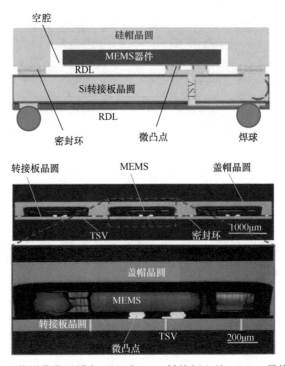

图 1.48　德国弗劳恩霍夫 IZM 在 TSV 转接板上的 MEMS 异构集成

1.17.4 美国帝时华公司的 MEMS 异构集成

美国帝时华公司（Discera）生产了带有 TSV 的 MEMS 谐振器，如图 1.49 所示。可以看出，MEMS 谐振器位于 ASIC 的正上方，ASIC 和 TSV 之间的连接是通过引线键合，以及 TSV 位于 MEMS 谐振器的硅基板中。

1.17.5 亚德诺半导体技术有限公司的 MEMS 异构集成

图 1.50 显示了 MEMS 器件和亚德诺（Analog Devices）半导体技术有限公司提供的 ASIC 的异构集成。它由芯片（MEMS）到（ASIC）晶圆键合组装而成。MEMS 和 ASIC 之间的互连通过 RDL 和 TSV 实现。

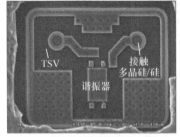 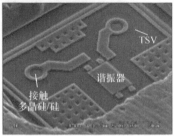

图 1.49 Discera 公司 ASIC 与具有 TSV 的 MEMS 异构集成

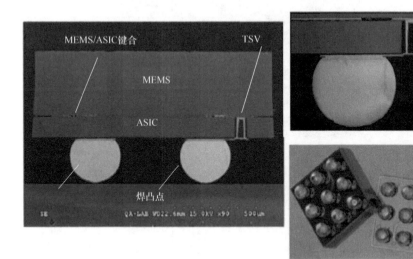

图 1.50 Analog Devices 公司具有 TSV 的 ASIC 与 MEMS 异构集成

1.17.6 IME 在 ASIC 上的 MEMS 异构集成

图 1.51 显示了 IME 的 MEMS 器件在 ASIC 上的异构集成,并由硅帽密封[154]。它是通过芯片(MEMS)到(ASIC)晶圆键合和(盖帽)晶圆到(载有 MEMS

的 ASIC）晶圆键合来组装的。有关 MEMS 封装的组装和特性，可参阅本书第 10 章。

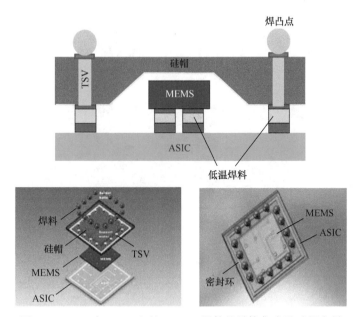

图 1.51　IME 在 ASIC 上的 MEMS 器件的异构集成及硅帽密封

1.17.7　安华高科技公司的 MEMS 异构集成

安华高科技（Avago）公司生产了薄膜体声谐振器（FBAR）MEMS 滤波器，ACMD-7612：UMTS Band I Duplexer，如图 1.52 所示[156, 157]。可以看出帽内有电路，TSV 在 Tx（发射器）芯片和 Rx（接收器）芯片中，TSV 的侧壁被金属化并且 TSV 没有被填充。有关 Avago FBAR MEMS 滤波器的更多详细信息，可参阅本书第 10 章。

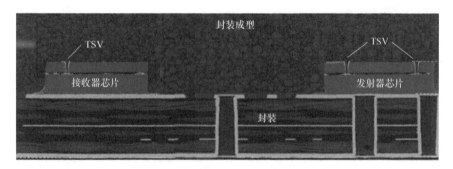

图 1.52　FBAR 密封封装的照片

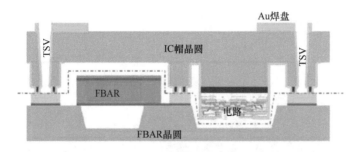

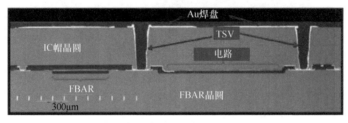

图 1.52　FBAR 密封封装的照片（续）

1.18　VCSEL 的异构集成

1.18.1　IME 的 VCSEL 异构集成

图 1.53 显示了 IME 使用传统 PCB 制造工艺[157-162]的带有嵌入式波导的单通道光电电路板（OECB）。OECB 由嵌入在 60μm 厚 BT 基板下的 4 个电层和 1 个光学层组成。通过形成 2 个直径为 100 μm 的光通孔来引导光束从垂直腔面发射激光器（VCSEL）到 45° 反射镜耦合器。同样，离开波导的光束从 45° 反射镜耦合器通过光通孔并被光电探测器接收。一个 10 厘米长嵌入式聚合物波导由一个 70μm×70μm 核心和一个 15μm 厚的顶部和底部覆层组成。使用 90° 金刚石切割刀片在波导的两个角落形成 2 个 45° 反射镜耦合器。这些反射镜将垂直路径上 VCSEL 发射的光束转换为平面方向，并进入波导。更多信息可参阅本书第 10 章。

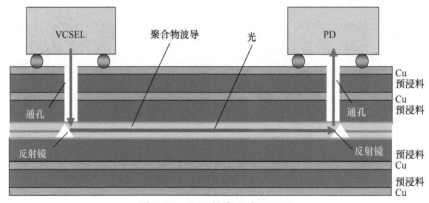

图 1.53　IME 的嵌入式 OECB

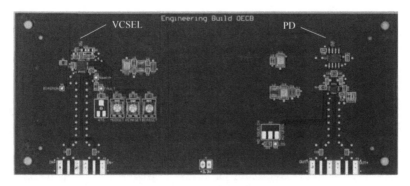

图 1.53　IME 的嵌入式 OECB（续）

1.18.2　中国香港科技大学的 VCSEL 异构集成

图 1.54 显示了香港科技大学（HKUST）用于光电互连的嵌入式混合 3D 异构集成 [75, 163, 164]。可以看出，VCSEL 是用任何材料进行倒装焊接在带有 TSV 的 VCSEL 驱动芯片上，而 TSV 是焊凸点倒装焊接在串行芯片上。当串行芯片是晶圆形式时，带有任何材料的较大凸点都安装在串行芯片上。切割 3D 混合 IC 芯片组后，将其放置在聚合物波导顶部的 PCB（或基板）上。可能需要使用特殊的密封剂（下填料）如透明聚合物来保护芯片组，这时通过热界面材料（TIM）可以将均热板或任何导电材料连接到串行芯片的背面，另外，也可以使用 TIM 将散热器连接到均热板的顶部。同样，光电探测芯片是焊凸点倒装焊接在 TIA 芯片上，而 TIA 芯片是焊凸点倒装焊接在解串器芯片上。热管理封装技术与 VCSEL 芯片组相同。更多信息可参阅本书第 10 章。

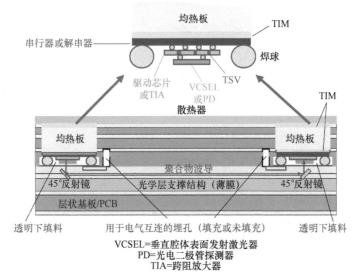

图 1.54　HKUST 的嵌入式 3D 光电互连的异构集成

1.19 总结和建议

一些重要的结果和建议如下：

1）本书中异构集成均被定义为采用封装技术将从不同IC设计公司、代工厂设计生产的不同晶圆尺寸、特征尺寸的不同材料、不同功能的芯片、光电器件、封装芯片集成在一个系统或不同基板的子系统中，也可以独立封装。

2）本书中异构集成分为有机基板上的异构集成、硅基板上的异构集成（TSV转接板）、硅基板（桥）上的异构集成、扇出RDL基板上的异构集成，以及陶瓷基板上的异构集成（本书不详述）。

3）有机基板、硅基板异构集成示例（TSV转接板）、硅基板（桥）和扇出RDL基板相关内容均有被提及。

4）在未来的几年中，无论是在商业化时长、性能、外形尺寸、功耗、信号完整性还是成本方面，更多更高水平的异构集成将涌现出来。

5）芯片到芯片、面对面、面对背、内存堆叠，PoP、AiP、LED、CIS、MEMS和VCSEL等均进行了简要介绍。

参考文献

[1] Martins, A., M. Pinheiro, A. Ferreira, R. Almeida, F. Matos, J. Oliveira, H. Santos, M. Monteiro, H. Gamboa, and R. Silva, "Heterogeneous Integration Challenges Within Wafer Level Fan-Out SiP for Wearables and IoT", *IEEE/ECTC Proceedings*, May 2018, pp. 1485–1492.

[2] Ko, CT, H. Yang, J. H. Lau, M. Li, M. Li, C. Lin, et al., "Chip-First Fan-Out Panel-Level Packaging for Heterogeneous Integration", *IEEE/ECTC Proceedings*, May 2018, pp. 355–363.

[3] Ko, CT, H. Yang, J. H. Lau, M. Li, M. Li, C. Lin, J. W. Lin, T. Chen, I. Xu, C. Chang, J. Pan, H. Wu, Q. Yong, N. Fan, E. Kuah, Z. Li, K. Tan, Y. Cheung, E. Ng, K. Wu, J. Hao, R. Beica, M. Lin, Y. Chen, Z. Cheng, S. Koh, R. Jiang, X. Cao, S. Lim, N. Lee, M. Tao, J. Lo, and R. Lee, "Chip-First Fan-Out Panel-Level Packaging for Heterogeneous Integration", *IEEE Transactions on CPMT*, September 2018, pp. 1561–1572.

[4] Hsu, F., J. Lin, S. Chen, P. Lin, J. Fang, J. Wang, and S. Jeng, "3D Heterogeneous Integration with Multiple Stacking Fan-Out Package", *IEEE/ECTC Proceedings*, May 2018, pp. 337–342.

[5] Lin, Y., S. Wu, W. Shen, S. Huang, T. Kuo, A. Lin, T. Chang, H. Chang, S. Lee, C. Lee, J. Su, X. Liu, Q. Wu, and K. Chen, "An RDL-First Fan-out Wafer Level Package for Heterogeneous Integration Applications", *IEEE/ECTC Proceedings*, May 2018, pp. 349–354.

[6] Lau, J. H., M. Li, M. Li, T. Chen, I. Xu, X. Qing, Z. Cheng, et al., "Fan-Out Wafer-Level Packaging for Heterogeneous Integration", *Proceedings of IEEE/ECTC*, May 2018, pp. 2354–2360.

[7] Lau, J. H., M. Li, M. Li, T. Chen, I. Xu, X. Qing, Z. Cheng, N. Fan, E. Kuah, Z. Li, K. Tan, Y. Cheung, E. Ng, P. Lo, K. Wu, J. Hao, S. Koh, R. Jiang, X. Cao, R. Beica, S. Lim, N. Lee, C. Ko, H. Yang, Y. Chen, M. Tao, J. Lo, and R. Lee, "Fan-Out Wafer-Level Packaging for Heterogeneous Integration", *IEEE Transactions on CPMT*, 2018, September 2018, pp. 1544–1560.

[8] Knickerbocker, J., R. Budd, B. Dang, Q. Chen, E. Colgan, L. W. Hung, S. Kumar, K. W. Lee, M. Lu, J. W. Nah, R. Narayanan, K. Sakuma, V. Siu, and B. Wen, "Heterogeneous Integration Technology Demonstrations for Future Healthcare, IoT, and AI Computing Solutions", *IEEE/ECTC Proceedings*, May 2018, pp. 1519–1522.

[9] Lau, J. H., "Fan-Out Wafer-Level Packaging for 3D IC Heterogeneous Integration", *Proceedings of CSTIC*, March 2018, pp. VII_1–6.
[10] Lau, J. H., "Heterogeneous Integration with Fan-Out Wafer-Level Packaging", *Proceedings of IWLPC*, October 2017, pp. 1–25.
[11] Panigrahi, A., C. Kumar, S. Bonam, B. Paul, T. Ghosh N. Paul, S. Vanjari, and S. Singh, "Metal-Alloy Cu Surface Passivation Leads to High Quality Fine-Pitch Bump-Less Cu-Cu Bonding for 3D IC and Heterogeneous Integration Applications", *IEEE/ECTC Proceedings*, May 2018, pp. 1555–1560.
[12] Faucher-Courchesne, C., D. Danovitch, L. Brault, M. Paquet, and E. Turcotte, "Controlling Underfill Lateral Flow to Improve Component Density in Heterogeneously Integrated Packaging Systems", *IEEE/ECTC Proceedings*, May 2018, pp. 1206–1213.
[13] Lau, J. H., "3D IC Heterogeneous Integration by FOWLP", *Chip Scale Review*, Vol. 22, January/February 2018, pp. 16–21.
[14] Hu, Y., C. Lin, Y. Hsieh, N. Chang, A. J. Gallegos, T. Souza, W. Chen, M. Sheu, C. Chang, C. Chen, K. Chen, "3D Heterogeneous Integration Structure Based on 40 nm- and 0.18 μm-Technology Nodes", *Proceedings of IEEE/ECTC*, May 2015, pp. 1646–1651.
[15] Bajwa, A., S. Jangam, S. Pal, N. Marathe, T. Bai, T. Fukushima, M. Goorsky, and S. S. Iyer, "Heterogeneous Integration at Fine Pitch (≤10 μm) using Thermal Compression Bonding", *IEEE/ECTC Proceedings*, May 2017, pp. 1276–1284.
[16] Dittrich, M., A. Heinig, F. Hopsch, and R. Trieb, "Heterogeneous Interposer Based Integration of Chips with Copper Pillars and C4 Balls to Achieve High Speed Interfaces for ADC Application", *Proceedings of IEEE/ECTC*, May 2017, pp. 643–648.
[17] Chuang, Y., C. Yuan, J. Chen, C. Chen, C. Yang, W. Changchien, C. Liu, and F. Lee, "Unified Methodology for Heterogeneous Integration with CoWoS Technology", *IEEE/ECTC Proceedings*, May 2013, pp. 852–859.
[18] Ko, C., H. Yang, J. H. Lau, M. Li, M. Li, et al., "Design, Materials, Process, and Fabrication of Fan-Out Panel-Level Heterogeneous Integration", *Proceedings of IMAPS Symposium*, October 2018, pp. TP2_1–7.
[19] Lau, J. H., M. Li, Y. Lei, M. Li, I. Xu, T. Chen, Q. Yong, Z. Cheng, et al., "Reliability of Fan-Out Wafer-Level Heterogeneous Integration", *Proceedings of IMAPS Symposium*, October 2018, pp. WA2_1–9.
[20] Beal, A., and R. Dean, "Using SPICE to Model Nonlinearities Resulting from Heterogeneous Integration of Complex Systems", *IMAPS Proceedings*, October 2017, pp. 274–279.
[21] Lau, J. H., M. Li, Y. Lei, M. Li, I. Xu, T. Chen, Q. Yong, Z. Cheng, et al., "Reliability of Fan-Out Wafer-Level Heterogeneous Integration", *IMAPS Transactions, Journal of Microelectronics and Electronic Packaging*, Vol. 15, Issue 4, October 2018, pp. 148–162.
[22] Ko, C. T., H. Yang, and J. H. Lau, "Design, Materials, Process, and Fabrication of Fan-Out Panel-Level Heterogeneous Integration", *IMAPS Transactions, Journal of Microelectronics and Electronic Packaging*, Vol. 15, Issue 4, October 2018, pp. 141–147.
[23] Hanna, A, A. Alam, T, Fukushima, S. Moran, W. Whitehead, S. Jangam, S. Pal, G. Ezhilarasu, R. Irwin, A. Bajwa, and S. Iyer, "Extremely Flexible (1 mm Bending Radius) Biocompatible Heterogeneous Fan-Out Wafer-Level Platform with the Lowest Reported Die-Shift (<6 μm) and Reliable Flexible Cu-Based Interconnects", *IEEE/ECTC Proceedings*, May 2018, pp. 1505–1511.
[24] Kyozuka, M., T. Kiso, H. Toyazaki, K. Tanaka, and T. Koyama, "Development of Thinner POP base Package by Die Embedded and RDL Structure", *IMAPS Proceedings*, October 2017, pp. 715–720.
[25] Yoon, S., J. Caparas, Y. Lin, and P. Marimuthu, "Advanced Low Profile PoP Solution with Embedded Wafer Level PoP (eWLB-PoP) Technology", *IEEE/ECTC Proceedings*, 2012, pp. 1250–1254.
[26] Yoon, S., P. Tang, R. Emigh, Y. Lin, P. Marimuthu, and R. Pendse, "Fanout Flipchip eWLB (Embedded Wafer Level Ball Grid Array) Technology as 2.5D Packaging Solutions", *IEEE/ECTC Proceedings*, 2013, pp. 1855–1860.
[27] Lin, Y., W. Lai, C. Kao, J. Lou, P. Yang, C. Wang, and C. Hseih, "Wafer Warpage Experiments and Simulation for Fan-Out Chip on Substrate", *IEEE/ECTC Proceedings*, May 2016, pp. 13–18.

[28] Lau, J. H., *Fan-Out Wafer-Level Packaging*. Springer Book Company, 2018.
[29] Lau, J. H., et al, "Apparatus Having Thermal-Enhanced and Cost-Effective 3D IC Integration Structure with Through Silicon via Interposer". US Patent No: 8,604,603, Date of Patent: December 10, 2013.
[30] Chiu, C., Z. Qian, and M. Manusharow, "Bridge Interconnect with Air Gap in Package Assembly". US Patent No. 8,872,349, 2014.
[31] Mahajan, R., R. Sankman, N. Patel, D. Kim, K. Aygun, Z. Qian, et al., "Embedded Multi-die Interconnect Bridge (EMIB)—A High-Density, High-Bandwidth Packaging Interconnect", *IEEE/ECTC Proceedings*, May 2016, pp. 557–565.
[32] Suk, K., S. Lee, J. Kim, S. Lee, H. Kim, S. Lee, P. Kim, D. Kim, D. Oh, and J. Byun, "Low Cost Si-less RDL Interposer Package for High Performance Computing Applications", *IEEE/ECTC Proceedings*, May 2018, pp. 64–69.
[33] Podpod, A., J. Slabbekoorn, A. Phommahaxay, F. Duval, A. Salahouedlhadj, M. Gonzalez, K. Rebibis, R. A. Miller, G. Beyer, and E. Beyne, "A Novel Fan-Out Concept for Ultra-High Chip-to-Chip Interconnect Density with 20-μm Pitch", *IEEE/ECTC Proceedings*, May 2018, pp. 370–378.
[34] Lau, J. H., C. Lee, C. Zhan, S. Wu, Y. Chao, M. Dai, R. Tain, H. Chien, et al., "Low-Cost Through-Silicon Hole Interposers for 3D IC Integration", *IEEE Transactions on CPMT*, Vol. 4, No. 9, September 2014, pp. 1407–1419.
[35] Souriau, J., O. Lignier, M. Charrier, and G. Poupon, "Wafer Level Processing Of 3D System in Package for RF and Data Applications", *IEEE/ECTC Proceedings*, 2005, pp. 356–361.
[36] Henry, D., D. Belhachemi, J-C. Souriau, C. Brunet-Manquat, C. Puget, G. Ponthenier, J. Vallejo, C. Lecouvey, and N. Sillon, "Low Electrical Resistance Silicon Through Vias: Technology and Characterization", *IEEE/ECTC Proceedings*, 2006, pp. 1360–1366.
[37] Khan, N., V. Rao, S. Lim, H. We, V. Lee, X. Zhang, E. Liao, R. Nagarajan, T. C. Chai, V. Kripesh, and J. H. Lau, "Development of 3-D Silicon Module With TSV for System in Packaging", *IEEE Proceedings of Electronic, Components & Technology Conference*, Orlando, FL, May 27–30, 2008, pp. 550-555. Also, *IEEE Transactions on CPMT*, Vol. 33, No. 1, March 2010, pp. 3–9.
[38] Lau, J. H., C.-J. Zhan, P.-J. Tzeng, C.-K. Lee, M.-J. Dai, H.-C. Chien, Y.-L. Chao, et al., "Feasibility Study of a 3D IC Integration System-in-Packaging (SiP) from a 300 mm Multi-Project Wafer (MPW)", *IMAPS International Symposium on Microelectronics*, October 2011, pp. 446–454. Also, *IMAPS Transactions, Journal of Microelectronic Packaging*, Vol. 8, No. 4, Fourth Quarter 2011, pp. 171–178.
[39] Zhan, C., P. Tzeng, J. H. Lau, M. Dai, H. Chien1, C. Lee, S. Wu, et al., "Assembly Process and Reliability Assessment of TSV/RDL/IPD Interposer with Multi-Chip-Stacking for 3D IC Integration SiP", *IEEE/ECTC Proceedings*, San Diego, CA, May 2012, pp. 548–554.
[40] Che, F., M. Kawano, M. Ding, Y. Han, and S. Bhattacharya, "Co-design for Low Warpage and High Reliability in Advanced Package with TSV-Free Interposer (TFI)", *Proceedings of IEEE/ECTC*, May 2017, pp. 853–861.
[41] Hou, S., W. Chen, C. Hu, C. Chiu, K. Ting, T. Lin, W. Wei, W. Chiou, V. Lin, V. Chang, C. Wang, C. Wu, and D. Yu, "Wafer-Level Integration of an Advanced Logic-Memory System Through the Second-Generation CoWoS Technology", *IEEE Transactions on Electron Devices*, October 2017, pp. 4071–4077.
[42] Selvanayagam, C., J. H. Lau, X. Zhang, S. Seah, K. Vaidyanathan, and T. Chai, "Nonlinear Thermal Stress/Strain Analysis of Copper Fill TSV (Through Silicon Via) and Their Flip-Chip Microbumps", *IEEE/ECTC Proceedings*, May 27–30, 2008, pp. 1073–1081.
[43] Selvanayagam, C., J. H. Lau, X. Zhang, S. Seah, K. Vaidyanathan, and T. Chai, "Nonlinear Thermal Stress/Strain Analyses of Copper Filled TSV (Through Silicon Via) and Their Flip-Chip Microbumps", *IEEE Transactions on Advanced Packaging*, Vol. 32, No. 4, November 2009, pp. 720–728.
[44] Lau, J. H., and G. Tang, "Thermal Management of 3D IC Integration with TSV (Through Silicon Via)", *IEEE/ECTC Proceedings*, May 2009, pp. 635–640.
[45] Lau, J. H., Y. S. Chan, and R. S. W. Lee, "3D IC Integration with TSV Interposers for High-Performance Applications", *Chip Scale Review*, Vol. 14, No. 5, September/October, 2010, pp. 26–29.
[46] Lau, J. H., "TSV Manufacturing Yield and Hidden Costs for 3D IC Integration", *IEEE/ECTC Proceedings,* May 2010, pp. 1031–1041.

[47] Zhang, X., T. Chai, J. H. Lau, C. Selvanayagam, K. Biswas, S. Liu, D. Pinjala, et al., "Development of Through Silicon Via (TSV) Interposer Technology for Large Die (21 × 21 mm) Fine-pitch Cu/low-k FCBGA Package", *IEEE Proceedings of ECTC*, May, 2009, pp. 305–312.

[48] Chai, T. C., X. Zhang, J. H. Lau, C. S. Selvanayagam, D. Pinjala, et al., "Development of Large Die Fine-Pitch Cu/low-*k* FCBGA Package with Through Silicon Via (TSV) Interposer", *IEEE Transactions on CPMT*, Vol. 1, No. 5, May 2011, pp. 660–672.

[49] Chien, H. C., J. H. Lau, Y. Chao, R. Tain, M. Dai, S. T. Wu, W. Lo, and M. J. Kao, "Thermal Performance of 3D IC Integration with Through-Silicon Via (TSV)", *IMAPS Transactions, Journal of Microelectronic Packaging*, Vol. 9, 2012, pp. 97–103.

[50] Chaware, R., K. Nagarajan, and S. Ramalingam, "Assembly and Reliability Challenges in 3D Integration of 28 nm FPGA Die on a Large High-Density 65 nm Passive Interposer", *IEEE/ECTC Proceedings*, May 2012, pp. 279–283.

[51] Banijamali, B., S. Ramalingam, K. Nagarajan, and R. Chaware, "Advanced Reliability Study of TSV Interposers and Interconnects for the 28 nm Technology FPGA", *IEEE/ECTC Proceedings*, May 2011, pp. 285–290.

[52] Banijamali, B., S. Ramalingam, H. Liu, and M. Kim, "Outstanding and Innovative Reliability Study of 3D TSV Interposer and Fine-Pitch Solder Micro-Bumps", *IEEE/ECTC Proceedings*, May 2012, pp. 309–314.

[53] Banijamali, B., C. Chiu, C. Hsieh, T. Lin, C. Hu, S. Hou, et al., "Reliability Evaluation of a CoWoS-Enabled 3D IC Package", *IEEE/ECTC Proceedings*, May 2013, pp. 35–40.

[54] Xie, J., H. Shi, Y. Li, Z. Li, A. Rahman, K. Chandrasekar, et al., "Enabling the 2.5D Integration", *Proceedings of IMAPS International Symposium on Microelectronics*, October 2012, pp. 254–267.

[55] Li, L., P. Su, J. Xue, M. Brillhart, J. H. Lau, P. Tzeng, C. Lee, C. Zhan, et al., "Addressing Bandwidth Challenges in Next Generation High Performance Network Systems with 3D IC Integration", *IEEE/ECTC Proceedings*, May 2012, pp. 1040–1046.

[56] Lau, J. H., P. Tzeng, C. Zhan, C. Lee, M. Dai, J. Chen, Y. Hsin, et al., "Large Size Silicon Interposer and 3D IC Integration for System-in-Packaging (SiP)", *Proceedings of the 45th IMAPS International Symposium on Microelectronics*, September 2012, pp. 1209–1214.

[57] Wu, S. T., J. H. Lau, H. Chien, Y. Chao, R. Tain, L. Li, P. Su, et al., "Thermal Stress and Creep Strain Analyses of a 3D IC Integration SiP with Passive Interposer for Network System Application", *Proceedings of the 45th IMAPS International Symposium on Microelectronics*, September 2012, pp. 1038–1045.

[58] Chien, H., J. H. Lau, T. Chao, M. Dai, and R. Tain, "Thermal Management of Moore's Law Chips on Both sides of an Interposer for 3D IC integration SiP", *IEEE ICEP Proceedings*, Japan, April 2012, pp. 38–44.

[59] Chien, H., J. H. Lau, T. Chao, M. Dai, R. Tain, L. Li, P. Su, et al., "Thermal Evaluation and Analyses of 3D IC Integration SiP with TSVs for Network System Applications", *IEEE/ECTC Proceedings*, San Diego, CA, May 2012, pp. 1866–1873.

[60] Ji, M., M. Li, J. Cline, D. Seeker, K. Cai, J. H. Lau, P. Tzeng, et al., "3D Si Interposer Design and Electrical Performance Study", *Proceedings of DesignCon*, Santa Clara, CA, January 2013, pp. 1–23.

[61] Wu, S. T., H. Chien, J. H. Lau, M. Li, J. Cline, and M. Ji, "Thermal and Mechanical Design and Analysis of 3D IC Interposer with Double-Sided Active Chips", *IEEE/ECTC Proceedings*, Las Vegas, NA, May 2013, pp. 1471–1479.

[62] Tzeng, P. J., J. H. Lau, C. Zhan, Y. Hsin, P. Chang, Y. Chang, J. Chen, et al., "Process Integration of 3D Si Interposer with Double-Sided Active Chip Attachments", *IEEE/ECTC Proceedings*, Las Vegas, NA, May 2013, pp. 86–93.

[63] Stow, D., Y. Xie, T. Siddiqua, and G. H. Loh, "Cost-Effective Design of Scalable High-Performance Systems Using Active and Passive Interposers", *Proceedings of IEEE/ACM International Conference on Computer-Aided Design*, November 2017, pp. 728–735.

[64] Hwang, T., D. Oh, E. Song, K. Kim, J. Kim, and S. Lee, "Study of Advanced Fan-Out Packages for Mobile Applications", *IEEE/ECTC Proceedings*, May 2018, pp. 343–348.

[65] Hong, J., K. Choi, D. Oh, S Park, S. Shao, H. Wang, Y. Niu, and V. Pham, "Design Guideline of 2.5D Package with Emphasis on Warpage Control and Thermal Management", *IEEE/ECTC Proceedings*, May 2018, pp. 682–692.

[66] You, S., S. Jeon, D. Oh, K. Kim, J. Kim, S. Cha, and G. Kim, "Advanced Fan-Out Package SI/PI/Thermal Performance Analysis of Novel RDL Packages", *IEEE/ECTC Proceedings*, May 2018, pp. 1295–1301.

[67] Miao, M., L. Wang, T. Chen, X. Duan, J. Zhang, N. Li, L. Sun, R. Fang, X. Sun, H. Liu, and Y. Jin, "Modeling and Design of a 3D Interconnect Based Circuit Cell Formed with 3D SiP Techniques Mimicking Brain Neurons for Neuromorphic Computing Applications", *IEEE/ECTC Proceedings*, May 2018, pp. 490–497.

[68] Borel, S., L. Duperrex, E. Deschaseaux, J. Charbonnier, J. Cledière, R. Wacquez, J. Fournier, J.-C. Souriau, G. Simon, and A. Merle, "A Novel Structure for Backside Protection against Physical Attacks on Secure Chips or SiP", *IEEE/ECTC Proceedings*, May 2018, pp. 515–520.

[69] Lee, E., M. Amir, S. Sivapurapu, C. Pardue, H. Torun, M. Bellaredj, M. Swaminathan, and S. Mukhopadhyay, "A System-in-Package Based Energy Harvesting for IoT Devices with Integrated Voltage Regulators and Embedded Inductors", *IEEE/ECTC Proceedings*, May 2018, pp. 1720–1725.

[70] Li, J., S. Ma, H. Liu, Y. Guan, J. Chen, Y. Jin, W. Wang, L. Hu, and S. He, "Design, Fabrication and Characterization of TSV Interposer Integrated 3D Capacitor for SiP Applications", *IEEE/ECTC Proceedings*, May 2018, pp. 1968–1974.

[71] Ki, W., W. Lee, I. Lee, I. Mok, W. Do, M. Kolbehdari, A. Copia, S. Jayaraman, C. Zwenger, and K. Lee, "Chip Stackable, Ultra-thin, High-Flexibility 3D FOWLP (3D SWIFT® Technology) for Hetero-Integrated Advanced 3D WL-SiP", *IEEE/ECTC Proceedings*, May 2018, pp. 580–586.

[72] Lee, J., C. Lee, C. Kim, and S. Kalchuri, "Micro Bump System for 2nd Generation Silicon Interposer with GPU and High Bandwidth Memory (HBM) Concurrent Integration", *IEEE/ECTC Proceedings*, May 2018, pp. 607–612.

[73] Lim, Y., X. Xiao, R. Vempati, S. Nandar, K. Aditya, S. Gaurav, T. Lim, V. Kripesh, J. Shi, J. H. Lau, and S. Liu, "High Quality and Low Loss Millimeter Wave Passives Demonstrated to 77-GHz for SiP Technologies Using Embedded Wafer-Level Packaging Platform (EMWLP)", *IEEE Transactions on Advanced Packaging*, Vol. 33, 2010, pp. 1061–1071.

[74] Manessis, D., L. Boettcher, A. Ostmann, R. Aschenbrenner, and H. Reichl, "Chip Embedding Technology Developments Leading to the Emergence of Miniaturized System-in-Packages", *Proceedings of IEEE/ECTC*, May 2010, pp. 803–810.

[75] Lau, J. H., M. S. Zhang, and S. W. R. Lee, "Embedded 3D Hybrid IC Integration System-in-Package (SiP) for Opto-Electronic Interconnects in Organic Substrates", *ASME Transactions, Journal of Electronic Packaging*, Vol. 133, September 2011, pp. 1–7.

[76] Lau, J. H., C.-J. Zhan, P.-J. Tzeng, C.-K. Lee, M.-J. Dai, H.-C. Chien, et al., "Feasibility Study of a 3D IC Integration System-in-Packaging (SiP) from a 300 mm Multi-Project Wafer (MPW)", *IMAPS Transactions, Journal of Microelectronic Packaging*, Vol. 8, No. 4, Fourth Quarter 2011, pp. 171–178.

[77] Lau, J. H., and G. Y. Tang, "Effects of TSVs (through-silicon vias) on Thermal Performances of 3D IC Integration System-in-Package (SiP)", *Journal of Microelectronics Reliability*, Vol. 52, Issue 11, November 2012, pp. 2660–2669.

[78] Ahmad, M., M. Nagar, W. Xie, M. Jimarez, and C. Ryu, "Ultra Large System-in-Package (SiP) Module and Novel Packaging Solution for Networking Applications", *Proceedings of IEEE/ECTC*, May 2013, pp. 694–701.

[79] Wu, H., D. S. Gardner, C. Lv, Z. Zou, and H. Yu, "Integration of Magnetic Materials into Package RF and Power Inductors on Organic Substrates for System in Package (SiP) Applications", *Proceedings of IEEE/ECTC*, May 2014, pp. 1290–1295.

[80] Qian, R., and Y. Liu, "Modeling for Reliability of Ultra-Thin Chips in a System in Package", *Proceedings of IEEE/ECTC*, May 2014, pp. 2063–2068.

[81] Hsieh, C., C. Tsai, H. Lee, T. Lee, H. Chang, "Fan-out Technologies for WiFi SiP Module Packaging and Electrical Performance Simulation", *Proceedings of IEEE/ECTC*, May 2015, pp. 1664–1669.

[82] Li, L., P. Chia, P. Ton, M. Nagar, S. Patil, J. Xue, J. DeLaCruz, M. Voicu, J. Hellings, B. Isaacson, M. Coor, and R. Havens, "3D SiP with organic interposer of ASIC and memory integration", *Proceedings of IEEE/ECTC*, May 2016, pp. 1445–1450.

[83] Tsai, M., A. Lan, C. Shih, T. Huang, R. Chiu, S. L. Chung, J. Y. Chen, F. Chu, C. Chang,

S. Yang, D. Chen, and N. Kao, "Alternative 3D Small Form Factor Methodology of System in Package for IoT and Wearable Devices Application", *Proceedings of IEEE/ECTC*, May 2017, pp. 1541–1546.

[84] Das, R., F. Egitto, S. Rosser, E. Kopp, B. Bonitz, and R. Rai, "3D Integration of System-in-Package (SiP) using Organic Interposer: Toward SiP-Interposer-SiP for High-End Electronics", *IMAPS Proceedings*, September 2013, pp. 531–537.

[85] Chien, H., C. Chien, M. Dai, R. Tain, W. Lo, Y. Lu, "Thermal Characteristic and Performance of the Glass Interposer with TGVs (Through-Glass Via)", *IMAPS Proceedings*, September 2013, pp. 611–617.

[86] Vincent, M., D. Mitchell, J. Wright, Y. Foong, A. Magnus, Z. Gong, S. Hayes, and N. Chhabra, "3D RCP Package Stacking: Side Connect, An Emerging Technology for Systems Integration and Volumetric Efficiency", *IMAPS Proceedings*, September 2013, pp. 447–451.

[87] Renaud-Bezot, N., "Size-Matters—Embedding as an Enabler of Next-Generation SiPs", *IMAPS Proceedings*, September 2013, pp. 740–744.

[88] Couderc, P., Noiray, J., and C. Val, "Stacking of Known Good Rebuilt Wafers for High Performance Memory and SiP", *IMAPS Proceedings*, September 2013, pp. 804–809.

[89] Lim, J., and V. Pandey, "Innovative Integration Solutions for SiP Packages Using Fan-Out Wafer Level eWLB Technology", *IMAPS Proceedings*, October 2017, pp. 263–269.

[90] Becker, K., M. Minkus, J. Pauls, V. Bader, S. Voges, T. Braun, G. Jungmann, H. Wieser, M. Schneider-Ramelow, and K.-D., "Non-Destructive Testing for System-in-Package Integrity Analysis", *IMAPS Proceedings*, October 2017, pp. 182–187.

[91] Lee, Y., and D. Link, "Practical Application and Analysis of Lead-Free Solder on Chip-On-Flip-Chip SiP for Hearing Aids", *IMAPS Proceedings*, October 2017, pp. 201–207.

[92] Milton, B., O. Kwon, C. Huynh, I. Qin, and B. Chylak, "Wire Bonding Looping Solutions for High Density System-in-Package (SiP)", *IMAPS Proceedings*, October 2017, pp. 426–431.

[93] Morard, A., J. Riou, and G. Pares, "Flip Chip Reliability and Design Rules for SiP Module", *IMAPS Proceedings*, October 2017, pp. 754–760.

[94] Yu, D., "Wafer-Level System Integration (WLSI) Technologies for 2D and 3D System-in-Package", *SEMIEUROPE* 2014.

[95] Lin, J., J. Hung, N. Liu, Y. Mao, W. Shih, and T. Tung, "Packaged Semiconductor Device With a Molding Compound and a Method of Forming the same". US Patent 9,000,584, Filed on December 28, 2011, Patented on April 7, 2015.

[96] Tseng, C., Liu, C., Wu, C., and D. Yu, "InFO (Wafer Level Integrated Fan-Out) Technology", *IEEE/ECTC Proceedings*, 2016, pp. 1–6.

[97] Lau, J. H., "TSV-Less Interposers", *Chip Scale Review*, Vol. 20, September/October 2016, pp. 28–35.

[98] Lau, J. H., *Chip On Board Technologies for Multichip Modules*. Van Nostrand Reinhold, New York, March 1994.

[99] Lau, J. H., *3D IC Integration and Packaging*. McGraw-Hill, New York, 2016.

[100] Lau, J. H., *Through-Silicon Via (TSV) for 3D Integration*. McGraw-Hill, New York, 2013.

[101] Lau, J. H., "Semiconductor and Packaging for Internet of Things", *Chip Scale Review*, Vol. 19, May/June 2015, pp. 25–30.

[102] Moore, G., "Cramming More Components onto Integrated Circuits", *Electronics*, Vol. 38, No. 8, April 19, 1965. Reprinted in *IEEE Solid-State Circuits Newsletter*, Vol. 11, No. 3, October 2006, pp. 33–35.

[103] Hubner, M., and J. Becker, *Multiporcessor System-on-Chip*. Springer, 2011.

[104] Lin, S., C. Hsu, Y. Hsu, F. Han, D. Ho, W. Wu, C. Chen, "De-sensitization Design and Analysis for Highly Integrated RFSoC and DRAM Stacked-Die Design", *IEEE/ECTC Proceedings*, May 2018, pp. 1310–1317.

[105] Mi, M., M. Moallem, J. Chen, M. Li, and R. Murugan, "Package Co-Design of a Fully Integrated Multimode 76–81 GHz 45 nm RFCMOS FMCW Automotive Radar Transceiver", *IEEE/ECTC Proceedings*, May 2018, pp. 1054–1061.

[106] Jokerst, N. M., "Hybrid Integrated Optoelectronics: Thin Film Devices Bonded to Host Substrates", *International Journal of High-Speed Electronics and Systems*, Vol. 8, No. 2, pp. 325–356, 1997.

[107] Vrazel, M., J. Chang, I. Song, K. Chung, M. Brooke, N. Jokerst, A. Brown, D. Wills, "Highly

Alignment Tolerant InGaAs Inverted MSM Photodetector Heterogeneously Integrated on a Differential Si CMOS Receiver Operating at 1 Gbps", *IEEE/ECTC Proceedings*, May 2001, pp. 1–6.

[108] Jokerst, N. M., M. A. Brooke, S. Cho, and S. Wilkinson, M. Vrazel, S. Fike, J. Tabler, Y. Joo, S. Seo, D. Wills, and A. Brown, "The Heterogeneous Integration of Optical Interconnections Into Integrated Microsystems", *IEEE Journal of Selected Topic in Quantum Electronics*, Vol. 9, No. 2, March 2003, pp. 350–360.

[109] Lau, J. H., P. Tzeng, C. Lee, C. Zhan, M. Li, J. Cline, et al., "Redistribution Layers (RDLs) for 2.5D/3D IC Integration", *Proceedings of IMAPS Symposium*, 2013, pp. 434–441.

[110] Lau, J. H., P. Tzeng, C. Lee, C. Zhan, M. Li, J. Cline, et al., "Redistribution Layers (RDLs) for 2.5D/3D IC Integration", *IMAPS Transactions, Journal of Microelectronic Packaging*, Vol. 11, No. 1, First Quarter 2014, pp. 16–24.

[111] Lau, J. H., M. Li, N. Fan, E. Kuah, Z. Li, K. Tan, T. Chen, et al., "Fan-Out Wafer-Level Packaging (FOWLP) of Large Chip with Multiple Redistribution-Layers (RDLs)", *Proceedings of IMAPS Symposium*, 2017, pp. 576–583.

[112] Lau, J. H., M. Li, N. Fan, E. Kuah, Z. Li, K. Tan, T. Chen, et al., "Fan-Out Wafer-Level Packaging (FOWLP) of Large Chip with Multiple Redistribution-Layers (RDLs)", *IMAPS Transactions Journal of Microelectronics and Electronic Packaging*, October 2017, pp. 123–131.

[113] Lau, J. H., M. Li, Q. Li, I. Xu, T. Chen, Z. Li, K. Tan, X. Qing, C. Zhang, K. Wee, R. Beica, C. Ko, S. Lim, N. Fan, E. Kuah, K. Wu, Y. Cheung, E. Ng, X. Cao, J. Ran, H. Yang, Y. Chen, N. Lee, M. Tao, J. Lo, and R. Lee, "Design, Materials, Process, and Fabrication of Fan-Out Wafer-Level Packaging", *IEEE Transactions on CPMT*, June 2018, pp. 991–1002.

[114] Li, M., Q. Li, J. H. Lau, N. Fan, E. Kuah, K. Wu, et al., "Characterizations of Fan-Out Wafer-Level Packaging", *Proceedings of IMAPS Symposium*, October 2017, pp. 557–562.

[115] Lim, S., Y. Liu, J. H. Lau, M. Li, "Challenges of ball-attach process using Flux for Fan-Out Wafer/Panel Level (FOWL/PLP) Packaging", *Proceedings of IWLPC*, October 2017, pp. S10_P3_1–7.

[116] Kuah, E., W. Chan, J. Hao, N. Fan, M. Li, J. H. Lau, K. Wu, et al., "Dispensing Challenges of Large Format Packaging and Some of its Possible Solutions", *IEEE/EPTC Proceedings*, December 2017, pp. S27_1–6.

[117] Hua, X., H. Xu, Z. Li, D. Chen, K. Tan, J. H. Lau, M. Li, et al., "Development of Chip-First and Die-Up Fan-Out Wafer-Level Packaging", *IEEE/EPTC Proceedings*, December 2017, pp. S23_1–6.

[118] Lau, J. H., M. Li, Y. Lei, M. Li, Q. Yong, Z. Cheng, T. Chen, I. Xu, et al., "Reliability of FOWLP with Large Chips and Multiple RDLs", *IEEE/ECTC Proceedings*, May 2018, pp. 1568–1576.

[119] Ma, S., J. Wang, F. Zhen, Z. Xiao, T. Wang, and D. Yu, "Embedded Silicon Fan-Out (eSiFO): A Promising Wafer Level Packaging Technology for Multi-chip and 3D System Integration", *IEEE/ECTC Proceedings*, May 2018, pp. 1493–1498.

[120] Chang, P., C. Hsieh, C. Chang, C. Chung, and C. Chiang, "Signal and Power Integrity Analysis of InFO Interconnect for Networking Application", *IEEE/ECTC Proceedings*, May 2018, pp. 1714–1719.

[121] Yu, C. K., W. S. Chiang, P. S. Huang, M. Z. Lin, Y. H. Fang, M. J. Lin, C. Peng, B. Lin, and M. Huang, "Reliability Study of Large Fan-Out BGA Solution on FinFET Process", *IEEE/ECTC Proceedings*, May 2018, pp. 1617–1621.

[122] Ravichandran, S., S. Yamada, G. Park, H. Chen, T. Shi, C. Buch, F. Liu, V. Smet, V. Sundaram, and R. Tummala, "2.5D Glass Panel Embedded (GPE) Packages with Better I/O Density, Performance, Cost and Reliability than Current Silicon Interposers and High-Density Fan-Out Packages", *IEEE/ECTC Proceedings*, May 2018, pp. 625–630.

[123] Kim, J., I. Choi, J. Park, J. Lee, T. Jeong, J. Byun, Y. Ko, K. Hur, D. Kim, and K. Oh, "Fan-Out Panel Level Package with Fine Pitch Pattern", *IEEE/ECTC Proceedings*, May 2018, pp. 52–57.

[124] Braun, T., K.-F. Becker, O. Hoelck, R. Kahle, M. Wöhrmann, L. Boettcher, M. Töpper, L. Stobbe, H. Zedel, R. Aschenbrenner, S. Voges, M. Schneider-Ramelow, and K.-D. Lang, "Panel Level Packaging—A View along the Process Chain", *IEEE/ECTC Proceedings*, May 2018, pp. 70–78.

[125] Lee, C., J. Su, X. Liu, Q. Wu, J. Lin, P. Lin, C. Ko, Y. Chen, W. Shen, T. Kou, S. Huang, A. Lin,

Y. Lin, and K. Chen, "Optimization of Laser Release Process for Throughput Enhancement of Fan-Out Wafer-Level Packaging", *IEEE/ECTC Proceedings*, May 2018, pp. 1818–1823.

[126] Braun, T., S. Voges, M. Töpper, M. Wilke, M. Wöhrmann, U. Maaß, M. Huhn, K.-F. Becker, S. Raatz, J.-U. Kim, R. Aschenbrenner, K.-D. Lang, C. O'Connor, R. Barr, J. Calvert, M. Gallagher, E. Iagodkine, T. Aoude, and A. Politis, "Material and Process Trends for Moving From FOWLP to FOPLP", *IEEE/EPTC Proceedings*, December 2015, pp. 424–429.

[127] Braun, T., S. Raatz, U. Maass, M. Dijk, H. Walter, O. Hölck, K.-F. Becker, M. Töpper, R. Aschenbrenner, M. Wöhrmann, S. Voges, M. Huhn, K.-D. Lang, M. Wietstruck, R. Scholz, A. Mai, M. Kaynak, "Development of a Multi-Project Fan-Out Wafer Level Packaging Platform", *IEEE/ECTC Proceedings*, May 2017, pp. 1–7.

[128] Braun, T., K.-F. Becker, S. Raatz, M. Minkus, V. Bader, J. Bauer, R. Aschenbrenner, R. Kahle, L. Georgi, S. Voges, M. Wöhrmann, K.-D. Lang, "Foldable Fan-Out Wafer Level Packaging", *IEEE/ECTC Proceedings*, May 2016, pp. 19–24.

[129] Cardoso, A., R. Pinto, E. Fernandes, S. Kroehnert, "Implementation of Wafer Level Packaging KOZ using SU-8 as Dielectric for the Merging of WL Fan Out to Microfluidic and Bio-Medical Applications", *IMAPS Proceedings*, October 2017, pp. 569–575.

[130] Ishibashi, D., and Y. Nakata, "Planar Antenna for Terahertz Application in Fan Out Wafer Level Package", *IMAPS Proceedings*, October 2017, pp. 599–603.

[131] Palesko, C., and A. Lujan, "Cost Comparison of Fan-out Wafer-Level Packaging to Embedded Die Packaging", *IMAPS Proceedings*, October 2017, pp. 721–726.

[132] Pendse, R., "Semiconductor Device and Method of Forming Extended Semiconductor Device with Fan-Out Interconnect Structure to Reduce Complexity of Substrate", filed on December 23, 2011, US 2013/0161833 A1, pub. date: June 27, 2013.

[133] Chen, N. C., T. Hsieh, J. Jinn, P. Chang, F. Huang, J. Xiao, A. Chou, B. Lin, "A Novel System in Package with Fan-out WLP for High Speed SERDES Application", *IEEE/ECTC Proceedings*, May 2016, pp. 1496–1501.

[134] Wang, C.-T., T.-C. Tang, C.-W. Lin, C.-W. Hsu, J.-S. Hsieh, C.-H. Tsai, K.-C. Wu, H.-P. Pu, and D. Yu, "InFO_AiP Technology for High Performance and Compact 5G Millimeter Wave System Integration", *IEEE/ECTC Proceedings*, May 2018, pp. 202–207.

[135] Lee, M., Yoo, M., Cho, J., Lee, S., Kim, J., Lee, C., Kang, D., Zwenger, C., and Lanzone, R., "Study of Interconnection Process for Fine Pitch Flip Chip", *IEEE/ECTC Proceedings*, May 25–28, 2009, pp. 720–723.

[136] Zhang, Z., J. H. Lau, C. S. Premachandran, S. Chong, L. Wai, V. Lee, T. C. Chai, V. Kripesh, et al., "Development of a Cu/Low-*k* Stack Die Fine Pitch Ball Grid Array (FBGA) Package for System in Package Applications", *IEEE Transactions on CPMT*, Vol. 1, No. 3, March 2011, pp. 299–309.

[137] Choi, W., C. Premachandran, C. Ong, L. Xie, E. Liao, A. Khairyanto, B. Ratmin, K. Chen, P. Thaw, and J. H. Lau, "Development of Novel Intermetallic Joints Using Thin Film Indium Based Solder by Low Temperature Bonding Technology for 3D IC Stacking", *IEEE/ECTC Proceedings*, May 2009, pp. 333–338.

[138] Yu, A., J. H. Lau, S. Ho, A. Kumar, W. Hnin, W. Lee, M. Jong, V. Sekhar, V. Kripesh, D. Pinjala, S. Chen, C. Chan, C. Chao, C. Chiu, C. Huang, and C. Chen, "Fabrication of High Aspect Ratio TSV and Assembly with Fine-Pitch Low-Cost Solder Microbump for Si Interposer Technology with High-Density Interconnects", *IEEE Transactions on CPMT*, Vol. 1, No. 9, September 2011, pp. 1336–1344.

[139] Yu, A., J. H. Lau, Ho, S., Kumar, A., Yin, H., Ching, J., Kripesh, V., Pinjala, D., Chen, S., Chan, C., Chao, C., Chiu, C., Huang, M., and Chen, C., "Three Dimensional Interconnects with High Aspect Ratio TSVs and Fine Pitch Solder Microbumps", *IEEE Proceedings of ECTC*, May 2009, pp. 350–354.

[140] Lim, S., V. Rao, H. Yin, W. Ching, V. Kripesh, C. Lee, J. H. Lau, J. Milla, and A. Fenner, "Process Development and Reliability of Microbumps", *IEEE/EPTC Proceedings*, December 2008, pp. 367–372.

[141] Lim, S., V. Rao, W. Hnin, W. Ching, V. Kripesh, C. Lee, J. H. Lau, J. Milla, and A. Fenner, "Process Development and Reliability of Microbumps", *IEEE Transactions on CPMT*, Vol. 33, No. 4, December 2010, pp. 747–753.

[142] Kagawa, Y., N. Fujii, K. Aoyagi, Y. Kobayashi, S. Nishi, N. Todaka, et al., "Novel stacked CMOS image sensor with advanced Cu2Cu hybrid bonding", *IEEE/IEDM Proceedings*,

December 2016, pp. 8.4.1–4.
[143] Sukegawa, S., T. Umebayashi, T. Nakajima, H. Kawanobe, K. Koseki, I. Hirota, et al., "A 1/4-inch 8Mpixel Back-Illuminated Stacked CMOS Image Sensor", *Proceedings of IEEE/ISSCC*, February 2013, pp. 484–484.
[144] Coudrain, P., D. Henry, A. Berthelot, J. Charbonnier, S. Verrun, R. Franiatte, N. Bouzaida, et al., "3D Integration of CMOS Image Sensor with Coprocessor Using TSV last and Micro-Bumps Technologies", *Proceedings of IEEE/ECTC*, Las Vegas, NV, May 2013, pp. 674–682.
[145] Zhang, R., R. Lee, D. Xiao, and H. Chen, "LED Packaging using Silicon Substrate with Cavities for Phosphor Printing and Copper-filled TSVs for 3D Interconnection", *Proceeding of IEEE/ECTC*, Orlando, FL, May 2011, pp. 1616–1621.
[146] Zhang, R., and R. Lee, "Moldless Encapsulation for LED Wafer Level Packaging using Integrated DRIE Trenches", *Journal of Microelectronics Reliability*, Vol. 52, 2012, pp. 922–932.
[147] Chen, D., L. Zhang, Y. Xie, K. Tan, and C. Lai, "A Study of Novel Wafer Level LED Package Based on TSV Technology", *IEEE Proceedings on ICEPT*, August 2012, pp. 52–55.
[148] Xie, Y., D. Chen, L. Zhand, K. Tan, and C. Lai, "A Novel Wafer Level Packaging for White Light LED", *IEEE Proceedings on ICEPT*, August 2013, pp. 1170–1174.
[149] Sekhar, V., J. Toh, J. Cheng, J. Sharma, S. Fernando, and B. Chen, "Wafer Level Packaging of RF MEMS Devices Using TSV Interposer Technology", *Proceedings of IEEE/EPTC*, Singapore, December 2012, pp. 239–243.
[150] Chen, B., V. Sekhar, C. Jin, Y. Lim, J. Toh, S. Fernando, and J. Sharma, "Low-Loss Broadband Package Platform With Surface Passivation and TSV for Wafer-Level Packaging of RF-MEMS Devices", *IEEE Transactions on CPMT*, Vol. 3, No. 9, September 2013, pp. 1443–1452.
[151] Pham, N., V. Cherman, B. Vandevelde, P. Limaye, N. Tutunjyan, R. Jansen, N. Hoovels, et al., "Zerolevel Packaging for (RF-)MEMS Implementing TSVs and Metal Bonding", *Proceedings of IEEE/ECTC*, May 2011, 1588–1595.
[152] Pham, N., V. Cherman, N. Tutunjyan, L. Teugels, D. Teacan, and H. Tilmans, "Process Challenges in 0-level Packaging Using 100 μm-thin Chip Cappin with TSV", *Proceedings of IMAPS International Symposium on Microelectronics*, September 2012, San Diego, CA, pp. 276–282.
[153] Zoschke, K., C.-A. Manier, M. Wilke, N. Jurgensen, H. Oppermann1, D. Ruffieux, J. Dekker, et al., "Hermetic Wafer Level Packaging of MEMS Components Using Through Silicon Via and Wafer to Wafer Bonding Technologies", *Proceedings of IEEE/ECTC*, May 2013, Las Vegas, NV, pp. 1500–1507.
[154] Premachandran, C. S., J. H. Lau, X. Ling, A. Khairyanto, K. Chen, and M. Pa, "A Novel, Wafer-level Stacking Method for Low-chip Yield and Non-uniform, Chip-size Wafers for MEMS and 3D SiP Applications", *IEEE/ECTC Proceedings*, Orlando, FL, May 27–30, 2008, pp. 314–318.
[155] Pang, W., R. Ruby, R. Parker, P. W. Fisher, M. A. Unkrich, and J. D. Larson, III, "A Temperature-Stable Film Bulk Acoustic Wave Oscillator," *IEEE Electron Device Letters*, Vol. 29, No. 4, April 2008, pp. 315–318.
[156] Small, M., R. Ruby, S. Ortiz, R. Parker, F. Zhang, J. Shi, and B. Otis, "Wafer-Scale Packaging For FBARBased Oscillators", *Proceedings of IEEE International Joint Conference of FCS*, 2011, pp. 1–4.
[157] Lau, J. H., Y. Lim, T. Lim, G. Tang, K. Houe, X. Zhang, P. Ramana, et al., "Design and Analysis of 3D Stacked Optoelectronics on Optical Printed Circuit Boards", *Proceedings of SPIE, Photonics Packaging, Integration, and Interconnects VIII*, Vol. 6899, San Jose, CA, January 19–24, 2008, pp. 07.1–07.20.
[158] Lim, T. G., B. Lee, T. Shioda, H. Kuruveettil, K. Li, K. Suzuki, J. H. Lau, et al., "Demonstration of High Frequency Data link on FR4 PCB Using Optical Waveguides", *IEEE Transactions of Advanced Packaging*, Vol. 32, May 2009, pp. 509–516.
[159] Chai, J., G. Yap, T. Lim, C. Tan, Y. Khoo, C. Teo, J. H. Lau, et al., "Electrical Interconnect Design Optimization for Fully Embedded Board-level Optical Interconnects", *IEEE/EPTC Proceedings*, December 2008, pp. 1126–1130.
[160] Lim, L, C. Teo, H. Yee, C. Tan, O. Chai, Y. Jie, J. H. Lau, et al., "Optimization and Characterization of Flexible Polymeric Optical Waveguide Fabrication Process for Fully Embedded Board-level Optical Interconnects", *IEEE/EPTC Proceedings*, December 2008, pp. 1114–1120.

[161] Teo, C., W. Liang, H. Yee, L. Lim, C. Tan, J. Chai, J. H. Lau, et al., "Fabrication and Optimization of the 45° Micro-mirrors for 3-D Optical Interconnections", *IEEE/EPTC Proceedings*, December 2009, pp. 1121–1125.
[162] Chang, C., J. Chang, J. H. Lau, A. Chang, T. Tang, S. Chiang, M. Lee, et al., "Fabrication of Fully Embedded Board-Level Optical Interconnects and Optoelectronic Printed Circuit Boards", *IEEE/EPTC Proceedings*, December 2009, pp. 973–976.
[163] Lau, J. H., S. W. Lee, M. Yuen, J. Wu, J. Lo, H. Fan, and H. Chen, "Apparatus Having an Embedded 3D Hybrid Integration for Optoelectronic Interconnects in Organic Substrate". US Patent No: 9,057,853, Date of Patent: June 16, 2015.
[164] Lau, J. H., M. S. Zhang, and S. W. R. Lee, "Embedded 3D Hybrid IC Integration System-in-Package (SiP) for Opto-Electronic Interconnects in Organic Substrates", *ASME Paper IMECE2010-40974*.

第 2 章

有机基板上的异构集成

2.1 引言

截至目前,在大批量制造(HVM)中,有机基板上 70% 的异构集成再布线层(RDL),金属导线的线宽与间距大于等于 10μm。这类异构集成实际上主要应用于系统级封装(SiP)中。而有机基板上仅有 5% 的异构集成 RDL 金属导线的线宽与间距小于 10μm。

在过去几年中,一般通过增加积层层数、在积层顶部制作薄膜层、缩小金属导线线宽与间距的尺寸、减小焊盘尺寸和节距、去除芯基板层[6-28]、采用引线凸点(BOL)[29-37] 以及层压嵌入式线路基板(ETS)[39-44] 等技术来提高传统低成本积层有机封装基板的性能[1-5]。本章将简要介绍积层有机封装基板的发展历史,并讨论积层顶部的超薄线路层制造技术。此外,本章还将介绍支撑异构集成的有机转接板(基板)。

首先简要介绍 Amkor 和 Apple/ASE 的常规 SiP 技术。

2.2 安靠科技公司的汽车 SiP

安靠科技公司的汽车 SiP 专注于自动驾驶、信息娱乐和高级驾驶辅助系统(ADAS)以及车内计算机。图 2.1a、b 显示了 Amkor 汽车 SiP 的几个示例。图 2.1a 中,42.5mm×42.5mm 信息娱乐功能有机基板上有中央处理器和 DDR(双数据速率)存储器。图 2.1b 中,55mm×72mm 有机基板上有网络交换机、专用集成电路(ASIC)和存储器。

第 2 章　有机基板上的异构集成

a) 42.5mm×42.5mm信息娱乐功能有机基板　　　b) 55mm×72mm有机基板

图 2.1　Amkor 的汽车 SiP

2.3　日月光半导体公司组装的 Apple Watch Ⅲ（SiP）

通过环隆电气股份有限公司（USI），日月光半导体公司（ASE）成为苹果定制设计的如图 2.2 所示 S3 SiP 模块的唯一后端供应商，S3 SiP 模块用于 Apple Watch Ⅲ。从图 2.2 可以看出，有机基板上有超过 40 个片式元器件，这些片式元器件包括电容器和电阻器等分立无源元件以及 ASIC、处理器、控制器、转换器、DRAM（动态随机存取存储器）、NAND、Wi-Fi、NFC、GPS 以及传感器等。

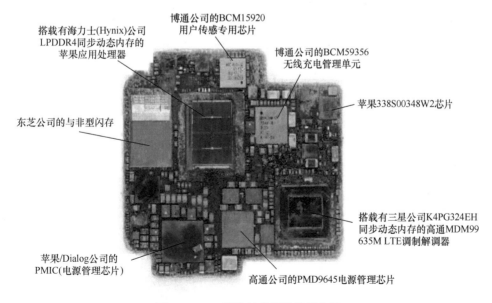

图 2.2　ASE 组装的苹果智能手表 Ⅲ

2.4 IBM 公司的 SLC 技术

25年以前，IBM日本野州公司发明了表面积层线路（SLC）技术，如图2.3所示[1-4]。SLC形成了当今非常流行的低成本有机封装基板的基础，其积层通过微通孔垂直连接，以支持倒装芯片等异构集成。SLC技术分为两部分：一是芯板的制作，另一部分是信号布线SLC。芯板由普通玻璃纤维环氧树脂板制成，而SLC层是由光敏环氧树脂制成的电介质层和镀铜的平面导电层等依次堆积而成，这属于半加成法（SAP）的技术范畴。通常，具有12层（如2层芯板和10层积层（5-2-5））和10μm线宽与间距的封装基板可以满足绝大多数芯片集成的需求。

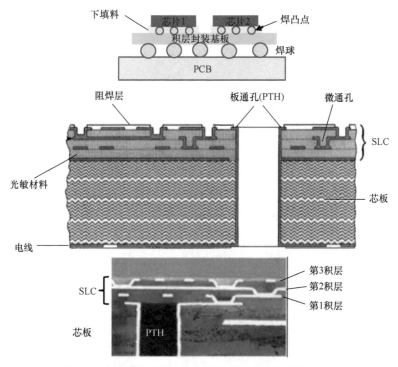

图 2.3 IBM 用于倒装芯片有机积层封装基板的 SLC

2.5 无芯基板

2006年，富士通（Fujitsu）公司首次提出了无芯基板。图2.4展示了通过积层方法制作的常规有机封装基板以及先进的有机无芯封装基板的剖面图。从图2.4可以看出，最大的区别在于无芯封装基板中没有芯板层，无芯封装基板的所有层都是通过堆积层进行叠层制作的[6-28]。

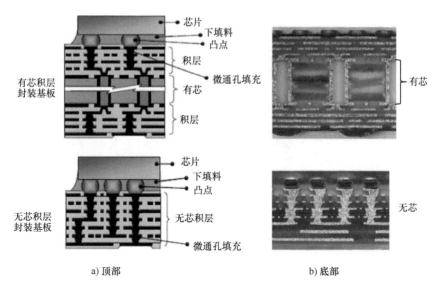

图 2.4 常规积层封装基板上的倒装芯片封装与无芯基板上的倒装芯片封装比较

无芯封装基板的优点：由于消除了芯板层，无芯基板的成本较更低；通过消除芯板层，可以实现更高的布线能力；高速传输特性改善，电气性能更好；更小的形变系数。另一方面[6-28]：由于消除了芯板层，无芯基板的翘曲较大；容易出现分层；基板刚度低，导致焊点成品率低；需要新的制造基础设施[6-29]。2010年，索尼为其 PlayStation 3 的单元处理器制造了第一块无芯封装基板[17]。

尽管无芯基板有许多优点，但由于翘曲控制问题并没有得到广泛应用。影响翘曲的关键因素之一是基板材料的热膨胀系数失配。因此，适当控制该因素将有助于减少无芯基板的翘曲问题。影响翘曲的另一个因素是封装技术。因此，对封装技术进行适当的改进（通过真空和压力）有利于控制无芯基板的翘曲问题。

图 2.5 显示了无芯基板的应用示例。可以看出，应用处理器芯片组的上部封装通过引线键合将手机 DRAM 封装在 3 层无芯有机封装基板上。

图 2.5 iPhone 7 中的 3 层无芯基板

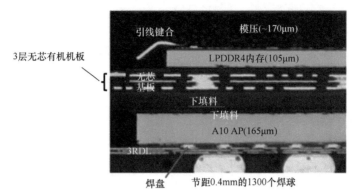

图 2.5　iPhone 7 中的 3 层无芯基板（续）

2.6　布线上凸点（BOL）

BOL 最早由星科金朋（STATS ChipPAC）公司[30-34]提出，并被高通[35]等其他公司[35-37]使用。图 2.6a 为倒装芯片有机基板上的常规捕捉焊盘凸点（BOC）或简单焊盘凸点（BOP）结构示意图。可以看出，在阻焊层（SR）确定的图形中，倒装芯片焊盘位于 210μm 的区域阵列节距上，焊盘凸点之间设计了一条布线，导致有效布线节距为 105μm。

如图 2.6b 所示，在 BOL 方法中，倒装芯片的焊盘仅仅为布线宽度或者比布线稍宽的设计即可实现。该方案可以极大地释放出焊接空间，允许在凸点之间设计更多的线路，形成高密度的布线结构。在不改变基板的设计规则（如布线线宽与间距）的情况下可以获得 70μm 的有效凸点节距。如图 2.6c 所示，在改进型 BOL 结构中，通过线路上设计稍加凸点焊盘没有任何阻焊限制，即开放式阻焊层[35]，线路可以直接作为凸点进行焊接[35]。

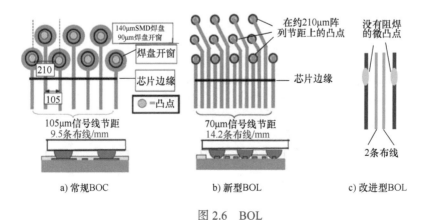

图 2.6　BOL

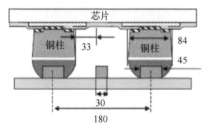
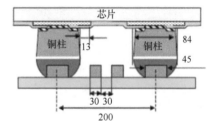

d) 采用铜柱FC(180μm节距)的改进型BOL   e) 采用铜柱FC(200μm节距)的改进型BOL

图 2.6　BOL（续）

参考文献 [35] 中 BOL 焊接中的铜柱示意图如图 2.6d、e 所示。可以看到，凸点节距为 180μm，中间有一条布线；凸点节距 200μm，中间可以轻松设计 2 条布线。图 2.7 显示了焊料 BOL 上铜柱的横截面图像，以及垂直和平行 BOL 方向的典型横截面图。

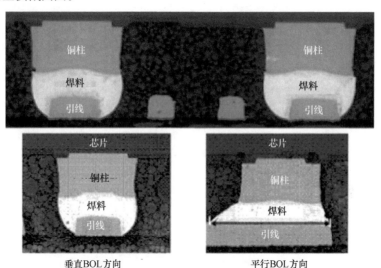

图 2.7　高通 BOL 上的铜柱典型横截面图像

2.7　嵌入式线路基板（ETS）

ETS 是一种无芯基板，具有精细线宽 / 间距布线，将顶层金属布线嵌入预浸料层[38-43]。如图 2.8a 所示为顶层制作 ETS 的工艺流程图。

其具体工艺流程为：在载板上制作一层超薄铜层（铜箔）并使用干膜制作掩模版，然后使用电镀铜方法形成第一层铜图形，在铜图形上层压半固化片。接着是激光通孔钻孔、化学镀铜、干膜层压、曝光和显影、第二层铜图形电镀、剥离和微蚀刻腐蚀。一旦完成了所有的铜图形层，将移除载板。因为铜箔连接到第一层铜图形，所以在 SR 涂覆之前需要微蚀刻腐蚀。在 SR 开口工艺之后，需要完成金属表面处理,如利用有机可焊性保护剂(OSP)等。图2.8b显示了矽品精密工业股份有限公司(SPIL)[41]组装的ETS组件上的铜柱倒装芯片横截面。另一个示例是英特尔为iPhone XS和XR制造的调制解调器芯片组的异构集成,如图2.9所示。目前使用的大多数ETS的线宽/间距为15μm/15μm。然而,韩国信泰电子公司(Simmtech)[43]正在生产13μm/13μm线宽/间距的ETS。

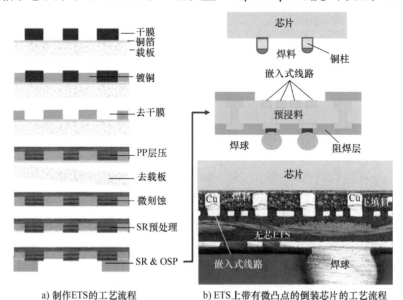

图2.8 顶层制作ETS的工艺流程图

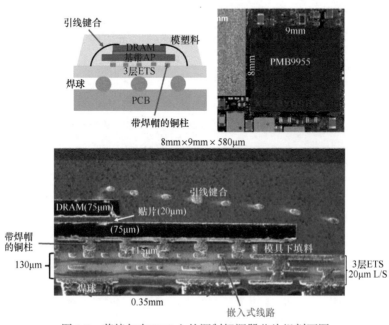

图2.9 英特尔在ETS上的调制解调器芯片组剖面图

2.8 新光公司的具有薄膜层的积层基板

自 2013 年、2014 年以来，新光（Shinko）一直建议在有机封装基板的积层顶部制作薄膜层（薄至 2μm），并将其称为 i-THOP 基板 [44, 45]，旨在用于高性能异构集成应用，如图 2.10 所示。

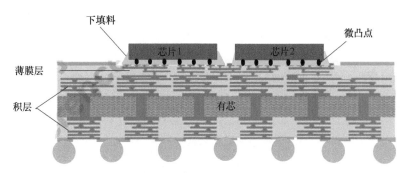

图 2.10　Shinko 用于异构集成的 i-THOP 基板

2.8.1　基板结构

图 2.11~图 2.13 显示了用于高性能应用的 Shinko 集成薄膜高密度有机封装（i-THOP）基板 [44]。它采用 4+（2-2-3）设计结构，即有 1 个 2 层金属芯板层，底部（PCB）侧有 3 个堆积金属层，顶部（芯片侧）有 2 个堆积金属层，第一个数字"4"表示顶部积层表面有 4 个铜薄膜布线层（RDL）。铜布线层的厚度、线宽和间距可以小到 2 μm，如图 2.12 所示。如图 2.11 所示，铜薄膜布线层通过一个 10μm 的通孔垂直连接。表面铜焊盘节距为 40μm，如图 2.13 所示，铜焊盘直径为 25μm，高度为 10~12μm。

2.8.2　制作工艺

Shinko 基板的制作工艺：首先在封装基板芯板（具有 100μm 的电镀通孔）的两侧堆积传统的积层，接着通过正常的半加成工艺形成铜金属积层，并且通过大约 50μm 的积层微通孔垂直连接 [44]（见图 2.11）。在背面对球栅阵列（BGA）焊球层涂覆（25μm 厚）阻焊膜后，通过化学机械抛光（CMP）对顶层表面进行抛光，以使激光钻孔和填铜的通孔变平，并使表面光滑，为制作精细布线绝缘树脂层做准备。

使用薄膜工艺沉积绝缘树脂层，应用正常工艺制备小直径通孔，并在树脂层上溅射 Ti/Cu 种子层，然后旋涂光致抗蚀剂（光刻胶），并通过光刻机曝光，以制作 2μm 布线轨迹（RDL）图案。布线厚度通过电解镀铜形成。最后，顶层的铜焊盘用有机可焊性保护剂（OSP）处理。

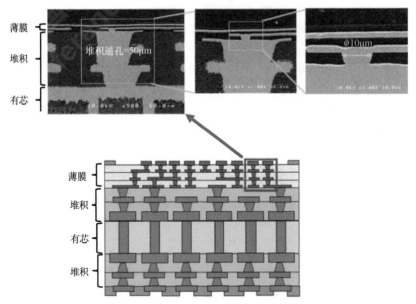

图 2.11 Shinko 的 i-THOP（积层上的薄膜布线层通过一个 10μm 通孔垂直连接）

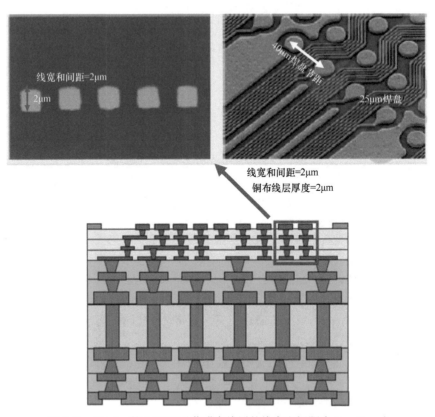

图 2.12 Shinko 的 iTHOP（薄膜布线层的线宽和间距为 2μm/2μm）

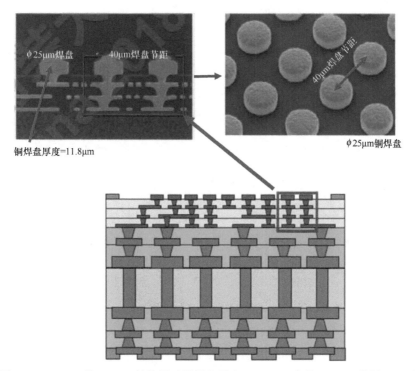

图 2.13 Shinko 的 iTHOP 结构图（铜焊盘厚度 11.8μm，直径 25μm，节距 40μm）

图 2.14 显示了 4+（2-2-3）i-THOP 基板（40mm×40mm）从室温（RT）到 260℃，然后再回到室温的温度变化下的翘曲测量结果。可以看出，对于所有温度，封装基板的变形形状（翘曲）都是凸形的，当基板被加热到无铅温度（260℃）时，翘曲仅增加至 10μm。总体而言，翘曲是稳定且可接受的。

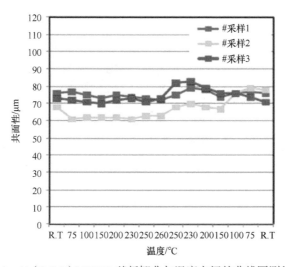

图 2.14 4+（2-2-3）i-THOP 基板翘曲与温度之间的曲线图测量结果

2.8.3 质量评价测试

对 i-THOP 基板结构进行了质量评价测试，其中包括：预处理，湿气敏感性等级（MSL）3A 和回流焊（至 260℃）3 次；热循环（-55℃~125℃）1000 次；高加速应力试验（HAST）130℃/85%RH/3.5V，持续 150h。i-THOP 基板通过了测试，并且没有观察到通孔分层等失效现象。

2.8.4 i-THOP 应用示例

2014 年，Shinko[46] 展示了将超细节距倒装芯片成功集成在 i-THOP 基板上。图 2.15 显示了 2 个芯片通过 2 个薄膜层的 2μm 线宽和间距布线层进行横向通信，这两个薄膜层构建在 1-2-2 堆积的有机基板的顶部，形成了 2+（1-2-2）结构。图 2.16 显示了 40μm 节距的微凸点（铜柱+NiSnAg）测试芯片和 40μm 节距倒装芯片焊盘（25μm 直径）图 2.16 显示了优化条件下倒装芯片组件横截面的典型图像。可以看出，集成组件的所有区域有良好的焊点特性。

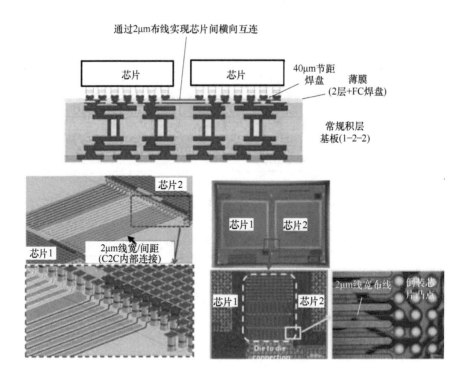

图 2.15 Shinko 的 i-THOP 测试样板结构（2 个薄膜层构建在 1-2-2 堆积的封装基板顶部）

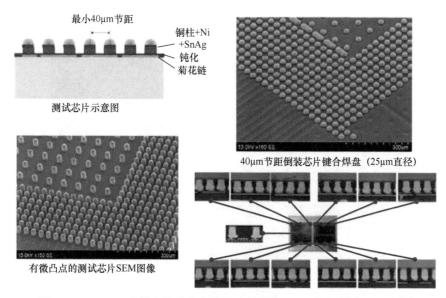

图 2.16 i-THOP 基板上的微凸点阵列（节距 40μm、凸点直径 25μm）图

2.9 思科公司的有机转接板

2.9.1 转接板层结构

图 2.17 展示了一个由思科（Cisco）公司设计和制造的 3D 异构集成结构[46]，其中设计有一个大尺寸的有机转接板。该转接板具有细间距和细线互连特性，尺寸为 38mm×30mm×0.4mm，如图 2.18 所示。有机转接板的正面和背面的最小线宽、间距和厚度相同，分别为 6μm、6μm 和 10μm。它是 10 层高密度有机转接板（基板），通孔尺寸为 20μm。图 2.18 中，高性能专用 IC（ASIC）芯片尺寸为 19.1mm×24mm×0.75mm，与 4 个高带宽存储器（HBM）动态随机存取存储器（DRAM）芯片堆叠一起安装在有机转接板的顶部，尺寸为 5.5mm×7.7mm×0.48mm 的 3D HBM 管芯堆叠包括 1 个基础缓冲管芯和 4 个 DRAM 核心管芯，它们通过 TSV（硅通孔）和具有焊帽凸点的细间距微凸点柱互连。有机转接板正面的焊盘尺寸和节距分别为 30μm 和 55μm。

2.9.2 制作工艺

制造有机转接板的主要步骤与制造有机积层封装基板的步骤相同，包括[46]芯层电镀通孔（PTH）形成和填铜、芯层的电路化及用 SAP 工艺在芯层的两侧构建铜布线层。

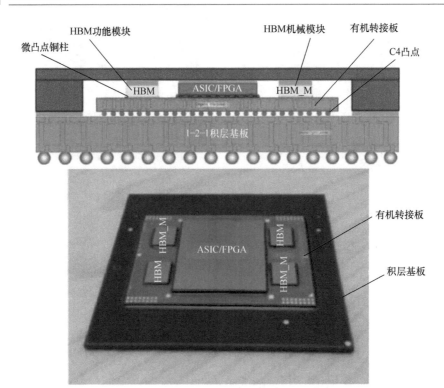

图 2.17 思科公司基于有机转接板的 3D 异构集成

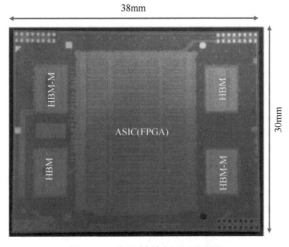

图 2.18 有机转接板的顶视图

HBM 和 HBM-M 芯片的翘曲以及有机转接板的翘曲通过阴影云纹法测量,结果如图 2.19 和图 2.20 所示。从图 2.19 可以看出,在室温至 280℃的温度范围内,HBM 芯片叠层的翘曲测量结果非常小(小于 8μm),并且不会随温度发生太大变化。从室温到 280℃，有机转接板 W/O 焊凸点的翘曲测量结果也非常小（约为 100μm），且随温度变化较小，如图 2.20 所示。

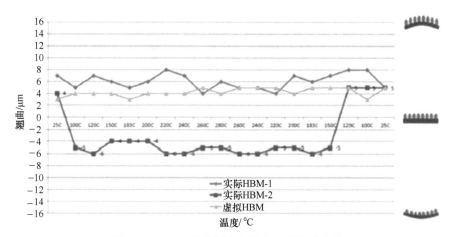

图 2.19　HBM 芯片翘曲与温度间的关系曲线

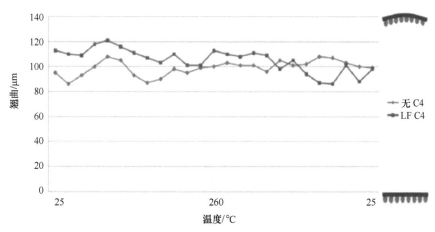

图 2.20　有机转接板翘曲与温度间的关系曲线

由于 HBM 芯片和有机转接板的翘曲很小，并且随温度变化小，所以 FPGA 管芯和 HBM 芯片被组装到转接板上，以首先形成 FPGA 和 HBM 子组件。底部填充封装用于保护由微凸点和规则凸点构成的接头。3D SiP 子组件的俯视图见图 2.17。图 2.21 显示了 HBM 芯片和有机转接板之间的微型铜柱焊点的横截面，没有异常。

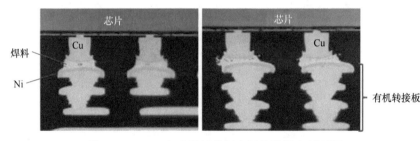

图 2.21 HBM 与有机转接板之间的微焊点剖面图

2.10 总结和建议

一些重要的结果和建议如下：

1）普通、高密度、基板状 PCB 和低成本有机封装基板的最新进展促进了更多异构集成（实际上是 SiP）应用。

2）此外，有机积层基板、具有薄膜层的有机积层基板、有机转接板、无芯基板、引线键合焊球凸点（BOL）和嵌入式线路基板（ETS）的发展进一步扩大了异构集成的应用范围。

3）有机基板上异构集成的 RDL 的趋势是以批量生产方式开发更为精细的金属线宽和间距（如 5μm）。

4）有机基板上的大多数异构集成组件（SiP）是通过表面安装技术设备实现的。

参考文献

[1] Tsukada, Y., S. Tsuchida, and Y. Mashimoto, "Surface Laminar Circuit Packaging", *42nd IEEE Electronic and Components Technology Conference*, San Diego, CA, May 1992, pp. 22–27.

[2] Tsukada, Y., and S. Tsuchida, "Surface Laminar Circuit, a Low Cost High Density Printed Circuit Board", *Surface Mount International Conference*, San Jose, CA, August 27–29, 1992, pp. 537–542.

[3] Tsukada, Y. "Solder Bumped Flip Chip Attach on SLC Board and Multichip Module", *in Chip on Board Technologies for Multichip Modules*, ed. J. H. Lau, 410–443. New York: Van Nostrand Reinhold, 1994.

[4] Tsukada, Y., Y. Maeda, and K. Yamanaka, "A Novel Solution for MCM-L Utilizing Surface Laminar Circuit and Flip Chip Attach Technology", *Proceedings of 2nd International Conference on Multichip Modules*, April 1993, pp. 252–259.

[5] Lau, J. H., and S. W. R. Lee, *Microvias for Low Cost, High Density Interconnects*. New York: McGraw-Hill, 2001.

[6] Koide, M., K. Fukuzono, H. Yoshimura, T. Sato, K. Abe, and H. Fujisaki, "High-Performance Flip-Chip BGA Technology Based on Thin-Core and Coreless Package Substrate", *IEEE 56th Electronic and Components Technology Conference*, San Diego, CA, May 2006, pp. 1869–1873.

[7] Sung, R., K. Chiang, Y. Wang, and C. Hsiao, "Comparative Analysis of Electrical Performance on Coreless and Standard Flip-Chip Substrate", *IEEE 57th Electronic and Components Technology Conference*, Reno, NV, May 2007, pp. 1921–1924.

[8] Chang, D., Y. Wang, and C. Hsiao, "High Performance Coreless Flip-Chip BGA Packaging Technology", *IEEE 57th Electronic and Components Technology Conference*, Reno, NV, May 2007, pp. 1765–1768.

[9] Savic, J., P. Aria, J. Priest, N. Dugbartey, R. Pomerleau, B. Shanker, M. Nagar, J. Lim, S. Teng, L. Li, and J. Xue, "Electrical Performance Assessment of Advanced Substrate Technologies for High Speed Networking Applications", *IEEE 59th Electronic and Components Technology Conference*, San Diego, CA, May 2009, pp. 1193–1199.

[10] Kurashina, M., D. Mizutani, M. Koide, and N. Itoh, "Precision Improvement Study of Thermal Warpage Prediction Technology for LSI Packages", *IEEE 59th Electronic and Components Technology Conference*, San Diego, CA, May 2009, pp. 529–534.

[11] Wang, J., Y. Ding, L. Liao, P. Yang, Y. Lai, and A. Tseng, "Coreless Substrate for High Performance Flip Chip Packaging", *IEEE 11th International Conference on Electronic Packaging Technology*, August 2010.

[12] Kimura, M., "Shinko Officially Announces Volume Production of Coreless Substrate", *Nikkei Electronics*, June 2011.

[13] Fujimoto, D., K. Yamada, N. Ogawa, H. Murai, H. Fukai, Y. Kaneato, and M. Kato, "New Fine Line Fabrication Technology on Glass-Cloth Prepreg Without Insulating Films for KG Substrate", *IEEE 61st Electronic and Components Technology Conference*, Lake Buena Vista, FL, May 2011, pp. 387–391.

[14] Kim, G., S. Lee, J. Yu, G. Jung, H. Yoo, and C. Lee, "Advanced Coreless Flip-Chip BGA Package With High Dielectric Constant Thin Film Embedded Decoupling Capacitor", *IEEE 61st Electronic and Components Technology Conference*, Lake Buena Vista, FL, May 2011, pp. 595–600.

[15] Nickerson, R., R. Olmedo, R. Mortensen, C. Chee, S. Goyal, A. Low, and C. Gealer, "Application of Coreless Substrate to Package on Package Architectures", *IEEE 62nd Electronic and Components Technology Conference*, San Diego, CA, May 2012, pp. 1368–1371.

[16] Kim, G., J. Yu, C. Park, S. Hong, J. Kim, G. Rinne, and C. Lee, "Evaluation and Verification of Enhanced Electrical Performance of Advanced Coreless Flip-Chip BGA Package With Warpage Measurement Data", *IEEE 62nd Electronic and Components Technology Conference*, San Diego, CA, May 2012, pp. 897–903.

[17] Nishitani, Y., "Coreless Packaging Technology for High-Performance Application", *IEEE/ECT/CMPT Seminar on Advanced Coreless Package Substrate and Material Technologies, Electronic Components and Technology Conference*, San Diego, CA, May 2012.

[18] Kurashina, M., D. Mizutani, M. Koide, M. Watanabe, K. Fukuzono, and H. Suzuki, "Low Warpage Coreless Substrate for Large-Size LSI Packages", *IEEE 62nd Electronic and Components Technology Conference*, San Diego, CA, May 2012, pp. 1378–1383.

[19] Kurashina, M., D. Mizutani, M. Koide, M. Watanabe, K. Fukuzono, N. Itoh, and H. Suzuki, "Low Warpage Coreless Substrate for IC Packages", *Transaction of the Japan Institute of Electronics Packaging*, October 2012, Vol. 5, Issue 1, pp. 55–62.

[20] Manusharow, M., S. Muthukumar, E. Zheng, A. Sadiq, and C. Lee, "Coreless Substrate Technology Investigation for Ultra-Thin CPU BGA Packaging", *IEEE 62nd Electronic and Components Technology Conference*, San Diego, CA, May 2012, pp. 892–896.

[21] Kim, J., S. Lee, J. Lee, S. Jung, and C. Ryu, "Warpage Issues and Assembly Challenges Using Coreless Package Substrate", *IPC APEX Expo*, San Diego, CA, February 28–March 1, 2012.

[22] Sakuma, K., E. Blackshear, K. Tunga, C. Lian, S. Li, M. Interrante, O. Mantilla, and J. Nah, "Flip Chip Assembly Method Employing Differential Heating/Cooling for Large Dies With Coreless Substrates", *IEEE 63rd Electronic and Components Technology Conference*, Las Vegas, NV, May 2013, pp. 667–673.

[23] Hahm, Y., M. Li, J. Yan, Y. Tretiakov, H. Lan, S. Chen, and S. Wong, "Analysis and Measurement of Power Integrity and Jitter Impacts on Thin-Core and Coreless Packages", *IEEE 66th Electronic Components and Technology Conference*, Las Vegas, NV, May 2016, pp. 387–392.

[24] Baloglu, B., W. Lin, K. Stratton, M. Jimarez, and D. Brady, "Warpage Characterization and Improvements for IC Packages With Coreless Substrate", *46th International Symposium on Microelectronics*, Orlando, FL, September 2013, pp. 260–284.

[25] Sun, Y., X. He, Z. Yu, and L. Wan, "Development of Ultra-Thin Low Warpage Coreless Substrate", *IEEE 63rd Electronic and Components Technology Conference*, Las Vegas, NV, May 2013, pp. 1846–1849.

[26] Lin, W., B. Baloglu, and K. Stratton, "Coreless Substrate With Asymmetric Design to Improve Package Warpage", *IEEE 64th Electronic and Components Technology Conference*, Orlando, FL, May 2014, pp. 1401–1406.

[27] Liu, W., G. Xiaa, T. Liang, T. Li, X. Wang, J. Xie, S. Chen, and D. Yu, "Development of High Yield, Reliable Fine Pitch Flip Chip Interconnects With Copper Pillar Bumps and Thin Coreless Substrate", *IEEE 65th Electronic and Components Technology Conference*, San Diego, CA, May 2015, pp. 1713–1717.

[28] Goodhue, N., D. Danovitch, J. Moussodji, B. Papineau, and E. Duchesne, "Warpage Control during Mass Reflow Flip Chip Assembly using Temporary Adhesive Bonding ", *Proceedings of IEEE/ECTC*, May 2018, pp. 703–711.

[29] Pendse, R., K. Kim, O. Kim, and K. Lee, "Bond-On-Lead: A Novel Flip Chip Interconnection Technology for Fine Effective Pitch and High I/O Density", *IEEE 56th Electronic and Components Technology Conference*, San Diego, CA, May 2006, pp. 16–23.

[30] Pendse, R., "Bump-On-Lead Flip Chip Interconnection", U.S. Patent No. 7,368,817, filed November 10, 2004 and issued May 6, 2008.

[31] Ouyang, E., M. Chae, S. Chow, R. Emigh, M. Joshi, R. Martin, and R. Pendse, "Improvement of ELK Reliability in Flip Chip Packages Using Bond-On-Lead (BOL) Interconnect Structure", *IMAPS 43rd International Symposium on Microelectronics*, Raleigh, NC, October 2010, pp. 197–203.

[32] Pendse, R, "Bump-On-Lead Flip Chip Interconnection. U.S. Patent No. 7,901,983, filed May 26, 2009 and issued March 8, 2011.

[33] Pendse, R., C. Cho, M. Joshi, K. Kim, P. Kim, S. Kim, S. Kim, H. Lee, K. Lee, R. Martin, A. Murphy, V. Pandey, and C. Palar, "Low CostmFlip Chip (LCFC): An Innovative Approach for Breakthrough Reduction in Flip Chip Package Cost", *IEEE 60th Electronic Components and Technology Conference*, Las Vegas, NV, June 2010.

[34] Movva, S., S. Bezuk, O. Bchir, M. Shah, M. Joshi, R. Pendse, E. Ouyang, Y. Kim, S. Park, H. Lee, S. Kim, H. Bae, G. Na, and K. Lee, "CuBOL (Cu-Column on BOL) Technology: A Low Cost Flip Chip Solution Scalable to High I/O Density, Fine Bump Pitch and Advanced Si-Nodes", *IEEE 61st Electronic and Components Technology Conference*, Lake Buena Vista, FL, May 2011, pp. 601–607.

[35] Lan, A., C. Hsiao, J. H. Lau, E. So, and B. Ma, "Cu Pillar Exposed-Die Molded FCCSP for Mobile Devices", *IEEE 62nd Electronic and Components Technology Conference*, San Diego, CA, May 2012, pp. 886–891.

[36] Kuo, F., J. Lee, F. Chien, R. Lee, C. Mao, and J. H. Lau, "Electromigration Performance of Cu Pillar Bump for Flip Chip Packaging With Bump on Trace by Using Thermal 64 2 Flip Chip Technology Versus FOWLP Compression Bonding", *IEEE 64th Electronic and Components Technology Conference*, Orlando, FL, May 2014, pp. 56–61.

[37] Li, M., D. Tian, Y. Cheung, L. Yang, and J. H. Lau, "A High Throughput and Reliable Thermal Compression Bonding Process for Advanced Interconnections", *IEEE 65th Electronic Components and Technology Conference*, San Diego, CA, May 26–29, 2015, pp. 603–608.

[38] Liu, F., C. Nair, V. Sundaram, and R. Tummala, "Advances in Embedded Traces for 1.5 μm RDL on 2.5D Glass Interposers", *IEEE 65th Electronic and Components Technology Conference*, San Diego, CA, May 2015, pp. 1736–1741.

[39] Chen, C., M. Lin, G. Liao, Y. Ding, and W. Cheng, "Balanced Embedded Trace Substrate Design for Warpage Control", *IEEE 65th Electronic and Components Technology Conference*, San Diego, CA, May 2015, pp. 193–199.

[40] Zhang, L., and G. Joseph, "Low Cost High Performance Bare Die PoP With Embedded Trace Coreless Technology and 'Coreless Cored' Build Up Substrate Manufacture Process", *IEEE 65th Electronic Components and Technology Conference*, San Diego, CA, May 2015, pp. 882–887.

[41] Lu, M., "Challenges and Opportunities in Advanced IC Packaging", *Chip Scale Review*, July/August 2014, Vol. 18, Issue 2, pp. 5–8.

[42] Lee, K., S. Cha, and P. Shim, "Form Factor and Cost Driven Advanced Package Substrates for Mobile and IoT Applications", *China Semiconductor Technology International Conference*, Shanghai, China, March 13–14, 2016.

[43] Chen, C., N. Kao, and D. Jiang, "Trend Plots for Different Mold-thick Selection on Warpage Design of MUF FCCSP with 4L ETS", *Proceedings of IEEE/ECTC*, May 2018, pp. 255–266.

[44] Shimizu, N., Kaneda, W., Arisaka, H., Koizumi, N., Sunohara, S., Rokugawa, A., and Koyama, T., "Development of Organic Multi Chip Package for High Performance Application", *IMAPS International Symposium on Microelectronics*, Orlando, FL, September 30–October 3, 2013, pp. 414–419.

[45] Oi, K., Otake, S., Shimizu, N., Watanabe, S., Kunimoto, Y., Kurihara, T., Koyama, T., Tanaka, M., Aryasomayajula, L., and Kutlu, Z., "Development of New 2.5D Package With Novel Integrated Organic Interposer Substrate With Ultra-Fine Wiring and High Density Bumps", *IEEE 64th Electronic and Components Technology Conference*, Orlando, FL, May 27–30, 2014, pp. 348–353.

[46] Li, L., P. Chia, P. Ton, M. Nagar, S. Patil, J. Xue, J. DeLaCruz, M. Voicu, J. Hellings, B. Isaacson, M. Coor, and R. Havens, "3D SiP with organic interposer of ASIC and memory integration", *Proceedings of IEEE/ECTC*, May 2016, pp. 1445–1450.

第 3 章

硅基板上的异构集成（TSV 转接板）

3.1 引言

在人工智能（AI）、机器学习（ML）以及 5G 的驱动下，中央处理器单元（CPU）、图形处理单元（GPU）、高带宽内存（HBM）、切片现场可编程门阵列（FPGA）等半导体器件的密度和 I/O 数增加，而焊盘节距减小。即使是第 2 章提到的 12 层（6-2-6）有机封装基板也不足以支撑切片芯片，因此，需要使用硅通孔（TSV）转接板[1-28]。台积电（TSMC）称这种结构为 CoWoS（chip-on-wafer-on substrate）[6-11]。莱蒂（Leti）公司[12,13]称它为 SoW（system-on-wafer）。

硅基板（TSV 转接板）上的异构集成指的是硅晶圆上的多芯片集成或 SoW。含有 TSV 的晶圆级倒装芯片的组装方法通常有两种：节距大于 50μm，通过回流焊实现（见文献 [29] 中的图 24a）；细小节距（小于 50μm）通过热压键合实现（见文献 [29] 中的图 24b~d）。一般来说，硅基板（TSV 转接板）上异构集成的再布线层（RDL）主要用来实现超细线宽和间距（低至亚微米）的应用。

本章将介绍 TSV 以及 TSV 转接板中 RDL 的加工，并讨论硅基板（TSV 转接板）上异构集成的热分析、热机械分析和可行性分析。下面简要介绍硅基板（TSV 转接板）上异构集成的原型和投入生产的 TSV 转接板。

3.2 莱蒂公司的 SoW

莱蒂（Leti）公司[12, 13]SoW 的早期应用之一如图 3.1 所示。其中，专用集成电路（ASIC）和存储器、电源管理集成电路（PMIC）和微机电系统（MEMS）等片上系统都集成在含有 TSV 的硅片上。切片后，硅基板上包含 RDL 和 TSV（即 TSV 转接板）的独立单元即为一个异构集成系统或子系统，可以贴装在有机基板上使用，也可以单独使用。

第 3 章 硅基板上的异构集成（TSV 转接板）

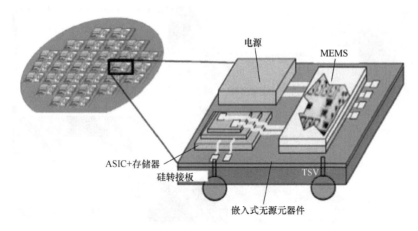

图 3.1 莱蒂公司的 SoW

3.3 台积电公司的 CoWoS 和 CoWoS-2

此后，台积电公司将 SoW 投入生产，并将 TSV 转接板尺寸为 800mm$^{2[7-10]}$ 的 SoW 称为 CoWoS，尺寸为 1200mm$^{2[6]}$ 的 SoW 称为 CoWoS-2，如图 3.2 所示。类似 CPU 或 GPU 的片上系统芯片（SoC）和通过 TSV 垂直互连由堆叠 DRAM 与底部逻辑芯片形成的 HBM 通过微凸点并排贴装在一块含有 RDL 的 TSV 转接板上。这块 TSV 转接板通过 Cu-C4 凸点贴装在积层封装基板上，而封装基板通过焊球连接在 PCB 上。

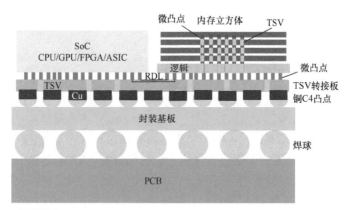

下填料用于：
➢ TSV 转接板和 SoC/逻辑芯片之间
➢ TSV 转接板和封装基板之间

图 3.2 台积电的 CoWoS 和 CoWoS-2

图 3.3a 是台积电和赛灵思（Xilinx）开发的切片 FPGA CoWoS[7-10]。可以看到 TSV（直径 10μm，深度 100μm）转接板上方有 4 层 RDL，即 3 层铜大马士

革层和 1 层铝层。在切片 FPGA 芯片之间超过 10000 个横向互连主要由转接板上最小节距为 0.4μm 的 RDL 连接。图 3.3b 是英伟达 Pascal 100 GPU[5]。它由台积电的 16nm 制程技术制造，并配备 4 个三星制造的 HBM2（16GB）存储器。每个 HBM2 由 4 个 DRAM 和 1 个底部逻辑芯片组成，它们通过铜柱、焊帽凸点以及垂直的 TSV 进行互连。每个 DRAM 芯片有超过 1000 个 TSV。该 GPU 和 HBM2 位于 TSV 转接板（1200mm^2）上方。此转接板是由台积电采用 64nm 制程技术制造的 CoWoS-2 硅芯片，并通过 C4 凸点与 5-2-5 有机封装基板连接。

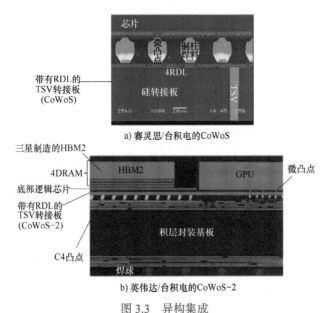

a) 赛灵思/台积电的 CoWoS

b) 英伟达/台积电的 CoWoS-2

图 3.3 异构集成

3.4 TSV 的制备

图 3.4 显示了制备 TSV 的关键工艺步骤[30, 31]。首先通过热氧化或者等离子体增强型化学气相沉积（PECVD）工艺沉积 SiN_x/SiO_x 绝缘层。在光刻之后，使用 Bosch 深反应离子刻蚀（DRIE）工艺刻蚀硅基底形成高深宽比（10.5）的通孔结构。然后在此结构表面通过亚常压化学气相沉积（SACVD）工艺沉积一层 SiO_x 绝缘层，并通过物理气相沉积（PVD）沉积一层钽阻挡层和一层铜种子层。接着通过铜的电化学沉积（ECD）工艺来填充 TSV 结构。最终的 TSV 盲孔顶部开口直径约为 10μm，深度约为 105μm，深宽比为 10.5。在这种高深宽比结构中，一般采用自底部向上的电镀方式以确保 TSV 没有缝隙，以及 TSV 之间合理的较低表面铜厚度。TSV 横截面的扫描电子显微镜（SEM）图像如图 3.5 所示。可以看到 TSV 的直径在底部略有减小，这从刻蚀工艺的角度来看是意料之中的。TSV 表面铜的厚度小于 5μm，电镀后在 400℃的温度下退火 30min。最后，使用化学机械抛光（CMP）以去除表面多余的铜。

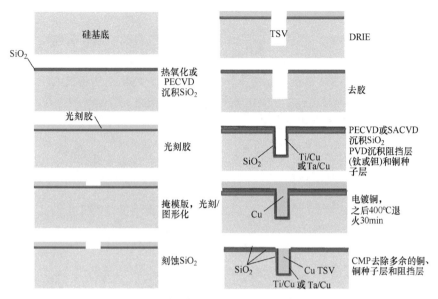

图 3.4 制备 TSV 的关键工艺步骤

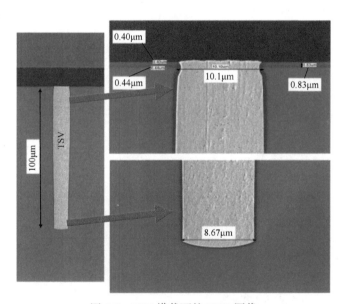

图 3.5 TSV 横截面的 SEM 图像

3.5 铜双大马士革工艺制备 RDL

图 3.6 给出了从 TSV 上开始制备 RDL 的关键工艺步骤。首先，通过 PECVD 工艺沉积 SiO_2 层，再通过步进式光刻技术（对准和曝光）在 SiO_2 层上方开孔，然后对 SiO_2 层进行反应离子刻蚀（RIE）形成通孔。应用步进式光刻

技术，刻出再分布布线的位置。然后通过 RIE 进一步刻蚀 SiO_2 层，去除光刻胶并溅射钛和铜。接着通过 ECD 在整个晶圆表面沉积铜，最后通过 CMP 去除表面多余的铜和钛，得到 RDL1。重复以上所有步骤获得 V_{12}（连接 RDL1 和 RDL2 的通孔）、RDL2 以及其他层。图 3.7 给出了在 TSV 上制备的 RDL 的截面图像，显示了 3 层 RDL，这些 RDL 层被称为无机 RDL 层，由 PECVD、铜双大马士革和 CMP 工艺制备得到。

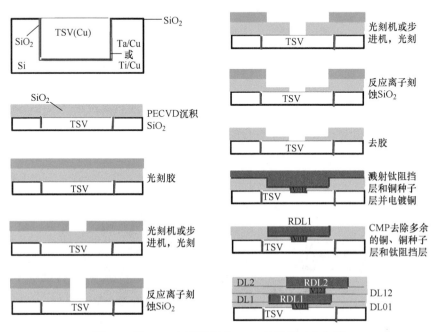

图 3.6 从 TSV 上开始制备 RDL 的关键工艺步骤

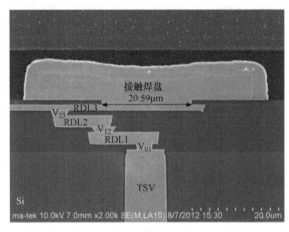

图 3.7 3 层 RDL 横截面的 SEM 图像

3.6 异构集成中的双面转接板

3.6.1 结构

图 3.8 给出了异构集成中双面转接板的示意图。图 3.9 展示了此转接板的布局。可以看到，图 3.9a 为 TSV 转接板顶部对应 2 个芯片的焊盘；图 3.9b 为底部芯片的焊盘和有机封装基板。封装基板的尺寸为 35mm × 35mm × 970μm。在芯片和转接板之间以及转接板和封装基板之间均使用了下填料。TSV 的直径为 10μm，节距为 150μm。芯片和转接板之间以及转接板和封装基板之间的焊凸点的直径为 90μm、节距为 125μm。封装基板和 PCB 之间的焊凸点的直径为 600μm、节距为 1000μm。表 3.1 总结了通过 TSV 转接板支撑双面芯片的异构集成的几何参数[24-28]。

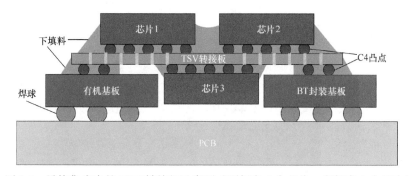

图 3.8 异构集成中的 TSV 转接板示意图（顶部有 2 个芯片，底部有 1 个芯片）

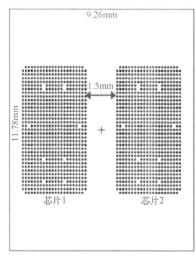
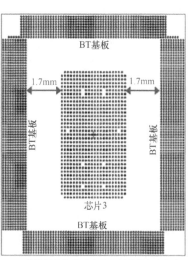

a) TSV 转接板顶部2个芯片的焊盘　　b) TSV 转接板底部芯片和有机封装基板的焊盘

图 3.9　TSV 转接板的布局

表 3.1　3D IC 异构集成模块的几何参数

	长	宽	高	直径	节距	注释
芯片 1~3	6.5mm	3.8mm	850μm			3 个芯片尺寸相同
转接板	11.9mm	9.4mm	100μm			
有机基板	35mm	35mm	970μm			
空腔	5.74mm	8.46mm				空腔在有机基板中
PCB	60mm	60mm	1600μm			
散热器	3.35mm	3.35mm	500μm			
C4 凸点				90μm	125μm	在芯片和转接板之间
C4 凸点				75μm	125μm	在转接板和有机基板之间
焊球				600μm	1000μm	在转接板和 PCB 之间
TSV				10μm	150μm	均匀分布在转接板中

3.6.2　热分析—边界条件

异构集成模型热分析的边界条件如图 3.10 所示，热载荷条件及边界条件见表 3.1 和表 3.2。可以看到，转接板上焊接的每个芯片的功耗为 5W，用来模拟安装在散热器上的假想散热片的冷却能力，热阻 R_{ca} 的变化范围为 0.1~−4.0℃/W，PCB 两端面的对流系数 h 为 20W/($m^2 \cdot$℃)。

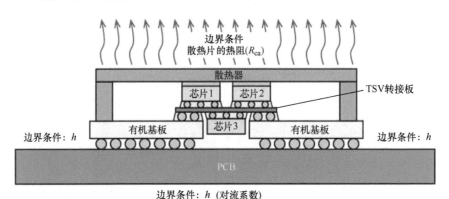

图 3.10　异构集成模型热分析的边界条件

表 3.2　分析中使用的热载荷及边界条件

热载荷	（芯片 1、芯片 2、芯片 3）每个芯片的功耗为 5W
边界条件	散热片的热阻（R_{ca}）可调（便于参数研究）
	PCB 两面散热系数均为 20W/($m^2 \cdot$K)
	假定其他表面绝热

3.6.3 热分析—TSV 等效模型

为了简化热仿真，可使用等效模型代替真实的 TSV 模型。在转换过程中，TSV、焊凸点和焊球的阵列可以被多个具有等效热导率的耦合等效区代替，如图 3.11 所示。特别地，为了准确起见，必须保留等效模型中转接板上的 SiO_2 层。如一个直径为 10μm、SiO_2 侧壁厚度为 0.2μm、高度为 100μm、节距为 50μm 的 2×2 TSV 阵列可以转化为 k_{xy}=144.56W/（m·K）、k_z=156.34W/（m·K）、面积为 100μm×100μm 的等效区域。

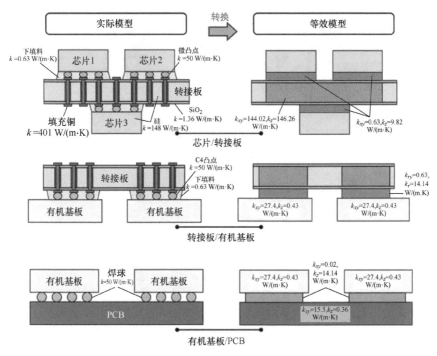

图 3.11 模型转换规则和计算得到的各等效区域的热导率

3.6.4 热分析—焊凸点/下填料等效模型

焊凸点和焊球的等效热导率应通过热阻串并联法确定。由于断路和大的节距，含有下填料的凸点阵列的等效热导率应在 x-y 方向上接近于下填料自身的等效热导率。同理，对于不含下填料的焊球阵列，其等效 k_{xy} 应非常小，几乎等于 0。图 3.11 为模型转换规则及计算得到的各等效区域的热导率。

3.6.5 热分析—结果

图 3.12 为各芯片（应用边界条件：散热器 R_{ca}=1.0℃/W）温度轮廓的顶视图，图 3.13 为异构集成模型的 3D 温度分布图。从温度分布轮廓可以看出，芯

片1和芯片2的温度分布相当不均匀，而芯片3的温度分布是均匀的。芯片1、2的最大温差约为4.7℃（63.2~58.5℃），而芯片3的最大温差只有0.4℃左右（70.6~70.2℃）。因此，芯片1、2在温度不均匀性方面比芯片3有更严重的热问题，可能会影响芯片的质量和可靠性。此外，从图3.12还可以看出散热片对于冷却异构集成模块来说是必要的，因为芯片1~3在自然对流且没附加散热器的条件下，分别达到410℃和417℃的高温。

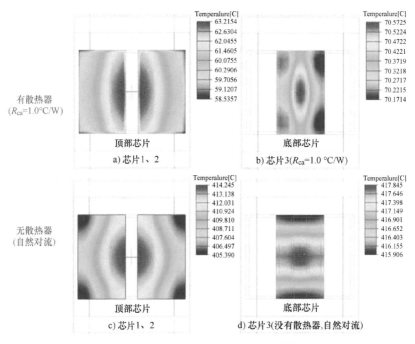

图3.12 温度分布图（4个温度条比例不同）

芯片平均温度和应用的散热器冷却能力（R_{ca}）之间的关系如图3.14所示。利用图3.14可以为散热器选择一个合适的散热片。如芯片的温度规范为100℃，那么，一个R_{ca}小于3.7℃/W的散热片对于冷却异构集成模块是必不可少的。另一方面，如果芯片温度必须低于60℃，那么必须选择一个R_{ca}小于0.3℃/W的大功率散热片。

图3.14显示了异构集成结构中一个严重的热问题，即连接在转接板底部的芯片温度总是比顶部的芯片温度高。这是因为顶部的芯片能直接与主冷却装置散热器和散热片接触，而底部的芯片不能。两种芯片平均温度的差值为9℃。芯片3的高温是不可避免的，因为芯片产生的热量主要经由散热器散去，而芯片1和芯片2的温度阻碍了芯片3到散热器的散热。如果不对异构集成模块进行结构上的改变，将很难解决这个严重的热问题。加厚散热器可以降低芯片的温度，但效果有限。文献[32]提出了一种实用而简单的技术，即从PCB端插入金属热

塞,并直接接触芯片 3 的背面。插入的金属热塞可以有效地排出芯片 3 的热量,也能显著降低芯片 1、2 的温度。

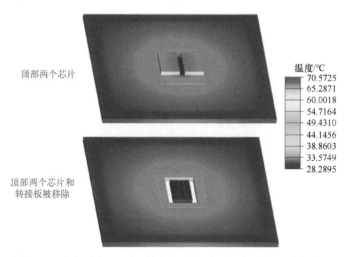

图 3.13 有机基板、TSV 转接板和芯片 1~3 集成的温度轮廓图
(上图芯片 3 被屏蔽;下图 TSV 转接板和芯片 1、2 被屏蔽)

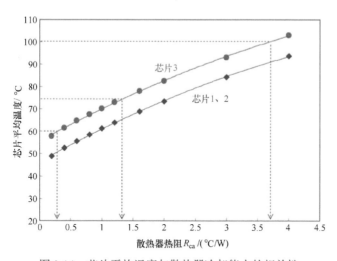

图 3.14 芯片平均温度与散热器冷却能力的相关性

3.6.6 热机械分析—边界条件

对于热机械学仿真,可使用非线性有限元建模和分析来计算结构的热应力/应变分布。同时,对该结构中焊点的可靠性进行评估。为了得到焊点有关温度、时间依赖的非线性蠕变曲线,在 −25~125℃ 的热循环条件下,对该结构进行了 5 次循环模拟,如图 3.15 所示。

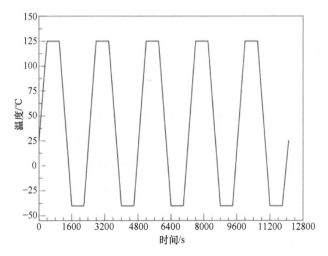

图 3.15　热循环边界条件（-25℃↔125℃）

3.6.7　热机械分析—材料特性

该结构中所使用材料的热机械材料特性如杨氏模量、泊松比和热膨胀系数（CTE）见表 3.3。由于温度和时间同时影响着无铅焊料的材料特性，因此采用以下非线性温度有关时间的本构方程[33]：

$$\frac{d\varepsilon}{dt} = C_1[\sinh(C_2\sigma)]^{C_3} \exp\left(-\frac{C_4}{T}\right)$$

$$E(\text{GPa}) = 49 - 0.07 \times T(\text{℃})$$

$$\alpha(\text{ppm/℃}) = 21.301 + 0.017 \times T(\text{℃})$$

其中，$C_1=50000$（1/s）；$C_2=0.01$（1/MPa）；$C_3=5$；$C_4=0.5$（K）。方程中，E 为杨氏模量，α 为热膨胀系数，ε 为蠕变应变，K 为开尔文温度，t 为时间。

表 3.3　用于有限元分析的材料特性

	杨氏模量（GPa）	泊松比	CTE/(ppm/K)
硅	130	0.28	2.8
BT 基板	X:26 Y:11	0.39	X:15 Y:52
FR4	X:22 Y:10	0.28	X:18 Y:70
下填料	9.07	0.3	40.75
合金焊料	取决于温度	0.35	取决于温度
电镀铜	70	0.34	18
低介电常数的铜焊盘	8	0.3	10
SiO_2	70	0.16	0.6
IMC(Cu_6Sn_5)	125	0.3	18.2
镍	131	0.3	13.4

在热机械分析中只进行了2D有限元建模。因为结构对称，所以只需对一半结构进行建模。对称轴是 y 轴，应用适当的位移和转动边界条件，热循环边界条件见图3.15，温度从 $-25℃ \leftrightarrow 125℃$ 循环60min，25℃时应力为0。

3.6.8 热机械分析—结果

采用ANSYS Mechanical R14的2D单元（PLANE182）构建半有限元模型，如图3.16、图3.17所示，对称轴为 y 轴。图3.16为转接板中TSV及周边区域的细致建模。芯片与转接板、转接板与封装基板之间的焊凸点、封装基板与PCB之间的焊球的细致建模如图3.17所示。整个模型经过如图3.15所示的热循环加载，其结果将在下一部分展示。

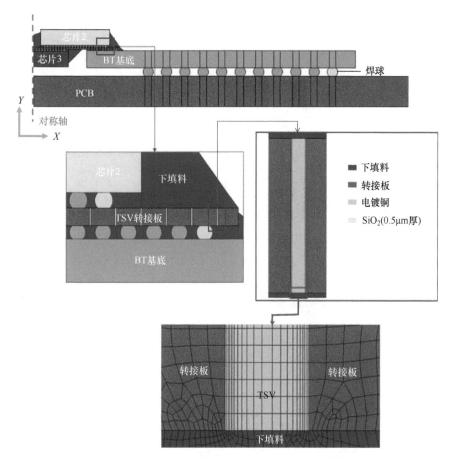

图3.16 异构集成的半有限元模型（展示了TSV及其周围区域的细节）

图3.18显示了125℃时TSV角落处的冯米斯应力轮廓图（25℃时应力为0）。可以看出，最大应力约为135MPa，临界位置在TSV的 $Cu-SiO_2$ 与下填料形成的界面附近。

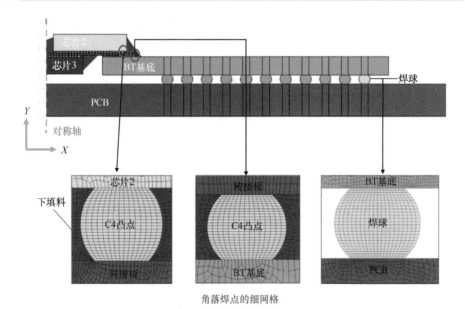

图 3.17　异构集成的半有限元模型（展示了芯片 2 与转接板、转接板与封装基板之间的焊凸点及封装基板与 PCB 之间的焊球细节（未按比例））

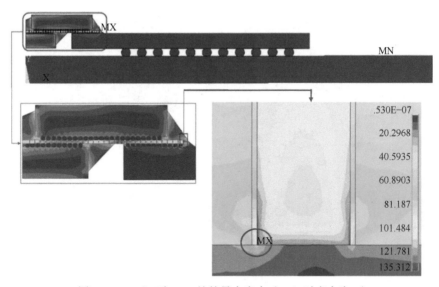

图 3.18　125℃ 时 TSV 处的最大应力（25℃时应力为 0）

滞回线稳定的时间对研究多次循环下的蠕变响应具有重要意义。图 3.19 为芯片 2 与转接板之间角落焊凸点处多个循环的剪切应力和剪切蠕变应变滞回线。可以看出，剪切蠕变应变 - 剪切应力滞回线在第三次循环后相当稳定。事实上，对于那些焊凸点，它们的滞回线在第一个周期之后就趋于稳定。图 3.19 显示了芯片 2 和转接板之间角落焊凸点的蠕变应变能密度发展历程。可以看出，芯片 2 与转接板之间角落焊凸点每次循环的蠕变应变能密度为 0.0107MPa。在 $-25℃\leftrightarrow125℃$ 循环 60min 的条件下，该数值太小，并不能造成焊点热疲劳可靠性问题。下填料在此处起到了作用！其他焊点位置的蠕变响应见参考文献 [27]。

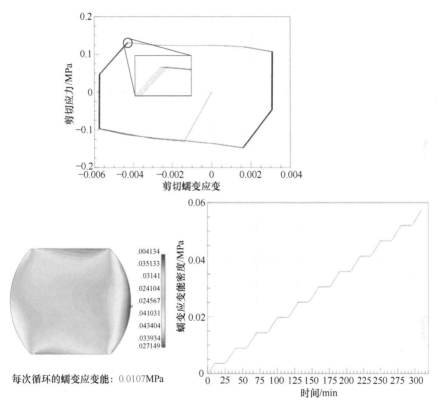

图 3.19 （上图）芯片 2 与转接板之间角落焊凸点的滞回线（5 次循环），（下图）芯片 2 与转接板之间角落焊凸点的蠕变应变能密度 – 时间曲线

3.6.9　TSV 加工

图 3.20 展示了 TSV/RDL 硅转接板的横截面结构。这是一种 5 层结构，正面是 3 层 RDL（TR1、TR2 和 TR3），背面是 2 层 RDL（BR1、BR2），5 层结构均由铜大马士革工艺制作。这种需要背面制作的结构给工艺带来了一定的挑战。在相同分辨率下，使用接触式光刻比使用步进式光刻的成本更低，因此本段中提到的光刻工艺均使用接触式光刻。表 3.4 列出了 TSV/RDL 转接板制作工艺流程。

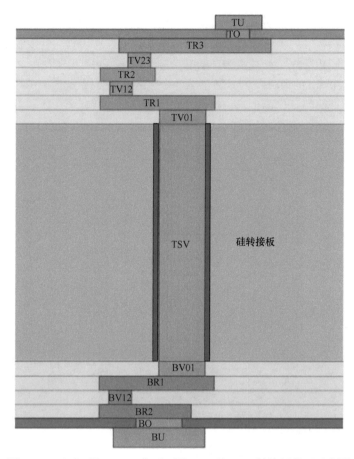

图 3.20　正面 3 层 RDL、背面 2 层 RDL 的 TSV 转接板截面示意图

表 3.4　TSV/RDL 转接板制作工艺流程

工艺阶段	
层间绝缘层（ILD）	
TSV	TSV 直径 =10μm TSV 深度 =105μm TSV 氧化硅绝缘层（SACVD）=0.5μm 钽阻挡层 =0.08μm
正面 3 层 RDL	铜大马士革工艺 最小线宽 =3μm 最小孔径 =3μm 金属层厚度 =2μm 和 1.5μm 钽阻挡层 =0.05μm
顶部钝化层开窗	开窗直径 =60μm

(续)

工艺阶段	
层间绝缘层（ILD）	
顶部凸点下金属层（UBM）	顶部 UBM 尺寸 =75μm 铜的厚度 =5μm 镍钯金 =2μm
正面临时键合至硅载体	键合胶厚度 =25μm
背面减薄	打磨 + 化学机械抛光（CMP）
硅凸槽 TSV 背面露头	干法刻蚀
背面绝缘层	等离子增强化学气相沉积（PECVD）氧化硅 温度 <200℃ 厚度 =1μm
CMP 除去硅通孔表面剩余的铜	背面绝缘层保留 >0.8μm
背面 2 层 RDL	铜双大马士革工艺 最小线宽 =5μm 最小孔径 =5μm 金属层厚度 =2μm 钽阻挡层 =0.05μm
底部钝化层开窗	开窗直径 =60μm
底部 UBM	UBM 尺寸 =75μm 铜的厚度 =5μm 镍钯金 =2μm

制作 TSV/RDL 转接板的第一步是使用等离子增强化学气相沉积（PECVD）方法制作 SiNx 或者 SiOx 绝缘层。进行光刻后，用 Bosch 工艺对硅衬底进行深反应离子刻蚀制作高深宽比 TSV，接着用正常压化学气相沉积（SACVD）方法在 TSV 侧壁制作 SiOx 绝缘层，然后用物理气相沉积（PVD）方法制作钽阻挡层和铜种子层，最后电镀铜填充 TSV，最终形成的盲孔 TSV 的开口直径约为 10μm，深度约为 105μm，深宽比为 10.5。在高深宽比通孔结构中，采用自底部向上的电镀方式可以得到表面铜厚度很薄的无缝隙 TSV，扫描电子显微镜截面图如图 3.21 所示，TSV 的直径在底部略有减小，从刻蚀工艺角度来看，这样的结果符合工艺规律。使用上述工艺形成的 TSV 之间的表面铜厚度小于 5μm。在 400℃下退火 30min，采用化学机械抛光（CMP）的方法去除表面剩余的铜覆盖层后，转接板中的 TSV 结构即制作完成。

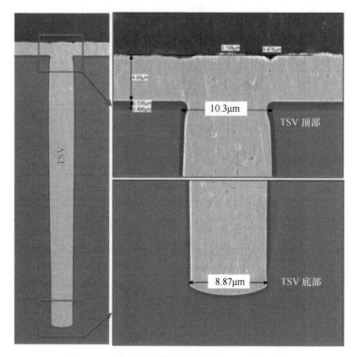

图 3.21　扫描电子显微镜下的 TSV 截面图

3.6.10　带有正面 RDL 的转接板加工

TSV 制作完成后,制作硅转接板需要使用铜大马士革工艺在其正面制作 3 层 RDL,这种工艺是从传统的铜后段工艺改进而来。受接触式光刻分辨率限制,本段中展示的 RDL 的临界尺寸(CD)为 3μm。3 层 RDL 制作完成后,制作倒装焊中与凸点接触的正面凸点下金属层(UBM)焊盘。UBM 由 Cu-UBM 焊盘和 Ni、Pd、Au 组成,这为有源芯片和转接板之间的下填料提供了空间,并且降低了键合失败的可能性。图 3.22 为扫描电子显微镜下转接板正面各层(TSV、3 层 RDL 和 UBM)截面图。图 3.23 为 TSV/RDL 转接板俯视图,互连区域为 2 个有源区域,其余区域为无源区域。

3.6.11　带有正面 RDL 铜填充转接板的 TSV 露头

正面工艺制作完成后,需要对晶圆进行减薄来进行临时键合。在对硅转接板背面进行研磨之前,需要将硅转接板晶圆与承载晶圆临时键合,然后进行 TSV 露头。如图 3.24 所示,硅转接板晶圆背面经过研磨后,剩余的硅转接板晶圆厚度分布非常均匀。接着沉积绝缘层,对晶圆背面进行干法刻蚀,露出 TSV 以便与背面 RDL 互连。由于临时键合剂最高工作温度的限制,背面工艺的温度应当控制在 200℃以下。

第 3 章 硅基板上的异构集成（TSV 转接板）

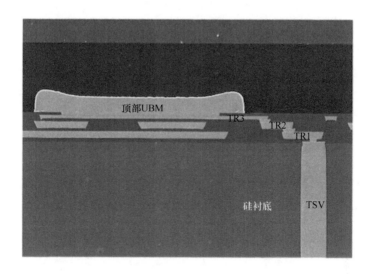

图 3.22　扫描电子显微镜转接板正面（TSV、3 层 RDL 和 UBM）截面图

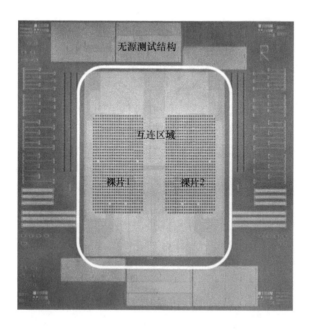

图 3.23　TSV/RDL 转接板俯视图

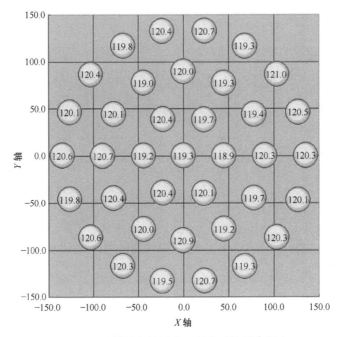

图 3.24 硅转接板晶圆背面研磨后的剩余厚度

3.6.12 带有背面 RDL 的转接板加工

考虑到掩模与接触式光刻的能力以及正面工艺和临时键合过程中发生的晶圆翘曲，背面 RDL 的最小 CD 应控制在 5μm。此外，为了减少工艺导致的已减薄晶圆的开裂或切屑，背面 RDL 使用铜双大马士革工艺制作（即 RDL+ 电镀、CMP 形成孔）。背面第一层与 TSV 背面互连的 RDL 的扫描电子显微镜截面图如图 3.25 所示。与正面工艺类似，2 层 RDL 制作完成后制作背面 UBM 焊盘，它们分别用于倒装焊中凸点和有机基板的互连。

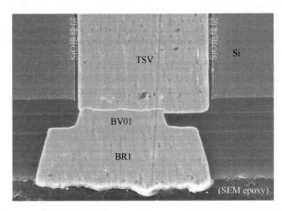

图 3.25 与 TSV 背面互连的铜 RDL（BR1 + BV01）扫描电子显微镜截面图

图 3.26 为转接板背面各层（TSV、2 层 RDL、BR1、BR2 和底部 UBM）扫描电子显微镜截面图，图 3.27 为 TSV 转接板俯视图。将减薄后的硅转接板晶圆与承载晶圆解键合之后，对硅转接板晶圆划片，划片后的转接板可用于异构集成，也可用于转接板和封装的无源电学表征。

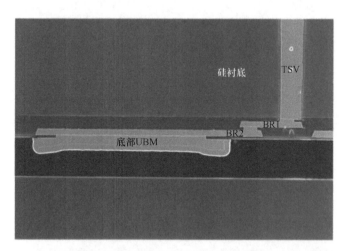

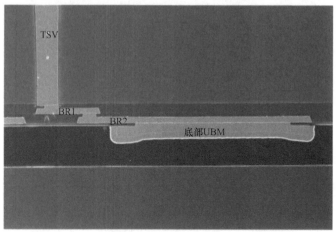

图 3.26 转接板背面各层扫描电子显微镜截面图
（TSV、2 层 RDL、BR1、BR2 和底部 UBM）

3.6.13 转接板的无源电学表征

文献 [24] 中报道了各种微带线、带线和共面波导线的仿真结果与测量结果。图 3.28 显示了 3 种不同情况下的（插入损耗（IL）和回波损耗（RL））。在使用公称设计尺寸时，测量结果与 3D 建模结果之间存在差异，但当使用实际制作尺寸时，如图3.29所示，仿真结果和测量结果非常较为吻合。

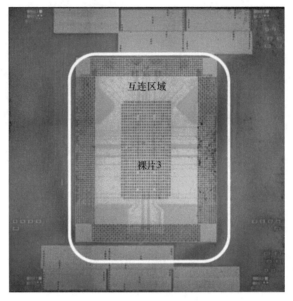

图3.27 TSV转接板俯视图

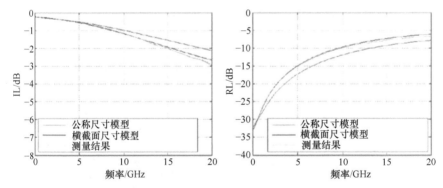

图3.28 3种不同情况下的插入损耗和回波损耗仿真和测量结果相关性

3.6.14 最终组装

图3.30为模块组装的最后工艺流程。首先将背面的芯片与转接板互连，回流并固化下填料，接着将转接板与有机封装基板正面相连，再次回流并固化下填料，然后将正面的2个芯片与转接板的正面互连，回流并固化下填料，最后如图3.30步骤7和步骤8所示，将焊球放置在封装基板底部并回流。

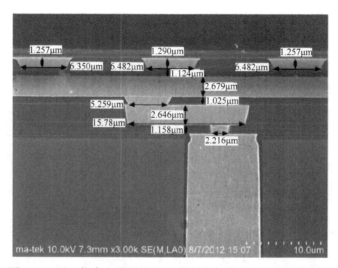

图 3.29 用于仿真和测量的 TSV 转接板扫描电子显微镜截面图

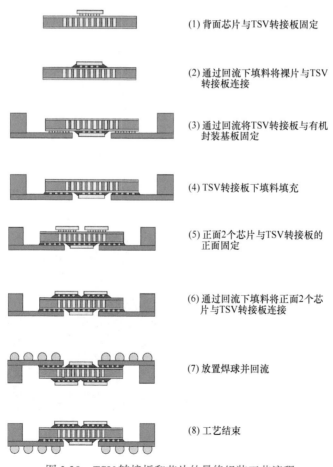

(1) 背面芯片与TSV转接板固定

(2) 通过回流下填料将裸片与TSV转接板连接

(3) 通过回流将TSV转接板与有机封装基板固定

(4) TSV转接板下填料填充

(5) 正面2个芯片与TSV转接板的正面固定

(6) 通过回流下填料将正面2个芯片与TSV转接板连接

(7) 放置焊球并回流

(8) 工艺结束

图 3.30 TSV 转接板和芯片的最终组装工艺流程

图 3.31 展示了一个组装完成的异构集成模块,它包括 1 个无源 TSV 转接板、2 个转接板正面的有源芯片(带下填料,如图 3.31a 所示),以及 1 个底部的有源芯片(带下填料,如图 3.31b 所示)。该转接板与芯片组被焊接在带空腔的封装基板正面。

a) 正面2个芯片

b) 背面1个芯片

图 3.31 组装完成的异构集成模块

图 3.32 为全组装模块横截面图。转接板通过底部填充的方式与 3 个芯片连接,并通过下填料的方式焊接到 4-2-4 封装基板上。图 3.33 为全组装的异构集成模块俯视图的 X 射线图像。可以看到,正面的 2 个芯片(未减薄)和底部的 1 个芯片很好地对齐,并且转接板背面周边区域有将转接板与封装 BT 基板互连的焊点。

图 3.32 全组装模块横截面图(3 个芯片、TSV 转接板、封装基板、下填料)

第 3 章 硅基板上的异构集成（TSV 转接板）

图 3.33 全组装的异构集成模块俯视图的 X 射线图像

3.7 总结和建议

1）人工智能（AI）的出现导致 CPU、GPU、HBM、FPGA 等半导体的密度和 I/O 数增加、引脚节距减小、芯片尺寸增大。即使是 12 层（6-2-6）的有机封装基板也不足以支持上述半导体芯片的集成，因此需要使用 TSV 转接板。

2）采用 PECVD、铜大马士革工艺、CMP 方法制备硅基板异构集成（TSV 转接板）RDL，可以使金属线宽度和 RDL 间距降到亚微米级别，这种异构集成可以使集成芯片获得最佳的性能。

3）然而在硅基板上进行异构集成（TSV 转接板）的成本过高，因此需要开发成本较低的 TSV 转接板。

4）目前，赛灵思、英伟达和超威半导体（AMD）等公司采用在硅基板上进行异构集成（TSV 转接板）工艺，台积电和联华电子公司制造了 TSV 转接板。

5）由于异构集成的双面转接板的正面和背面都可以集成芯片，转接板的尺寸可以做得更小，或者可以在相同尺寸的转接板上集成更多的芯片。而且由于芯片之间的互连是面对面而不是边对边的，这种集成方式的电学性能更优。TSV 转接板是一个真正的 3D IC 集成。

参考文献

[1] Shao, S., Y. Niu, J. Wang, R. Liu, S. Park, H. Lee, G. Refai-Ahmed, and L. Yip, "Comprehensive Study on 2.5D Package Design for Board-Level Reliability in Thermal Cycling and Power Cycling", *Proceedings of IEEE/ECTC*, May 2018, pp. 1662–1669.

[2] McCann, S., H. Lee, G. Refai-Ahmed, T. Lee, and S. Ramalingam, "Warpage and Reliability Challenges for Stacked Silicon Interconnect Technology in Large Packages", *Proceedings of IEEE/ECTC*, May 2018, pp. 2339–2344.

[3] Lai, C., H. Li, S. Peng, T. Lu, and S. Chen, "Warpage Study of Large 2.5D IC Chip Module", *Proceedings of IEEE/ECTC*, May 2017, pp. 1263–1268.

[4] Hong, J., K. Choi, D. Oh, S. Park, S. Shao, H. Wang, Y. Niu, and V. Pham, "Design Guideline of 2.5D Package with Emphasis on Warpage Control and Thermal Management", *Proceedings of IEEE/ECTC*, May 2018, pp. 682–692.

[5] Lee, J., C. Lee, C. Kim, and S. Kalchuri, "Micro Bump System for 2nd Generation Silicon Interposer with GPU and High Bandwidth Memory (HBM) Concurrent Integration", *Proceedings of IEEE/ECTC*, May 2018, pp. 607–612.

[6] Hou, S., W. Chen, C. Hu, C. Chiu, K. Ting, T. Lin, W. Wei, W. Chiou, V. Lin, V. Chang, C. Wang, C. Wu, and D. Yu, "Wafer-Level Integration of an Advanced Logic-Memory System Through the Second-Generation CoWoS Technology", *IEEE Transactions on Electron Devices*, October 2017, pp. 4071–4077.

[7] Chaware, R., K. Nagarajan, and S. Ramalingam, "Assembly and Reliability Challenges in 3D Integration of 28 nm FPGA die on a Large High-Density 65 nm Passive Interposer", *IEEE/ECTC Proceedings*, May 2012, pp. 279–283.

[8] Banijamali, B., S. Ramalingam, K. Nagarajan, and R. Chaware, "Advanced Reliability study of TSV Interposers and Interconnects for the 28 nm Technology FPGA", *IEEE/ECTC Proceedings*, May 2011, pp. 285–290.

[9] Banijamali, B., S. Ramalingam, H. Liu, and M. Kim, "Outstanding and Innovative Reliability study of 3D TSV Interposer and Fine-Pitch Solder Micro-Bumps", *IEEE/ECTC Proceedings*, May 2012, pp. 309–314.

[10] Banijamali, B., C. Chiu, C. Hsieh, T. Lin, C. Hu, S. Hou, et al., "Reliability evaluation of a CoWoS-enabled 3D IC package", *IEEE/ECTC Proceedings*, May 2013, pp. 35–40.

[11] Xie, J., H. Shi, Y. Li, Z. Li, A. Rahman, K. Chandrasekar, et al., "Enabling the 2.5D integration", *Proceedings of IMAPS International Symposium on Microelectronics*, October 2012, pp. 254–267.

[12] Souriau, J., O. Lignier, M. Charrier, and G. Poupon, "Wafer Level Processing Of 3D System in Package for RF and Data Applications", *IEEE/ECTC Proceedings*, 2005, pp. 356–361.

[13] Henry, D., D. Belhachemi, J-C. Souriau, C. Brunet-Manquat, C. Puget, G. Ponthenier, J. Vallejo, C. Lecouvey, and N. Sillon, "Low Electrical Resistance Silicon Through Vias: Technology and Characterization", *IEEE/ECTC Proceedings*, 2006, pp. 1360–1366.

[14] Yu, A., J. H. Lau, S. Ho, A. Kumar, W. Hnin, W. Lee, M. Jong, V. Sekhar, V. Kripesh, D. Pinjala, S. Chen, C. Chan, C. Chao, C. Chiu, C. Huang, and C. Chen, "Fabrication of High Aspect Ratio TSV and Assembly with Fine-Pitch Low-Cost Solder Microbump for Si Interposer Technology with High-Density Interconnects", *IEEE Transactions on CPMT*, Vol. 1, No. 9, September 2011, pp. 1336–1344.

[15] Yu, A., J. H. Lau, Ho, S., Kumar, A., Yin, H., Ching, J., Kripesh, V., Pinjala, D., Chen, S., Chan, C., Chao, C., Chiu, C., Huang, M., and Chen, C., "Three Dimensional Interconnects with high Aspect ratio TSVs and Fine Pitch Solder Microbumps", *IEEE Proceedings of ECTC*, May 2009, pp. 350–354.

[16] Selvanayagam, C., J. H. Lau, X. Zhang, S. Seah, K. Vaidyanathan, and T. Chai, "Nonlinear Thermal Stress/Strain Analysis of Copper Fill TSV (Through Silicon Via) and Their Flip-Chip Microbumps", *IEEE/ECTC Proceedings*, May 27–30, 2008, pp. 1073–1081.

[17] Selvanayagam, C., J. H. Lau, X. Zhang, S. Seah, K. Vaidyanathan, and T. Chai, "Nonlinear Thermal Stress/Strain Analyses of Copper Filled TSV (Through Silicon Via) and Their Flip-Chip Microbumps", *IEEE Transactions on Advanced Packaging*, Vol. 32, No. 4, November 2009, pp. 720–728.

[18] Lau, J. H., and G. Tang, "Thermal Management of 3D IC Integration with TSV (Through

Silicon Via)", *IEEE/ECTC Proceedings*, May 2009, pp. 635–640.
[19] Lau, J. H., Y. S. Chan, and R. S. W. Lee, "3D IC Integration with TSV Interposers for High-Performance Applications", *Chip Scale Review*, Vol. 14, No. 5, September/October 2010, pp. 26–29.
[20] Lau, J. H., "TSV Manufacturing Yield and Hidden Costs for 3D IC Integration", *IEEE/ECTC Proceedings*, May 2010, pp. 1031–1041.
[21] Zhang, X., T. Chai, J. H. Lau, C. Selvanayagam, K. Biswas, S. Liu, D. Pinjala, et al., "Development of Through Silicon Via (TSV) Interposer Technology for Large Die (21 × 21 mm) Fine-pitch Cu/low-k FCBGA Package", *IEEE Proceedings of ECTC*, May 2009, pp. 305–312.
[22] Chai, T. C., X. Zhang, J. H. Lau, C. S. Selvanayagam, D. Pinjala, et al., "Development of Large Die Fine-Pitch Cu/low-*k* FCBGA Package with through Silicon via (TSV) Interposer", *IEEE Transactions on CPMT*, Vol. 1, No. 5, May 2011, pp. 660–672.
[23] Chien, H. C., J. H. Lau, Y. Chao, R. Tain, M. Dai, S. T. Wu, W. Lo, and M. J. Kao, "Thermal Performance of 3D IC Integration with Through-Silicon Via (TSV)", *IMAPS Transactions, Journal of Microelectronic Packaging,* Vol. 9, 2012, pp. 97–103.
[24] Ji, M., M. Li, J. Cline, D. Seeker, K. Cai, J. H. Lau, P. Tzeng, et al., "3D Si Interposer Design and Electrical Performance Study", *Proceedings of Design Con*, Santa Clara, CA, January 2013, pp. 1–23.
[25] Lau, J. H., P. Tzeng, C. Lee, C. Zhan, M. Li, J. Cline, K. Saito, et al., "Redistribution Layers (RDLs) for 2.5D/3D IC Integration", *IMAPS Transactions, Journal of Microelectronic Packaging*, Vol. 11, No. 1, First Quarter 2014, pp. 16–24.
[26] Lau, J. H., P. Tzeng, C. Lee, C. Zhan, M. Li, J. Cline, K. Saito, et al., "(Redistribution Layers (RDLs) for 2.5D/3D IC Integration", *Proceedings of the 46th IMAPS International Symposium on Microelectronics*, Orlando, FL, October 2013, pp. 434–441.
[27] Wu, S. T., H. Chien, J. H. Lau, M. Li, J. Cline, and M. Ji, "Thermal and Mechanical Design and Analysis of 3D IC Interposer with Double-Sided Active Chips", *IEEE/ECTC Proceedings*, Las Vegas, NA, May 2013, pp. 1471–1479.
[28] Tzeng, P. J., J. H. Lau, C. Zhan, Y. Hsin, P. Chang, Y. Chang, J. Chen, et al., "Process Integration of 3D Si Interposer with Double-Sided Active Chip Attachments", *IEEE/ECTC Proceedings*, Las Vegas, NA, May 2013, pp. 86–93.
[29] Lau, J. H., "Recent Advances and New Trends in Flip Chip Technology", *ASME Transactions, Journal of Electronic Packaging*, September 2016, Vol. 138, Issue 3, pp. 1–23.
[30] Lau, J. H., *3D IC Integration and Packaging*, McGraw-Hill, New York, 2016.
[31] Lau, J. H., *Through-Silicon Via (TSV) for 3D Integration*, McGraw-Hill, New York, 2013.
[32] Chien, H., J. H. Lau, T. Chao, M. Dai, and R. Tain, "Thermal Management of Moore's Law Chips on Both Sides of an Interposer for 3D IC Integration SiP", *IEEE ICEP Proceedings*, Japan, April 2012, pp. 38–44.
[33] Lau, J. H., *Reliability of RoHS Compliant 2D & 3D IC Interconnects*, McGraw-Hill, New York, 2011.

第 4 章

硅基板（桥）上的异构集成

4.1 引言

硅通孔（TSV）转接板是非常昂贵的[1-10]。硅桥是支持硅衬底异构集成的一种形式。基本上，硅桥是一块带有 RDL（再布线层）和触点的挡片，但没有 TSV，即无 TSV 转接板。通常，RDL 和接触焊盘被制造在一个挡片上，然后切割成单独的硅桥。本章将介绍英特尔的 EMIB（嵌入式多模互连桥）的制造方式[11-13]，并简要讨论 IMEC 用于异构集成通信中的硅桥[14]，最后介绍 ITRI 的 TSH（through-silicon hole）转接板（桥）技术[15-18]。

4.2 无 TSV 转接板技术：英特尔的 EMIB 技术

通常，TSV 转接板是非常昂贵的[1-10]。英特尔提出了一种嵌入式多模互连桥（EMIB）来取代 TSV 转接板[11, 12]。芯片之间的横向连通将由嵌入式硅桥和电源/地完成，一些信号将通过有机封装基板，如图 4.1 和图 4.2 所示。2015 年 11 月 9 日，阿尔特拉（Altera）/英特尔宣布了行业首个异构集成器件，将 SK 海力士（Hynix）的堆叠 HBM 与高性能 Stratix®10 FPGA 和 SoC 集成在一起，如图 4.1 所示。可见，TSV 转接板已经消失，取而代之的是英特尔的 EMIB。用 EMIB 制作有机封装基板有两个主要任务：一是制作 EMIB，另一个任务是用 EMIB 制作基板。

第 4 章 硅基板（桥）上的异构集成

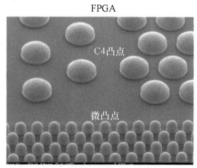
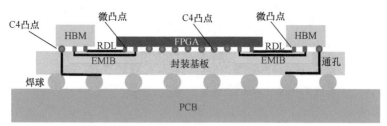

图 4.1 使用英特尔 EMIB 进行 HBM 和 FPGA 互连的异构集成

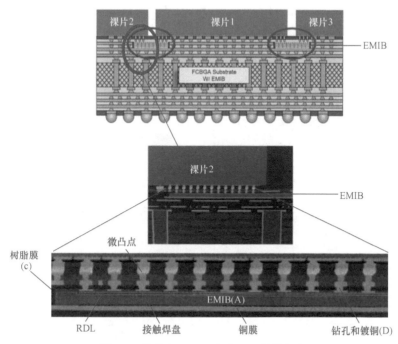

图 4.2 英特尔 EMIB（无 TSV 转接板）

4.3 EMIB 的制造

要制造 EMIB，必须首先在硅晶圆上构建 RDL（包括接触焊盘）。制作 RDL 的方式取决于 RDL 布线的线宽/间距。最后，将硅晶圆的非 RDL 侧连接到黏附芯片的薄膜（DAF），然后将硅晶圆分割成单独的桥。

4.3.1 利用精细 RDL 制造硅桥

对于 RDL 金属的精细线宽和间距（>2μm），有机 RDL（聚合物和 ECD Cu+ 蚀刻）应足够。如果线宽和间距为超细（≤2μm 甚至亚微米级尺度），则应使用无机 RDL（PECVD 和 双层铜镶嵌 +CMP）。

制造用于精细线宽和间距的有机 RDL 的关键工艺步骤如图 4.3 所示。可以看出，对于 PI（聚酰亚胺）显影，整个硅晶圆都旋涂有光敏 PI。随后应用曝光机掩模对准器或步进式光刻机（以获得更高的产量），然后使用光刻技术对准、曝光和制造 PI 的通孔。最后，PI 在 200°C 下固化 1h——这将形成 5μm 厚的 PI 层。然后在 175°C 下通过 PVD 在整个硅晶圆上溅射钛（Ti）和铜（CU）。应用光刻胶和掩模对准器或步进器，并使用光刻技术打开再分布迹线的位置。在室温下通过 ECD 在光刻胶开口中的钛/铜上电镀铜，然后剥离光刻胶并蚀刻掉钛/铜，从而获得 RDL1。最后，重复上述所有步骤以获取其他 RDL。如图 4.4 所示 RDL 就是用这种方法制造的。RDL 的金属线宽和间距可低至 5μm，以保证高产量。

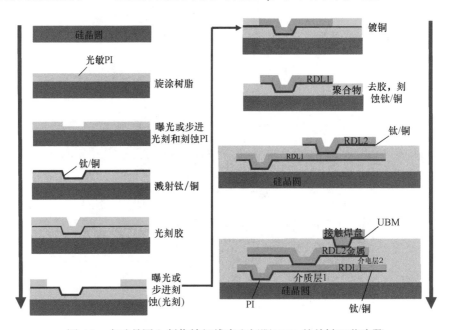

图 4.3 在硅晶圆上制作精细线宽和间距 RDL 的关键工艺步骤

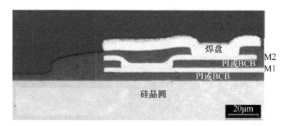

图 4.4　有机 RDL 法在硅晶圆上的双层 RDL 的横截面

4.3.2　利用超精细 RDL 制造硅桥

制造具有超精细线宽和间距 RDL 的关键工艺步骤如图 4.5 所示。这是最古老的后端半导体工艺。该工艺使用 SiO_2 或 SiN 作为介电层，使用 ECD 在整个晶圆上沉积铜。然后使用 CMP（化学机械抛光）去除覆盖层的铜和种子层，以制造 RDL 的铜导体层。首先，使用等离子增强化学气相沉积（PECVD）在裸硅晶圆上形成一层全面覆盖的 SiO_2（或 SiN）薄层，然后使用旋涂机层压光刻胶。在这些步骤之后，使用光刻机在聚合物阻挡层开口并使用反应离子蚀刻（RIE）去除 SiO_2。然后，利用光刻机使聚合物阻挡层开口更宽，并使用 RIE 蚀刻更多的 SiO_2。接下来，去除聚合物阻挡层，溅射钛/铜，并在整个晶圆上对铜进行 ECD。完成这些步骤后，利用 CMP 去除覆盖层铜和钛/铜，即获得 RDL1 和 V01（连接硅和 RDL1 的通孔），如图 4.6 所示。这种方法称为双层铜大马士革法[19]。最后，重复所有过程以获取其他 RDL。RDL 的金属线宽和间距可以达到 ≤ 2μm 的水平并低至亚微米级。

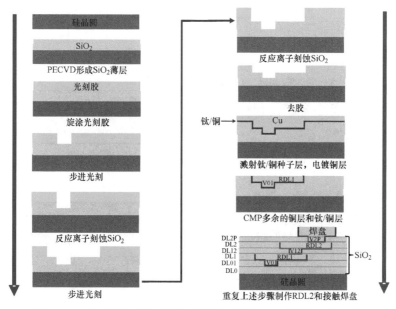

图 4.5　在硅晶圆上制作超精细线宽和间距 RDL 的关键工艺步骤

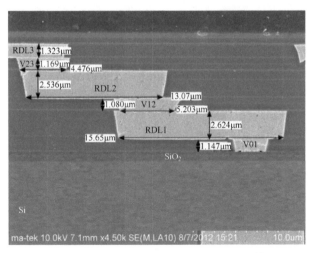

图 4.6 无机 RDL 法在硅晶圆上的 3 层 RDL 横截面

4.4 英特尔 EMIB 有机基板的制作

为了制作带有 EMIB 的基板，首先在基板空腔中的铜箔顶部放置带有薄膜的单个 EMIB，如图 4.7a 所示。然后在整个有机封装基板上层压树脂膜。（在环氧树脂上）钻孔和镀铜以填充孔（通孔），连接到 EMIB 的接触焊盘。如图 4.7b 所示，继续镀铜进行基板的横向连接。然后，在整个基板上层压另一层树脂膜，（在树脂上）钻孔和镀铜以填充孔并制作接触焊盘，如图 4.7c 所示。更小节距上的较小焊盘用于微凸点，而更大节距上的较大焊盘用于常规凸点，如图 4.7d 所示，具有 EMIB 的有机封装基板已经可以用于芯片的键合。在 EMIB 顶部的有机基板的堆积也可以通过半加成工艺制造，具体见第 2 章。

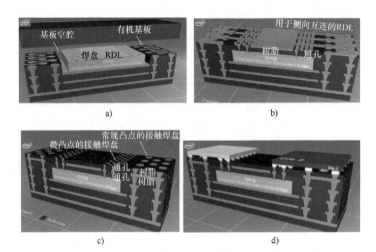

图 4.7 制作英特尔 EMIB 有机基板的关键工艺步骤

第 4 章 硅基板（桥）上的异构集成

4.5 总体组装

值得注意的是，为了使用 EMIB，芯片将具有不同类型/尺寸的凸点（见图 4.1），即 C4 凸点和微凸点（铜+焊帽）。晶圆碰撞和倒装芯片组装可能具有挑战性，需要在组装开发中进行更多的工作。图 4.8 显示了英特尔的 CPU 和 AMD 的 GPU+HBM 的异构集成，GPU 和 HBM 在具有 EMIB 和堆积层的有机基板上。图 4.9 显示了英特尔未来的 CPU、AMD 的 GPU 和海力士的 HBM 的异构集成[13]。它们通过 EMIB 连接。

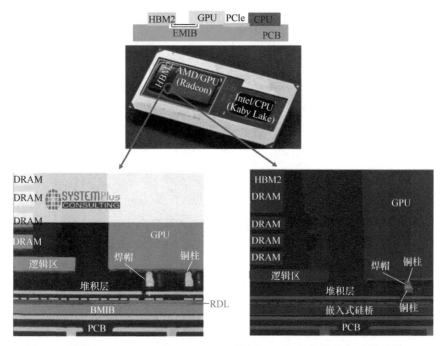

图 4.8 英特尔 CPU、AMD GPU+HBM（具有 EMIB）的异构集成用于其 NB

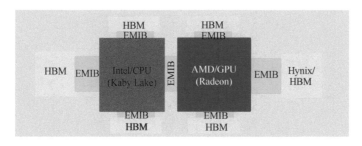

图 4.9 英特尔未来将 CPU、AMD GPU+HBM 与 EMIB 异构集成用于其 HPC

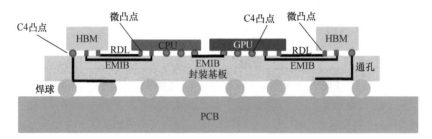

图 4.9　英特尔未来将 CPU、AMD GPU+HBM 与 EMIB 异构集成用于其 HPC（续）

4.6　IMEC 桥的异构集成

自从英特尔提出使用 EMIB 作为异构集成系统中芯片之间的高密度互连的提议后，"桥"就非常流行。如最近，IMEC 提出[14]使用桥转接板 + 扇出晶圆级封装（FOWLP）技术互连逻辑芯片、宽 I/O DRAM 和闪存，如图 4.10 所示，目的是不再对所有器件芯片都使用 TSV。

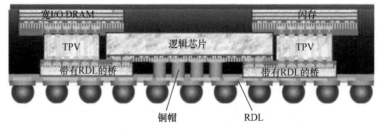

➤ 无TSV的芯片
➤ TPV是带有TSV的硅晶圆

图 4.10　IMEC 的逻辑芯片、存储器和闪存与桥的异构集成

4.7　IMEC 与桥的异构集成的组装工艺步骤

图 4.11 显示了组装图 4.10 异构集成系统的关键过程步骤。首先，在所有芯片、桥和 TPV 的微晶片植上凸点，在桥上制作 RDL。随后将逻辑芯片和 TPV 连接到载板 1 上，将桥接在逻辑芯片和 TPV 上，压缩成型和背面研磨以暴露逻辑芯片的凸点。然后，使 RDL 形成逻辑芯片的铜柱，将重组的晶圆连接到另一个载板 2 并移除载板 1，将内存芯片连接到逻辑芯片和 TPV。最后，移除载板 2，连接焊球，并将重组后的晶圆单独异构集成封装分离。

第4章 硅基板（桥）上的异构集成

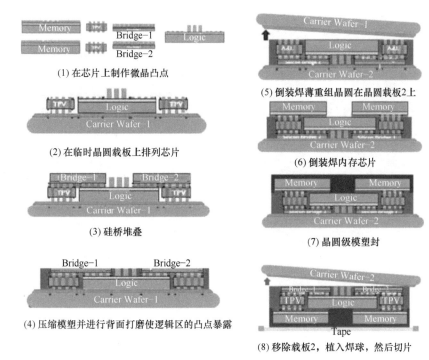

图 4.11 制作 IMEC 桥异构集成的关键过程步骤

4.8 用于异构集成的低成本 TSH 转接板（桥）

如第 3 章所述，制作一个 TSV 通常有 6 个关键步骤，造价昂贵。因此，如何制作低成本的无 TSV 转接板是异构集成的重要研究课题之一。本节提出并开发了一类具有硅通孔（TSH）并在其两侧均带有芯片的非常低成本的转接板（桥）[15-18]。

4.8.1 结构

图 4.12 示意性地显示了 TSH 转接板（桥）的异构集成，该转接板在其顶部和底部集成了几个芯片。TSH 中间层的关键特征是孔中没有金属化。因此，不需要介电层、阻挡层和种子层、通孔填充、CMP（用于去除多铜）和铜露出。与 TSV 中间层相比，TSH 中间层只需要在一片硅片上打孔（通过激光或 DRIE）。就像所有的桥一样，TSH 中间层也需要 RDL。

TSH 转接板可用于集成其顶部和底部的芯片。孔可以让底部芯片的信号通过铜柱和焊料传输到顶部芯片（反之亦然）。同侧的芯片可以与 TSH 转接板的 RDL 相互连通。物理上，顶部芯片和底部芯片通过铜柱和微焊点连接。此外，所有芯片的外围都焊接到 TSH 转接板，以实现结构的完整性，从而抵抗振

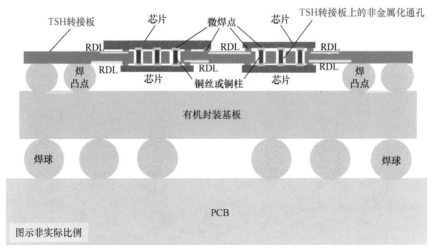

图 4.12 由一个 TSH 转接板（桥）组成的异构集成，转接板顶部集成的芯片有铜柱，底部集成的芯片有焊凸点

动和发热环境。另外，TSH 转接板底部的外围具有常规焊凸点，这些焊凸点附接到封装基板。

本节制作了一种非常简单的试验件，以证明这种与 TSH 转接板（桥）技术异构集成的可行性，介绍了带有铜柱的顶部芯片、带有焊凸点的底部芯片和 TSH 转接板的设计、材料和工艺。异构集成试验件的最终组装由芯片、桥、封装基板和 PCB 构成。将进行冲击和热循环测试，以证明异构集成结构的完整性。下面首先简要介绍上述结构的电学仿真结果以及与具有 TSV 的结构进行比较的结果。

4.8.2 电学仿真

图 4.13 显示了仿真环境下 TSV 和 TSH 的材料、几何形状和尺寸。可以看出，使用了 3D 轴对称结构，有限元代码为 Ansys-HFSS，仿真频率高达 20GHz。两个转接板的厚度相同（100μm）。考虑两个不同的节距，一个是 100μm，另一个是 200μm。TSV 的直径为 15μm，并且考虑两种不同的 SiO_2 厚度，即 0.2μm 和 0.5μm。考虑两种不同尺寸的空气孔，即 7.5μm 和 15μm。

100μm 和 200μm 的通孔节距的仿真结果分别如图 4.14 和图 4.15 所示。可以看出，对于两种节距，TSH 中考虑的两种气孔尺寸的 S21 差异很小，这在不影响电气性能的情况下为机械设计提供了更大的自由度，并且对于这两种节距，TSH 的 S21 都比 TSV 好得多，这意味着 TSH 在高频信号传输方面的插入损耗比 TSV 小得多。

因此，正如预期的那样，新设计的 TSH 的电气性能优于传统 TSV[17, 18]。

第 4 章 硅基板（桥）上的异构集成

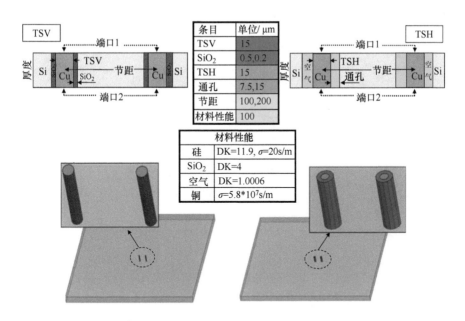

图 4.13　TSV 和 TSH（桥）转接板的电学仿真结构

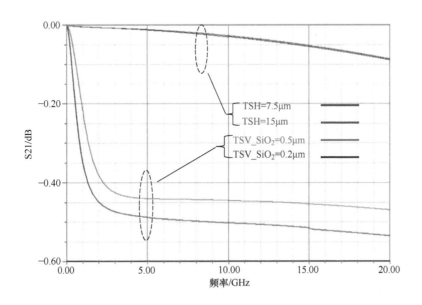

图 4.14　TSH（桥）和 TSV 转接板 S21 仿真结果对比（节距 =100μm）

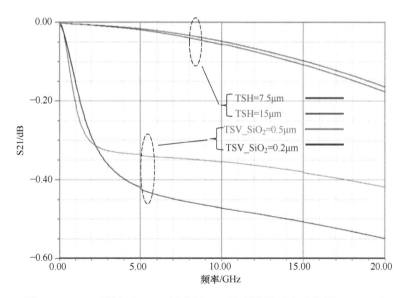

图 4.15 TSH（桥）和 TSV 转接板 S21 仿真结果对比（节距 =200μm）

4.8.3 试验件

试验件如图 4.16 所示。可以看出，它由一个 TSH（桥）组成，该桥支持一个带有铜柱的顶部芯片和一个带有 UBM/ 焊料的底部芯片。桥连接到封装基板，然后连接到 PCB。

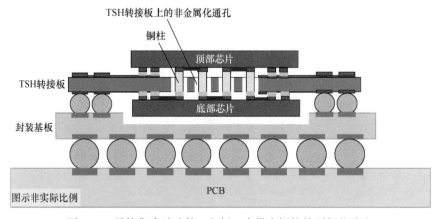

图 4.16 异构集成试验件，包括一个带有铜柱的顶部芯片和一个带有 TSH 转接板（桥）上焊点的底部芯片

顶部芯片的尺寸为 5mm × 5mm × 725μm。该芯片的中心部分有（16 × 16 = 256）个铜柱，外围有两排（176）铜 UBM/ 焊盘。铜柱的直径为 50μm，高为 100μm，节距为 200μm，如图 4.17 和表 4.1 所示。外围铜 UBM/ 焊盘的厚度为

9μm，采用化学镀（2μm）镍（Ni）和浸没（0.05μm）金（Au）(ENIG）。底部芯片的尺寸也是5mm×5mm×725μm。该芯片有432个铜UBM/焊盘（4μm）并涂有锡（Sn）焊料（5μm）。中心部分256个通孔则用于顶部芯片的铜柱的互连。

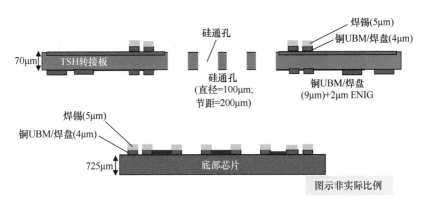

图4.17 顶部芯片、底部芯片和TSH转接板（桥）的几何形状、尺寸和互连

表4.1 异构集成结构的关键要素及其尺寸

带铜柱的顶部芯片	
尺寸	5mm×5mm×725μm
铜柱	直径=50μm；高度=100μm；节距=200μm
铜UBM/焊盘	9μm+2μm（ENIG）
无铜柱的底部芯片	
尺寸	5mm×5mm×725μm
铜UBM/焊盘	4μm
焊锡	5μm
TSH转接板	
尺寸	10mm×10mm×70μm
TSH	直径=100μm；节距=200μm；深度=70μm
顶部：铜UBM/焊盘和焊锡	4μm（铜）和5μm（锡）
底部：铜UBM/焊盘	9μm+2μm（ENIG）
封装基板（腔体）	15mm×15mm×1.6mm（6mm×6mm×0.9mm）
PCB	132mm×77mm×1.6mm

桥（TSH 转接板）的尺寸为 10mm×10mm×70μm，如图 4.17 和图 4.18 所示，中心部分有 256 个通孔，供铜柱穿过。通孔的直径为 100μm，节距为 200μm。在桥的顶部有两排（176）外围铜 UBM/焊盘（4μm），上面涂有锡焊料（5μm），用于顶部芯片的互连。另一方面，在桥的底部或底部芯片的互连处有两排（176）外围铜 UBM/焊盘（9μm）和 ENIG（2μm）。

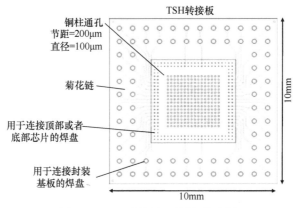

图 4.18　TSH 转接板（桥）的布局

有机封装基板的尺寸为 15mm×15mm×1.6mm。在基板的顶部有一个空腔（6mm×6mm×0.5mm），用于底部芯片。PCB 的尺寸为 132mm×77mm×1.6mm，这是 JEDEC（JESD22-B111）规范[20]的标准尺寸。

4.8.4　带 UBM/焊盘和铜柱的顶部芯片

制作顶部芯片铜柱的工艺流程如图 4.19 所示。由于这不是一个器件芯片，而是一个硅晶片，因此要首先制作菊花链（RDL）。用 PECVD 在 200℃下在硅晶片上沉积 SiO_2，然后溅射（0.1μm）钛和（0.3μm）铜。接下来，旋涂光刻胶并使用光刻技术的掩模图形化（对齐和曝光）。电镀铜（2μm）RDL 层。然后，剥离光刻胶，刻蚀钛/铜，用 PECVD 在整个晶片上沉积 0.5μm 的 SiO_2。再次进行光刻胶和图形化。然后，用 RIE（反应离子刻蚀法）对 SiO_2 进行干法刻蚀。接下来，剥离光刻胶，下面准备进行晶圆凸点制作工艺。

首先溅射（0.1μm）钛和（0.3μm）铜种子层。然后，对所有 432 个焊盘进行 9.5μm 光刻胶和图形化，电镀铜 UBM/焊盘（9μm）。剥离光刻胶，刻蚀钛/铜种子层。然后，在整个晶片上硬压叠 HD Microsystems 提供的正型光刻胶（100μm），仅对中心 256 个焊盘进行图形化，并在室温下镀铜（90μm）。晶圆片的顶面被 DISCO 的研磨机弄平。随后剥离光刻胶和刻蚀种子层钛/铜。最后化学镀镍（2μm）和浸金（0.05μm）。图 4.20 显示了顶部芯片上的铜柱、铜 UBM/焊盘和铜 RDL（菊花链）的 SEM（扫描电子显微镜）图像。铜柱顶部的直径较小（45μm），原因是干膜光刻胶顶部的开口较小。

第 4 章 硅基板（桥）上的异构集成

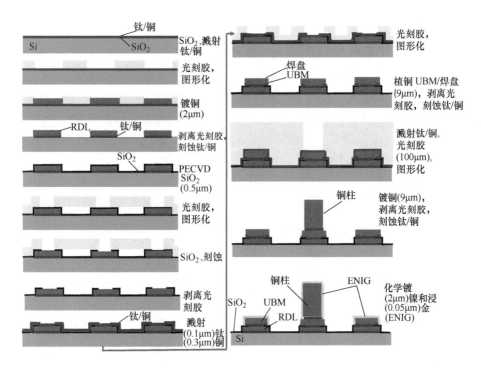

图 4.19 顶部芯片 RDL、铜 UBM/ 焊盘和铜柱的工艺流程

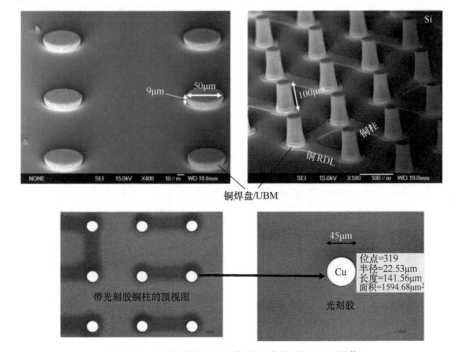

图 4.20 芯片上铜 UBM/ 焊盘和铜柱的 SEM 图像

4.8.5 带 UBM/焊盘/焊料的底部芯片

用 UBM/焊盘和焊料制作底部芯片的工艺流程如图 4.21 所示。可以看出，铜 RDL 的制作工艺与顶部芯片的制作工艺相同。对于大多数晶圆凸点工艺，除了光刻胶厚度和焊料之外，它们是相同的。在光刻胶（9.5μm）和所有（432）焊盘图形化之后，接着电镀铜 UBM/焊盘（4μm）和电镀锡焊料（5μm）。剥离光刻胶，刻蚀掉钛/铜种子层。底部芯片上的 RDL、铜 UBM/焊盘和锡焊料帽的照片图像如图 4.21 所示。

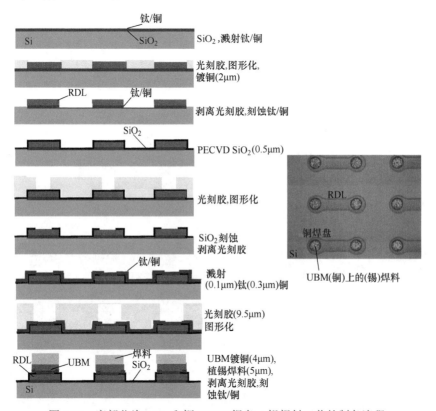

图 4.21　底部芯片 RDL 和铜 UBM/焊盘＋锡焊料工艺的制备流程

4.8.6 桥的制备

图 4.22 是顶部为铜 UBM/焊盘＋锡焊料、底部为铜 UBM/焊盘＋锡焊料的电桥制作工艺流程。可以看出，首先制作转接板底部的 RDL 和铜 UBM/焊盘，其工艺与顶部芯片基本相同。在剥离光刻胶和刻蚀掉较厚的 UBM/焊盘 9μm 的种子层（钛/铜）后，进行 ENIG（2μm 镍 -0.05μm 金）工艺。然后，用黏合剂将具有 UBM/焊盘的桥晶片的底侧暂时黏到 750μm 厚的硅支撑晶圆（载板）上。接着将仿制桥晶片的顶部减薄到 70μm。然后，重复前面提到的制造底部芯片的

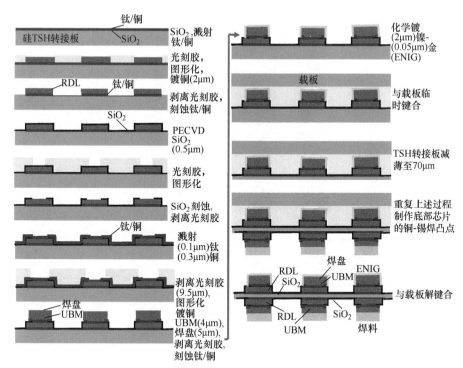

图 4.22 顶部为铜 UBM/ 焊盘 + 锡焊料、底部为铜 UBM/ 焊盘的
TSH 转接板（桥）制作工艺流程

UBM/ 焊盘 + 锡焊料的所有工艺步骤。最后，在室温下将载板晶圆与桥晶片解键合。在此阶段，桥晶片在其顶部具有 176 个外围 UBM/ 焊盘 + 锡焊料、底部具有 176 个外围 UBM/ 焊盘，如图 4.21 和图 4.22 所示。256 个孔是用 UV 激光打孔制造的，功率为 3400mW。图 4.23 显示了具有 RDL 的 70μm 厚桥晶片、封装基板的焊盘、芯片的外围焊盘和孔直径为 100μm、节距为 200μm 的照片图像。

4.8.7 总装

图 4.24 给出了异构集成试验装置的总装工艺流程。首先，176 个外围 UBM/ 焊盘的顶部芯片通过热压（TC）与桥顶部外围 UBM/ 焊盘 + 锡焊料相键合（铜柱穿过桥上的通孔）。TC 键合条件如图 4.25 所示。可以看出，最大键合力为 1600g（1g=9.8N/kg），卡盘的最高温度为 150℃，封头的最高温度为 250℃，循环时间为 120s。

然后，将底部芯片上的 432 个 UBM/ 焊盘 + 锡焊料全部 TC 键合到 256 个中心铜柱的顶端和桥底部 176 个外围 UBM/ 焊盘。除了键合力降低到 800g 外，键合条件与顶部芯片基本相同。焊料（Sn-3wt%Ag-0.5wt% 铜）凸点（直径 350μm）安装在桥底部。然后，沿着顶部芯片的两个相邻侧面填充 50% 填料含量（平均填料尺寸 =0.3μm，最大填料尺寸 =1μm）的毛细管型下填料。下填料

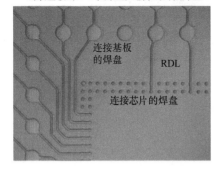

图 4.23　TSH 转接板（桥）晶圆，激光钻孔（右）和芯片用 RDL 和焊盘（左）

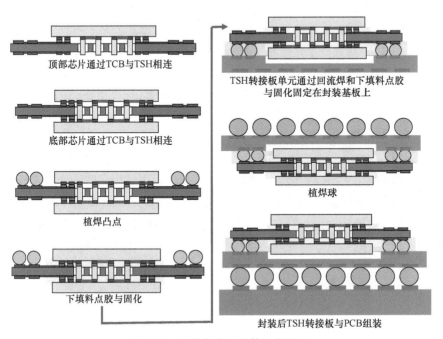

图 4.24　异构集成的总装工艺流程

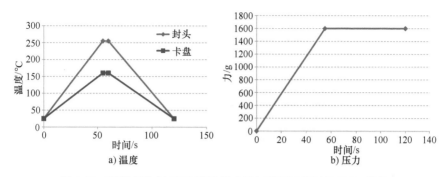

图 4.25 顶部芯片与 TSH 转接板（桥）顶部连接时的 TCB 条件

填充顶部芯片与 TSH 转接板之间的间隙，流过 TSH 转接板的通孔，填充底部芯片与桥层之间的间隙后，在 150℃下固化 30min。

整个 TSH 转接板（桥）模块以标准的无铅温度曲线在封装基板上回流，最高温度为 240℃。为了提高焊点的可靠性，在桥层和有机封装基板之间使用下填料。接着是，分布在封装基板底部植焊球（Sn-3wt%Ag-0.5wt%Cu，直径 450μm）。最后，整个异构集成封装在采用上述相同的无铅回流温度曲线回流焊在 PCB 上。最终组装而成的异构集成试验件如图 4.26 所示。可以看到，PCB 支撑着封装基板，封装基板支撑着桥层，桥层支撑着顶部芯片。底部芯片被桥层挡住看不见。

图 4.26 最终组装而成的异构集成系统

图 4.27 显示了异构集成组装完成的试验件的 X 射线图像。可以看出，铜柱不接触桥的侧壁，铜柱几乎在桥的中心。图 4.28 显示了异构集成试验件横截面的 SEM 图像，其中包括所有关键元素，如顶部芯片、TSH 转接板（桥）、底部

芯片、封装基板、PCB、微凸点、焊凸点、焊球和铜柱。通过 X 射线和 SEM 图像可以看出，SiP 结构的关键元素制作得当。

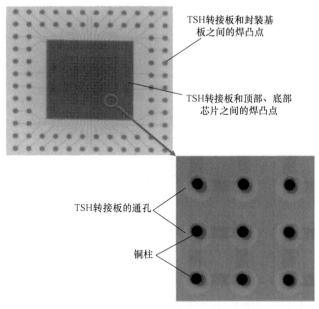

图 4.27　由顶部芯片、TSH 转接板（桥）、底部芯片、
封装基板和 PCB 组成的异构集成 TSH 和铜柱位置的 X 射线图像显示

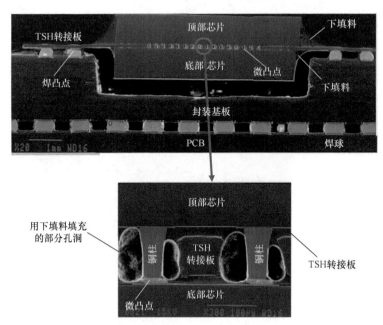

图 4.28　由顶部芯片、TSH 转接板（桥）、底部芯片、封装基板和 PCB 组成的
异构集成的横截面 SEM 图像

4.8.8 可靠性评估—冲击（坠落）试验及结果

验证异构集成装配结构和热完整性的可靠性评估包括坠落试验和热循环试验。坠落测试板和设置基于 JESD22-B111[19]。测试过程中，异构集成组件朝上。夹具上的 4 个支座为 PCB 在冲击期间的偏转提供了支撑和空间，如图 4.29 所示。坠落高度为 460mm，导致加速度为 1500g（1g=9.8m/s^2），如图 4.30 所示。10 次坠落后无失效，即菊花链电阻无变化，无明显失效。

图 4.29 异构集成系统坠落试验装置的搭建

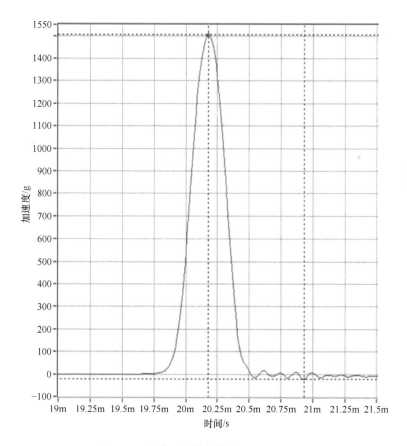

图 4.30 坠落试验结果符合 JESD22-B111

4.8.9 可靠性评估—热循环试验及结果

热循环试验条件为 $-55\sim125℃$，循环 1h（上升和下降各 15min，停留 15 分钟）。图 4.31 显示的试验结果表明，对于中位水平（50%），威布尔分布斜率为 2.52，样本的特征寿命（63.2% 失效时）为 1175 次。（对于给定的试验条件，如果特征寿命大于 1000 次，则被认为是可以接受的）菊花链焊点的样本平均寿命定义为平均失效时间（MTTF）=$1175\Gamma(1+1/2.52)$=1036 次，其中 Γ 为伽马函数。此平均寿命发生在 $F(1036)=1-\exp[-(1036/1175)^{2.52}]=0.52$，即失效概率为 52%。

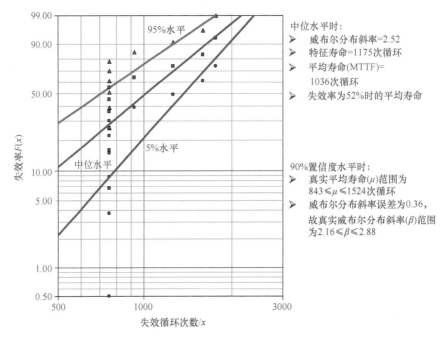

图 4.31 菊花链焊点在热循环试验中的威布尔图（$-55\sim125℃$，1h 循环（15min 上升和下降，各停留 15min），置信度为 90%）

图 4.31 还显示了在 90% 置信度水平下的测试结果，即 90% 的情况下，希望找出菊花链焊点的真实威布尔斜率和真实平均寿命。可以看出，焊点的真实平均寿命（在 100 种情况中的 90 种情况，其他 10 种情况下，没有人知道）将不小于 843 次循环，但不大于 1524 次循环。真实威布尔分布的斜率（β）在 $2.16 \leq \beta \leq 2.88$ 区间内。

其中一种典型的失效模式如图 4.32 所示。可见有机封装基板的桥层和焊盘之间的焊凸点有裂纹。失效（裂纹）位置在 TSH 转接板的焊料与 UBM 界面附近的焊凸点处。它在 1764 次循环时失效，失效准则是电阻无限变化。

第4章 硅基板（桥）上的异构集成

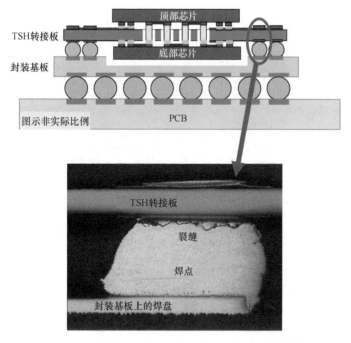

图4.32 热循环试验导致的失效（焊点开裂）模式

4.9 总结和建议

一些重要的结果和建议总结如下：

1）无TSV转接板，如桥，是制作异构集成的RDL非常有效的方法。它不仅减少了加工步骤，而且潜在地降低了成本；

2）桥的尺寸通常小于TSV转接板的尺寸。

3）桥的金属线宽和间距最多等于TSV转接板的线宽和间距。

4）低成本的TSH转接板在硅基板（桥）异构集成方面具有潜在优势：① TSH转接板（桥）的电性能优于TSV转接板；② TSH转接板的设计、材料和工艺的可行性；③试验件通过了热循环和冲击试验。

参考文献

[1] Hou, S., W. Chen, C. Hu, C. Chiu, K. Ting, T. Lin, W. Wei, W. Chiou, V. Lin, V. Chang, C. Wang, C. Wu, and D. Yu, "Wafer-Level Integration of an Advanced Logic-Memory System Through the Second-Generation CoWoS Technology," *IEEE Transactions on Electron Devices*, October 2017, pp. 4071–4077.

[2] Lee, J., C. Lee, C. Kim, and S. Kalchuri, "Micro Bump System for 2nd Generation Silicon Interposer with GPU and High Bandwidth Memory (HBM) Concurrent Integration," *Proceedings of IEEE/ECTC*, May 2018, pp. 607–612.

[3] Xie, J., H. Shi, Y. Li, Z. Li, A. Rahman, K. Chandrasekar, et al., "Enabling the 2.5D integration," *Proceedings of IMAPS International Symposium on Microelectronics*, October 2012, pp. 254–267.

[4] Shao, S., Y. Niu, J. Wang, R. Liu, S. Park, H. Lee, G. Refai-Ahmed, and L. Yip, "Comprehensive Study on 2.5D Package Design for Board-Level Reliability in Thermal Cycling and Power Cycling," *Proceedings of IEEE/ECTC*, May 2018, pp. 1662–1669.

[5] McCann, S., H. Lee, G. Refai-Ahmed, T. Lee, and S. Ramalingam, "Warpage and Reliability Challenges for Stacked Silicon Interconnect Technology in Large Packages," *Proceedings of IEEE/ECTC*, May 2018, pp. 2339–2344.

[6] Lau, J. H., *3D IC Integration and Packaging*, McGraw-Hill, New York, 2016.

[7] Lau, J. H., *Through-Silicon Via (TSV) for 3D Integration*, McGraw-Hill, New York, 2013.

[8] Lau, J. H., *Reliability of RoHS compliant 2D & 3D IC Interconnects*, McGraw-Hill, New York, 2011.

[9] Lau, J. H., C. K. Lee, C. S. Premachandran, and Yu Aibin, *Advanced MEMS Packaging*, McGraw-Hill, New York, 2010.

[10] Chai, T. C., X. Zhang, J. H. Lau, C. S. Selvanayagam, D. Pinjala, et al., "Development of Large Die Fine-Pitch Cu/low-*k* FCBGA Package with through Silicon via (TSV) Interposer," *IEEE Transactions on CPMT*, Vol. 1, No. 5, May 2011, pp. 660–672.

[11] Chiu, C., Z. Qian, and M. Manusharow, "Bridge Interconnect with Air Gap in Package Assembly," US Patent No. 8,872,349, 2014.

[12] Mahajan, R., R. Sankman, N. Patel, D. Kim, K. Aygun, Z. Qian, et al., "Embedded multi-die interconnect bridge (EMIB)—a high-density, high-bandwidth packaging interconnect," *IEEE/ECTC Proceedings*, May 2016, pp. 557–565.

[13] http://www.digitimes.com.tw/tw/dt/n/shwnws.asp?CnlID=1&Cat=10&id=493987&query=%A6%B3%A4F%AD%5E%AFS%BA%B8%A5%5B%AB%F9%A1A%B6W%B7L%B1N%B1o%A8%EC%A7%F3%A4j%AA%BA%A7U%A4O%B9%EF%A7%DCNVIDIA (2/20/2017).

[14] Podpod, A., J. Slabbekoorn, A. Phommahaxay, F. Duval, A. Salahouedlhadj, M. Gonzalez, K. Rebibis, R.A. Miller, G. Beyer, and E. Beyne, "A Novel Fan-Out Concept for Ultra-High Chip-to-Chip Interconnect Density with 20-μm Pitch," *IEEE/ECTC Proceedings*, May 2018, pp. 370–378.

[15] Wu, S., J. H. Lau, H. Chien, R. Tain, M. Dai, and Y. Chao, "Chip Stacking Structure and Fabricating Method of the Chip Stacking Structure," US Patent No.: 8,519,524, Date of Patent: August 27, 2013.

[16] Wu, S., J. H. Lau, H. Chien, J. Hung, M. Dai, Y. Chao, R. Tain, et al., "Ultra Low-Cost Through-Silicon Holes (TSHs) Interposers for 3D IC Integration SiPs," *IEEE ECTC Proceedings*, San Diego, CA, May 2012, pp. 1618–1624.

[17] Lau, J. H., C. Lee, C. Zhan, S. Wu, Y. Chao, M. Dai, R. Tain, H. Chien, C. Chien, R. Cheng, Y. Huang, Y. Lee, Z. Hsiao, W. Tsai, P. Chang, H. Fu, Y. Cheng, L. Liao, W. Lo, and M. Kao, "Low-Cost TSH (Through-Silicon Hole) Interposers for 3D IC Integration," *Proceedings of IEEE/ECTC*, Orlando, FL, May 2014, pp. 290–296.

[18] Lau, J. H., C. Lee, C. Zhan, S. Wu, Y. Chao, M. Dai, R. Tain, et al. Through-Silicon Hole Interposers for 3D IC Integration. *IEEE Transactions on CPMT*, 2014, 4 (9): 1407–1418.

[19] Lau, J. H., P. Tzeng, C. Lee, C. Zhan, M. Li, J. Cline, et al. 2014. Redistribution Layers (RDLs) for 2.5D/3D IC Integration, *IMAPS Transactions, Journal of Microelectronic Packaging*, 11 (1), First Quarter 2014, pp. 16–24.

[20] JESD22–B111, *Board Level Drop Test Method of Components for Handheld Electronic Products*, JEDEC Standard, July 2003.

第 5 章

异构集成的扇出晶圆级／板级封装

5.1 引言

英飞凌（Infineon）科技公司于 2001 年 10 月 31 日申请了第一个扇出晶圆级封装（FOWLP）美国专利[1, 2]，并且英飞凌及其商业合作伙伴长濑（Nagase）等还在 IEEE/ECTC 2006 和 IEEE/EPTC 2006 期间发表了第一篇技术论文[3, 4]。当时，他们称之为嵌入式晶圆级球栅阵列（eWLB）。该技术不使用引线键合或晶圆凸点的方法，不使用引线框架或封装基板，并有可能实现吞吐量更高、成本更低、性能更好和体积更小的封装[2]。该技术需要一个临时（重组）载体，用于已知合格芯片（KGD）、环氧树脂模塑料（EMC）、压缩或层压成型以及再布线层（RDL）的制造。

需要强调的是，FOWLP 的概念最早由英飞凌[1]提出。尽管该技术的一些知识已获得通用电气（GE）公司[5, 6]和 EPIC[7]的专利，但英飞凌的专利[1]明确指出使用 RDL 从芯片的金属焊盘扇出电路晶圆和焊球到 PCB 上的金属焊盘，如图 5.1 所示。英飞凌[1]还特别指出，一些 RDL 有一部分超出（扇出）芯片边缘。这些是英飞凌专利参考文献 [1] 中主要的权利要求，GE 和 EPIC 的专利[2, 5-7]没有要求。

如图 5.2a、b 所示 FOWLP 与 fcPBGA（倒装芯片塑料球栅阵列）封装相比的优势在于[2, 8]成本更低、外形更小、消除了基板、消除了晶圆凸点、消除了倒装芯片回流、省去了助焊剂清洗、消除了下填料以及更好的电气性能、更好的热性能、更容易实现异构集成和 3D IC 封装[9-11]。

如图 5.2c 所示，FOWLP 相对于 WLCSP（晶圆级芯片级封装）的优势为[2, 8]使用已知合格芯片（KGD）、更好的晶圆级良率、使用最好的硅、单芯片或多芯片、嵌入式集成无源元件，以及更多 RDL、更多引脚数、更好的热性能、更易于进行 SiP 和 3D IC 封装[9-11]、更高的 PCB 级可靠性。

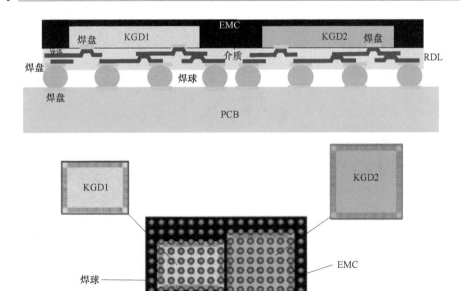

图 5.1 含有 2 个芯片的扇出晶圆 / 面板级封装

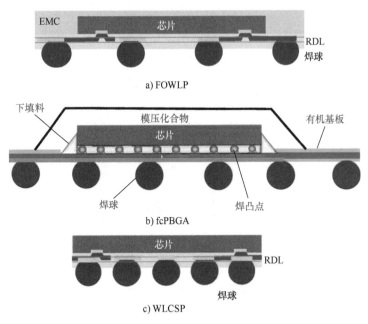

图 5.2 FOWLP、fcPBGA 和 WLCSP

在 ECTC 2007 期间,飞思卡尔半导体(即现在的恩智浦半导体)提出了一种类似的技术,并将其称为再分配芯片封装[12]。IME 将 FOWLP 技术扩展到 3D 设计的多芯片和堆叠多芯片,并在 ECTC 2008 期间展示[13]。在 ECTC 2009 期间,

IME 发表了 4 篇论文：一种预测压缩成型过程中模具移位的新方法[14]；横向放置和垂直堆叠的薄芯片[15]；3D FOWLP 的可靠性[16]；在 FOWLP 上验证的高质量、低损耗毫米波无源器件[17]。参考文献 [1，3，4，12-17] 中使用了芯片先置和芯片面朝下扇出晶圆级工艺[18]。

在 IEEE/ESTC 2010 和 ECTC 2011 期间，日本电气股份有限公司（NEC，即现在的瑞萨电子）发表了几篇关于晶圆级封装系统（SiWLP）[19]和 RDL 先置 FOWLP[20]的论文。这些论文基于其用于芯片间宽带数据传输[21, 22]和集成在逻辑器件上的 3D 堆叠存储器[23-27]的 SMAFTI（与馈通转接板连接的 SMArt 芯片）封装技术。SMAFTI 中使用的馈通转接板（FTI）是一种具有超细线宽和间距 RDL 的薄膜基板。FTI 的电介质通常是 SiO_2 或聚合物，RDL 的布线导体是铜。FTI 不仅支撑芯片下方的 RDL，还提供芯片边缘以外的支撑。面阵列焊球安装在 FTI 的底部，与 PCB 连接。EMC 内嵌芯片并支撑 RDL 和焊球。这项技术除了制备 RDL 外，还需要晶圆凸点、助焊剂、倒装芯片组装、清洁以及下填料点胶和固化，因此成本非常高。其潜在应用是非常高密度和高性能的产品，如超级计算机、高端服务器、电信和网络系统。该技术采用的工艺是芯片后置（chip-last）或再布线层先置（RDL-first）的 FOWLP 工艺[18]。

在 ECTC 2012 期间，星科金朋有限公司提出了一种采用 FOWLP 技术的应用处理器（AP）芯片组堆叠式封装（PoP）[28]。

台积电（TSMC）[29, 30]在 ECTC 2016 期间发表了两篇关于 FOWLP 的论文，一篇是集成扇出型（InFO）晶圆级封装，用于容纳移动应用中的最先进的应用处理器[31]，另一篇是集成扇出型封装技术与传统的基于封装基板的倒装焊芯片技术之间的热性能和电性能比较[32]。2016 年 9 月，台积电将采用集成扇出型晶圆级封装技术的 iPhone 应用处理器（AP）的堆叠式封装（PoP）投入量产。这具有很重要的意义，因为这意味着 FOWLP 不仅可用于封装小型芯片，如基带、电源管理 IC、射频（RF）开关/收发器、RF 雷达、音频编解码器、微控制器单元、连接 IC 等，还可用于封装高性能和大型（>120mm^2）的系统级芯片（SoC），如 AP。该技术使用了芯片先置和芯片面朝上的 FOWLP 工艺[18]。

最近，支持多个倒装芯片的无硅通孔（TSV）硅转接板[33]是半导体封装中非常热门的话题。在 ECTC 2013 期间，星科金朋有限公司[34]提出使用嵌入式晶圆级球栅阵列（eWLB）扇出倒装芯片来制作芯片的 RDL，以执行大部分横向通信。在 ECTC 2016 期间，日月光半导体公司（ASE）[35]和联发科技股份有限公司（Media Tek）[36]表明使用类似的技术制造了 FOWLP 的 RDL 布线电路，省去了 TSV 硅转接板、晶圆凸点、助焊剂、芯片到晶圆键合、清洁以及下填料点胶和固化，即无 TSV 硅转接板。

前面所有提到关于扇出的论文中的封装过程都采用 200mm 或 300mm 晶圆（基于现有的器件晶圆制造设备）作为重组载体，用于支撑 KGD 和制造模具、

RDL 等。为了增加吞吐量，提出了扇出板级封装（FOPLP）。如从 EPTC 2011 开始，J-Devices 就展示了他们的 FOPLP（320 mm×320mm），称为 WFOP™（宽条扇出封装）[37-39]。从 ECTC 2013 开始，弗劳恩霍夫陶瓷技术和系统研究所一直在展示他们对大面积 FOPLP（610mm×457mm）压缩成型的评估结果[40-42]。在 ECTC 2014 上，矽品（SPIL）发表了两篇关于 FOPLP 的论文，称为 P-FO，一篇是关于开发和表征 370mm×470mm 尺寸的 P-FO[43]，另一篇是测量其翘曲[44]。FOPLP 技术的瓶颈之一是板级设备的能力，因为缺乏面板尺寸的相关标准，如旋涂、物理气相沉积（PVD）、图形化/光刻、电化学沉积、刻蚀、背磨、植球和用于制造模具、RDL、封装的切割等。因此，潜在的 FOPLP 用户一致呼吁面板尺寸的行业标准诞生。本章将检查、讨论和更新 FOW/PLP 异构集成的以下重要主题：①封装形式，如芯片先置且芯片面朝下、芯片先置且芯片面朝上及芯片后置（chip-last）或再布线层先置（RDL-first）；② RDL 制造，如有机 RDL、无机 RDL、混合 RDL 和 LDI/PCB RDL；③翘曲；④热性能；⑤临时晶圆与面板载体；⑥ PCB 上扇出封装在温度循环下的可靠性。

5.2 FOW/PLP 的形式

FOW/PLP 有很多形式，但基本上可以分为三大类，即芯片先置（芯片面朝下）、芯片先置（芯片面朝上）和芯片后置或再布线层先置。

5.3 芯片先置（芯片面朝下）

采用芯片先置且芯片面朝下工艺的 FOW/PLP 实际上是由英飞凌[1, 2]首先提出、由星科金朋、日月光、意法半导体、英飞凌、英特尔和 NANIUM（现在的 Amkor）等量产（HVM）的 eWLB。这是制造 FOW/PLP 的最常规方法，当今制造的大多数 FOW/PLP 产品都使用这种方法。

5.3.1 芯片先置（芯片面朝下）工艺

如图 5.3 所示是芯片先置且芯片面朝下的 FOW/PLP 的一般制造工艺流程[3, 4, 12-18, 45-73]。首先，对器件晶圆进行检测找出已知合格裸片（KGD），然后将其分割成单个裸片。拿起 KGD 并将它们正面朝下放在一个如图 5.3a 所示的带有双面热剥离胶带的临时载板（可以是金属、硅、玻璃或有机物）上，该载体可以是圆形（圆片）或矩形（面板），如图 5.3b 所示。最常用的胶带是由日东电工（Nitto Denko）提供的 REVALPHA，如图 5.4 所示。然后，通过如图 5.5 所示压缩方法 + PMC（模后固化）或层压方法 + 后退火等工艺后，将带有 KGD 的重组（临时）载板进行 EMC 模压成型，再去除载板并剥离双面胶带，如图 5.3d 所示。最常用的 EMC 是由 Nagase 生产的液态材料 R4507，见表 5.1。在

EMC 之后，接下来是为 KGD 的铝或铜焊盘建立信号、能量、接地的 RDL（将在 5.6 节中详细介绍），如图 5.3e 所示。最后，安装焊球并将整个重组载板（带有 KGD、RDL 和焊球）切割成单独的封装，如图 5.3f 所示。

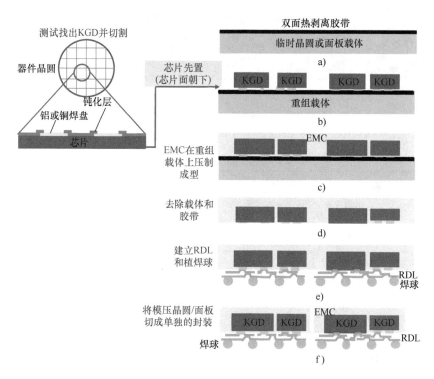

图 5.3 芯片先置且芯片面朝下型 FOW/PLP 的关键工艺过程

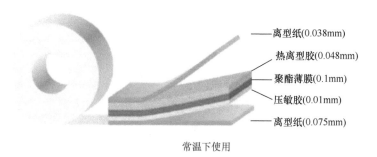

图 5.4 Nitto Denko 的 REVALPHA 双面胶

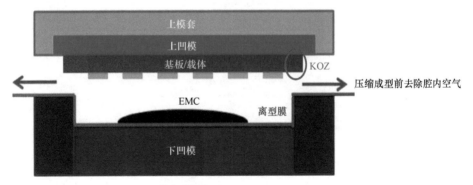

图 5.5 模压成型

表 5.1 EMC（Nagase R4507）的材料性能

性能	R4507
填料含量（%）	85
填料切割 /μm	25
填料平均尺寸 /μm	8
比重	1.96
黏度 / Pa·s	250
弯曲模量 /Gpa	19
玻璃化温度 T_g（DMA）/℃	150
热膨胀系数 CTE1/（ppm/K）	10
热膨胀系数 CTE2/（ppm/K）	41

注：1. 具有的高流动性适用于大表面积和薄膜模具。
2. 常温液态、可分配、无尘，适合洁净室内环境。
3. 可低温成型（125℃）。
4. 在低应力设计的大表面积模具中提供低反射。
5. 高可靠性。
6. 高纯度。

5.3.2　带有晶圆载板的芯片先置（芯片面朝下）

常用的 FOWLP 测试芯片如图 5.6a、b 所示，分别用于测试大芯片（5mm×5mm×150μm）和小芯片（3mm×3mm×150μm）。大芯片有 160 个节距为 100μm（内部行距）的焊盘，小芯片有 80 个节距为 100μm 的焊盘。对于这两种芯片，铝焊盘的 SiO_2 钝化开口为 50μm×50μm，铝焊盘的尺寸为 70μm×70μm，如图 5.6c 所示。

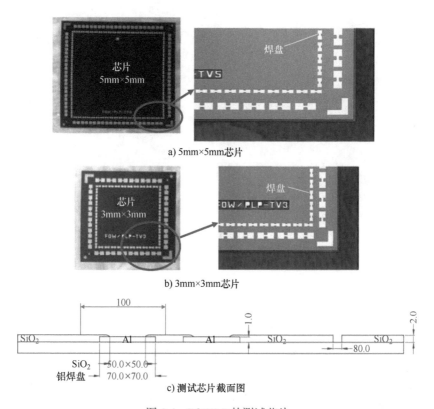

图 5.6 FOWLP 的测试芯片

常用的 10mm×10mm 的封装如图 5.7a 所示，由 1 个 5mm×5mm 的芯片、3 个 3mm×3mm 芯片和 4 个（0402）电容器组成。大芯片和小芯片的间距只有 100μm。该封装的一个实际应用是应用处理器芯片组，即大芯片可以是处理器，小芯片可以是存储器。

如图 5.7b 所示为测试封装的横截面示意图。可以看出有两个 RDL，RDL1 的金属层厚度为 3μm，RDL2 的金属层厚度为 7.5μm。RDL1 的线宽和间距为 10μm，RDL2 的线宽和间距为 15μm。DL1 和 DL2 的介电层厚度为 5μm，DL3 的介电层厚度为 10μm。钝化层（DL3）的开孔为 180μm。焊球尺寸为 200μm，球节距为 0.4mm。

图 5.8 所示是一个带有 629 个封装（10mm×10mm）的 300mm 重组晶圆载体 [45-47]。每个封装有 4 个芯片（1 个 5mm×5mm 和 3 个 3mm×3mm）和 4 个电容器（0402）。大芯片与小芯片的间距为 100μm。每个封装有两个 RDL。需要强调的是，FOWLP 是一个非常高吞吐量的过程。在这种情况下，一个过程可以产生 629 个 10mm×10mm 的异构集成封装。

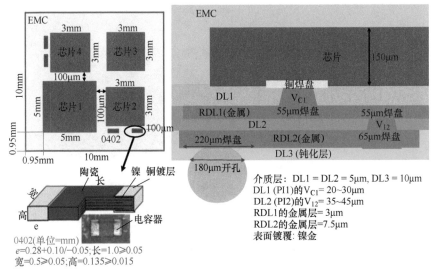

a) FOWLP的俯视图(1个5mm×5mm芯片、3个3mm×3mm 芯片和4个0402电容器)

b) FOWLP的横截面视图(两个RDL)

图 5.7 FOWLP 测试封装

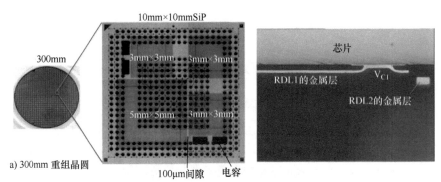

a) 300mm 重组晶圆

b) 测试封装的X射线图像(1个5mm×5mm芯片、3个3mm×3mm芯片和4个0402电容器)

c) 测试封装的横截面图

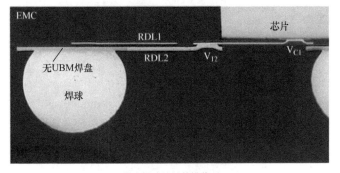

d) 带有焊球的封装横截面

图 5.8 FOWLP

5.3.3 带有面板载体的芯片先置（芯片面朝下）

图 5.9 是带有面板载体的芯片先置且芯片面朝下的关键工艺过程[48-50]。由于面板载体涉及 PCB 工艺，因此必须通过在图 5.9 左侧的铝焊盘顶部电镀一个 8μm 铜焊盘才能在器件晶圆上完成工作。铜焊盘的作用是阻止激光将铝焊盘钻透。此外，与图 5.3 所示的工艺不同，该工艺使用有机载体，如图 5.9a 所示，将 KGD 面朝下拾取并放置在载体上，如图 5.9b 所示，然后进行 EMC 层压或压缩成型，如图 5.9c 所示，并形成一个 ECM 面板（ECM-panel）。在本节中，将嵌入 KGD 的 EMC 称为 ECM 面板。然后用环氧树脂将 EMC 面板连接到核心基板的两侧，如图 5.9d 所示，形成 5 层的 PCB 叠层封装。接着移除载体并剥离双面胶带，如图 5.9e 所示，并制作 RDL，这将在 5.6 节中介绍。

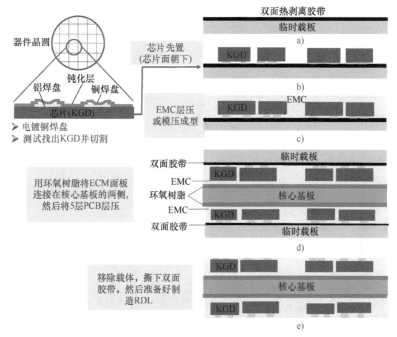

图 5.9 芯片先置且芯片面朝下型 FOWLP 的关键工艺过程

面板载体上的 10mm×10mm 测试封装也是由 4 个芯片组成，如图 5.10 和图 5.11 所示。芯片尺寸也为 5mm×5mm 和 3mm×3mm。但是，大芯片上有 88 个节距为 180μm 的焊盘，小芯片有 48 个节距为 180μm 的焊盘。铝焊盘的尺寸为 130μm×130μm，铝焊盘上的 SiO_2 钝化层开孔为 110μm×110μm。铜接触焊盘的直径为 110μm，距离铝焊盘的高度为 8μm。图 5.11b 为测试封装的横截面示意图。可以看出有两个 RDL，RDL1 和 RDL2 的金属层厚度都是 10μm。RDL1 的线宽和线间距为 20μm，RDL2 的线宽和线间距为 25μm。DL1、DL2、DL3 的介质层厚度为 20μm。钝化层（DL3）的开孔尺寸为 180μm。焊球尺寸为 200μm，球节距为 0.4mm。

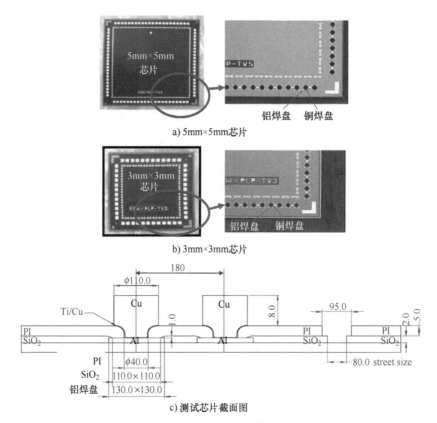

图 5.10 FOWLP 的测试芯片

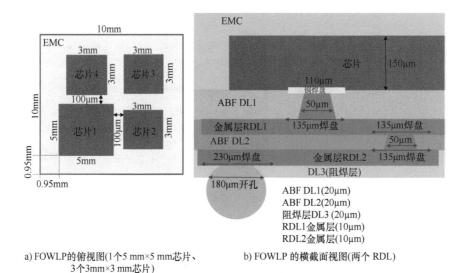

图 5.11 FOWLP 测试封装

图 5.12 为一个带有 378 个异构集成封装（10mm×10mm）的尺寸为 340mm×340mm 的面板载体[48, 49]。图 5.13 为一个具有 1512 个异构集成封装（10mm×10mm）的尺寸为 508mm×508mm 的面板载体[50]。可以看出，通过 C 模式扫描声学显微镜检查，即使在大芯片和小芯片之间的 100μm 间隙的 EMC 中也没有任何缺陷。

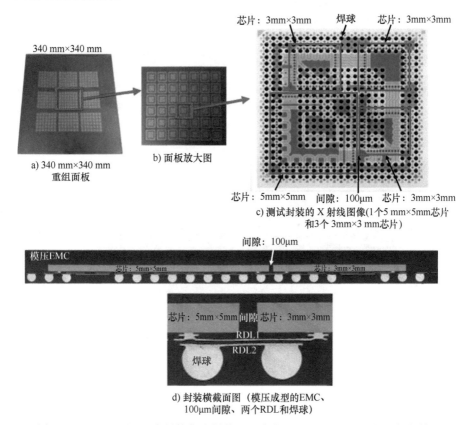

图 5.12　FOPLP（378 个异构集成封装，尺寸为 340mm×340mm 的面板载体）

5.3.4　芯片先置（芯片面朝下）封装组件的热循环试验

如图 5.8、图 5.12 和图 5.13 所示的封装组装在具有 305（Sn-3wt%Ag-0.5wt%Cu）焊点的 6 层 PCB 上[47]。热循环试验的样本大小等于 60。焊点（无下填料）可靠性的热循环试验结果如图 5.14 所示。热循环试验在完成 1300 次循环后停止。可以看出，威布尔曲线的特征寿命（63.2% 失效）为 1070 次循环，这对于智能手机和平板计算机等移动产品的预期寿命（通常小于 3 年）来说已经绰绰有余。失效模式和位置如图 5.15 所示。可以看出，焊点在焊料和封装接触焊盘之间的界面附近出现裂纹，裂纹发生在靠近封装角的芯片角下面——最长的 DNP（到中性点的距离）[74]。

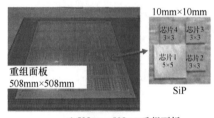

a) 508mm×508mm重组面板

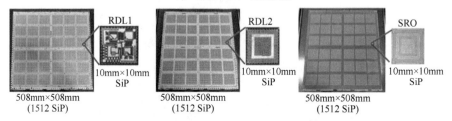

b) 具有 RDL1、RDL2 和阻焊层开孔的重组面板

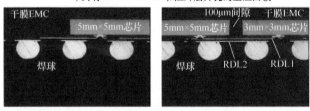

c) 封装横截面图(干膜叠压EMC、100μm 间隙、两个RDL和焊球)

图 5.13　FOPLP（1512 个异构集成封装、尺寸为 508mm × 508mm 的面板载体）

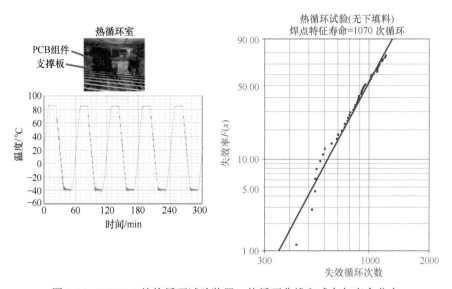

图 5.14　FOWLP 的热循环试验装置、热循环曲线和威布尔寿命分布

第 5 章 异构集成的扇出晶圆级/板级封装

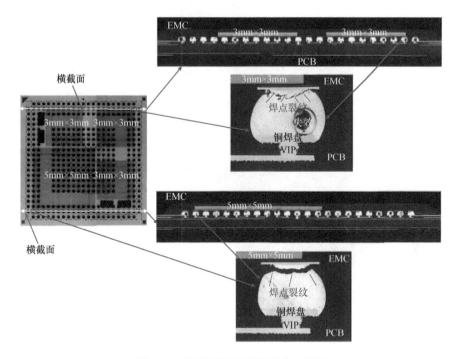

图 5.15 热循环试验失效模式和位置

SiP 扇出的 PCB 组件如图 5.8、图 5.12 和图 5.13 所示，沿该组件的对角线进行 3D 建模，如图 5.16 所示。该模型仅使用六面体实体单元，可以捕捉封装焊点的精确形状和潜在的 DNP 效应，同时保持比全八分圆模型显著的计算效率。尽管条带模型中的单元具有整体经济性，使用选择性网格细化可将高度细化的单元集中在预计会出现故障的焊点中。在目前的 PCB 组装中，预计会在具有最大 DNP（封装角）和芯片角附近的焊点出现失效，如图 5.16 所示。因此，这些焊点应使用高度精细的网格，其他焊点可以使用粗网格。该模型使用的是 ABAQUS 6.12（C3D8R）。假设 Sn-3wt%Ag-0.5wt%Cu 遵循广义 Garofalo 蠕变方程[9]，即 $d\varepsilon/dt = 500000 \sinh^5(\sigma/1\times10^8) \exp(-5807/T(K))$，其中 ε 为应变，σ 为应力（单位 Pa），T 为温度（单位 K）。其他材料性能见表 5.2。

PCB 组件按如图 5.14 所示的温度曲线，经过 5 个热循环，如图 5.16 所示，最大的累积蠕变应变发生在 3mm×3mm 芯片角和 5mm×5mm 芯片角下的焊点处。该位置位于封装底部和焊料之间的界面处。因此，在该位置可能会发生失效。这与图 5.15 所示的热循环试验失效模式和位置非常吻合。相关冲击测试和模拟结果，见参考文献 [47]。

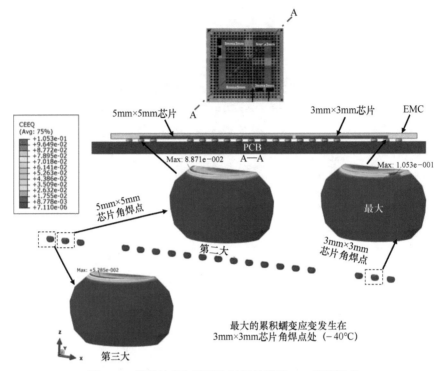

图 5.16 模拟结果与温度和时间的关系——累积蠕变

表 5.2 温度循环试验的材料特性

	热膨胀系数 /(ppm/℃)	杨氏模量 /GPa	泊松比
铜	16.3	121	0.34
PCB	α_1=18,α_2=18,α_3=70	E_1=22,E_2=22,E_3=10	0.28
硅	2.8	131	0.278
锡	21.3+0.017t	49−0.07T	0.3
聚酰亚胺	35	3.3	0.3
EMC	10（<150℃）	19	0.25

5.3.5 芯片先置（芯片面朝下）FOW/PLP 的应用

芯片先置且芯片面朝下的大多数应用是针对小芯片和引脚数不多的封装。此外，RDL 的金属线宽度和间距也不小，如 10～15μm 或更大。可封装的半导体包括基带、电源管理 IC、射频（RF）开关 / 收发器、RF 雷达、音频编解码器、微控制器单元和连接 IC。随着 SiP 或异构集成的普遍应用，扇出（可以处理多个裸片和分立元件）的方式将被更广泛地使用，因为扇入 WLCSP[75] 只能处理单个芯片。

5.4 芯片先置（芯片面朝上）

如图 5.17 所示是芯片先置且芯片面朝上的 FOW/PLP 的组装工艺[18, 29-32, 76-87]。与芯片先置且芯片面朝下（见图 5.3 和图 5.11）不同，基本上必须在器件晶圆（图 5.17a、b）和重组晶圆（图 5.17c～h）上完成工作。

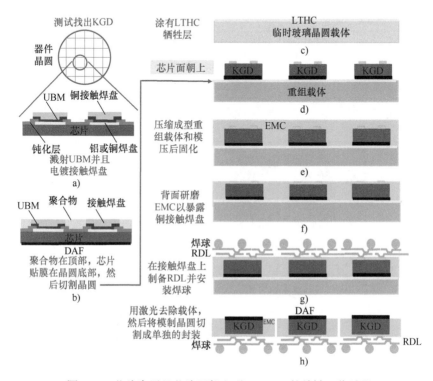

图 5.17 芯片先置且芯片面朝上型 FOWLP 的关键工艺过程

5.4.1 芯片先置（芯片面朝上）工艺

首先，必须通过溅射钛/铜作为凸点下金属层（UBM）的底层修改器件晶圆，UBM 是在铝（或铜）焊盘上进行物理气相沉积（PVD）形成，然后在 UBM 上通过电镀制造铜接触焊盘，如图 5.17a 所示。这个过程是芯片先置且芯片面朝上工艺所特有的，铜接触焊盘是为构建 RDL 做准备。如图 5.17b 所示，这个过程之后是在器件晶圆顶部旋涂聚合物，并（在约 70℃时）在器件晶圆的底部压一层（约 20μm）芯片贴膜（DAF）。如图 5.18 所示是日立（Hitachi）公司生产的一种 DAF。然后对器件晶圆进行测试并切割成单独的 KGD。在本节中，芯片尺寸为 10mm×10mm，如图 5.19 所示，封装尺寸为 13.42mm×13.42mm，如图 5.20 所示。

异构集成技术

项目			单位	FH-9011	测试方法
黏合剂厚度			μm	10 20 25 40	
DC 胶带特性	曝光能量		mJ/cm²	150~400	
	DCT 和 DBF 之间的黏合强度	紫外线前	N/25mm	1.4	
		紫外线后	N/25mm	<0.1	
	晶圆贴合温度		℃	60~80	
贴片条件	温度		℃	100~160	
	加载		MPa	0.05~0.2	
弹性模量（35℃）			MPa	200	DMA
玻璃化温度			℃	180	TMA
模具剪切强度（260℃）			N/芯片	>100	5mm × 5mm 芯片

晶圆背面层压：60~70℃
固化：125℃/1h

图 5.18 日立 FH-9011 芯片贴膜

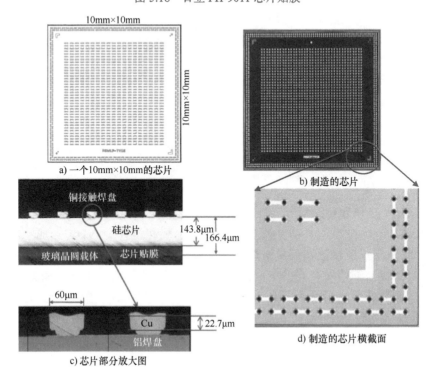

a) 一个10mm×10mm的芯片

b) 制造的芯片

c) 芯片部分放大图

d) 制造的芯片横截面

图 5.19 FOWLP 测试芯片

第 5 章 异构集成的扇出晶圆级/板级封装

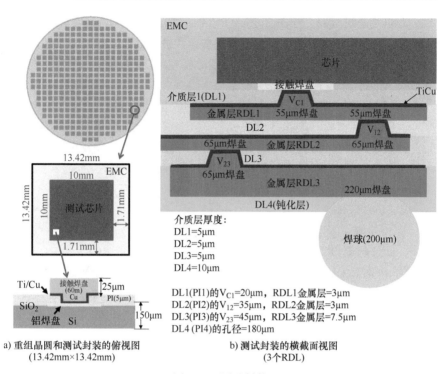

图 5.20 测试封装

如图 5.17c 所示,在重组的玻璃载板上,旋涂一层光热转换(LTHC)层(约 1μm),如图 5.21 是 3M 公司生产的一种 LTHC 层。如图 5.17d 所示,拾取单个 KGD 并将其正面朝上放置在 LTHC 载体上。为了制备 DAF,应使用与温

产品概述
3M™ 光热转换剥离型涂层(LTHC)胶是一种溶剂型涂层,使用旋涂方法进行涂敷。该涂层在 3M™ 晶圆支撑系统的玻璃基板上形成光热转换层。
特点和优势
实现黏合剂的无应力、室温剥离
典型特性说明
以下技术信息和数据仅具有代表性或典型性,不应用于规范目的
基础树脂:丙烯酸
颜色:黑色触变性液体
比重:1.00
固含量:11%
溶剂:丙二醇甲醚醋酸酯,乙二醇丁醚
闪点:45℃
一般信息
标准容量为 20L 不锈钢桶(UN:1A1/X/250)。
储存
将本产品在 5~35℃ 的正常条件下储存在原始容器中,可以获得最长的储存寿命
保质期
从 3M 发货后 6 个月
使用方法
使用前,产品必须以 600~700r/min 搅拌/混合 12h
产品在使用时必须不断搅拌

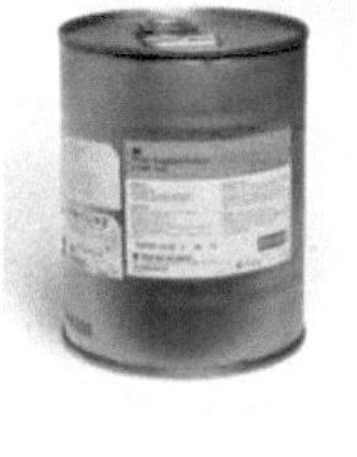

图 5.21 3M 光热转换剥离型涂层胶

135

度和压力相关的黏合剂。DAF 工艺在 120℃（键合头和键合阶段）下进行，键合力为 2kg，持续 2s。因此，重组载板将在芯片被拾取和放置期间发生热膨胀。然而，在 RDL 图形化 / 光刻期间，重组载体是在室温下操作的。因此，需要补偿由 DAF 受热引起的节距变化[76-78]。

参考文献 [76-78] 的研究中使用的 EMC 是一种液体状材料（Nagase R4507）。EMC 涂布后，接着是模压成型。经过几次实验，最佳的模压成型参数为：温度 =125℃，压力 =45kg/cm^2，时间 =10min，并在模压成型前去除空腔内的空气。接下来是 PMC，温度 =150℃，时间 ≥ 60min，加载 = 15kg，这样可以更好地控制翘曲。由于 DAF 的存在，模压成型引起的模具位移非常小（ ≤ ±4μm）[76-78]。

制造 RDL 的过程将在 5.6 节中讨论。电介质层 DL1、DL2 和 DL3 的厚度为 5μm，DL4 的厚度为 10μm，如图 5.20 所示。再分布层 RDL1 和 RDL2 的金属厚度为 3μm，RDL3 的金属厚度为 7.5μm。RDL3 的较厚金属用于较厚的无 UBM 铜焊盘，以阻挡焊球回流及在操作期间对铜的消耗。RDL 的线宽和线间距为 RDL1 为 5μm，RDL2 为 10μm，RDL3 为 15μm。RDL 如图 5.22d 所示。总体而言，可以看出所有 RDL 都已正确完成。

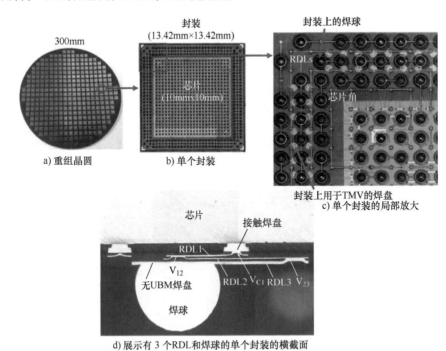

图 5.22　制备的测试封装

焊球安装有两种不同的模板：一种用于印刷助焊剂的模板，另一种用于安装焊球的模板。使用的焊球（焊料 Sn-3wt%Ag-0.5wt%Cu，直径 200μm）来自

Indium 公司。回流焊的峰值温度为 245 ℃。如图 5.22b 所示是一个独立封装，如图 5.22c 所示是封装上焊球的特写图。

如图 5.17h 所示，玻璃载板的剥离是通过从玻璃载板侧扫描激光（使用 355nm DPSS Nd:YAG UV 激光源）进行的。激光光斑尺寸为 240μm，扫描速度为 500mm/s，扫描节距为 100μm。当 LTHC 层"看到"激光时，它会转化为粉末，玻璃载板很容易被移除。然后通过化学清洗清理。

5.4.2 芯片先置（芯片面朝上）封装的热循环试验

热循环测试设置、数据采集系统和温度曲线与图 5.14 左图完全相同，样本量也是 60 个。热循环测试在 1100 次循环时停止，有 14 个失效（包括 1 个在 58 次循环时早期失效）。可以看出，威布尔图的特征寿命（63.2% 失效）为 2382 次循环，这对于智能手机和平板计算机等移动产品的预期寿命（通常小于 3 年）来说绰绰有余[87]。失效模式如图 5.23 右图所示。可以看出，焊点的裂纹出现在焊料与 RDL3 之间的界面处，并且发生在封装角附近。

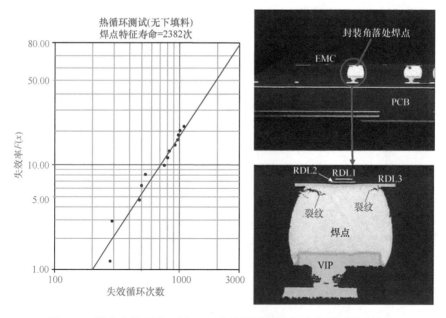

图 5.23 没有底填料的 6 层 PCB 上的测试封装的温度循环测试结果，故障位置和故障模式

图 5.24 所示模拟的基本原理与异构集成情况（见图 5.16）相同，但几何结构不同，边界（温度和运动学）条件相同。模拟（蠕变应变）结果如图 5.24 所示。可以看出，在封装角和芯片角附近的焊点的应变最大，失效模式是封装和焊料界面附近的焊点开裂。这与实际的实验观察现象非常吻合。

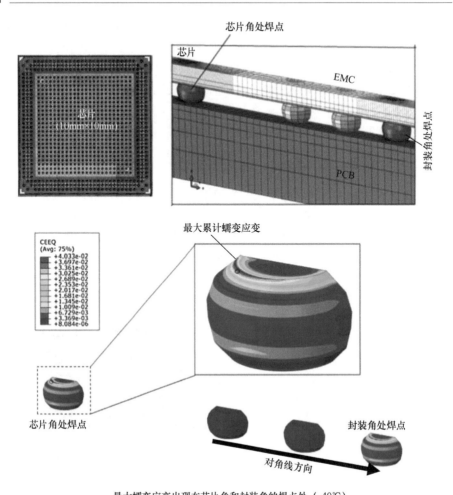

最大蠕变应变出现在芯片角和封装角的焊点处（-40℃）

图 5.24 模拟结果与温度和时间的关系——累积蠕变产生的失效位置和失效模式

5.4.3 芯片先置（芯片面朝上）封装的热性能

图 5.25 所示是图 5.22 FOWLP 结构的热分析的顶视图和横截面图[79, 80]。可以看出，芯片尺寸为 10mm×10mm，具有各种厚度（10μm、25μm、50μm、100μm、150μm、200μm、250μm 和 300μm）。芯片顶部被 100μm 的 EMC 覆盖，从芯片底部到扇出电路之间有一层 40μm 的 RDL。该封装有 1024 个（直径 0.2mm）焊球，节距 0.4mm，通过回流焊焊在 PCB 上。PCB 的尺寸为 25mm×25mm×0.8mm。

假设环境温度为 25℃，PCB 顶面和底面与芯片顶面的边界条件为对流换热系数 $h=10W/m^2·K$，模拟自然对流条件。芯片的散热量为 5W。不同厚度的 10mm×10mm 芯片的结-空气热阻（R_{ja}）如图 5.25 所示。可以看出，芯片越薄，R_{ja}越高（即热性能越差）。这是因为较薄芯片的散热能力较差。如图5.25所示，当芯片厚度低于100μm时，热性能迅速下降。典型的温度等值线分布如图5.25所示，最高温度为101.5℃，最低温度为89.9℃，R_{ja}为15.3℃/W。

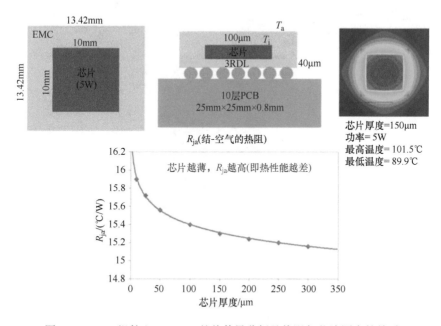

图5.25　PCB组件上FOWLP的热传导分析及热阻与芯片厚度的关系

5.4.4　芯片先置（芯片面朝上）FOW/PLP的应用

图5.26所示是PoP（堆叠式封装）的横截面示意图和SEM图像，此PoP装有iPhone的Apple应用处理器（AP）A10和动态随机存取存储器（DRAM）。这个PoP是由台积电使用其集成扇出（InFO）WLP技术制造的[29-32]。从底部的封装可以看出，消除了晶圆凸点、助焊剂、倒装芯片组装、清洁、下填料点胶和固化，以及积层封装基板（见参考文献[89]图3中AP A9）等，取而代之的是RDL，如图5.26所示A10。这种技术使得封装的成本更低、性能更高、外形更小。InFO是芯片先置且芯片面朝上型FOWLP。

上述几点非常重要，因为苹果和台积电是潮流的引领者，一旦他们使用了这项技术，很可能会有许多其他生产商效仿。此外，这意味着FOWLP不仅可用于封装基带、电源管理集成电路、射频（RF）开关/收发器、音频编解码器、微控制器、RF雷达、连接IC等，还可用于封装高性能和大尺寸（>120mm^2）SoC，如AP。此外，随着SiP或异构集成的普及，芯片先置且芯片面朝上型扇出将更多地用于精细线宽和间距的RDL（如5μm）。

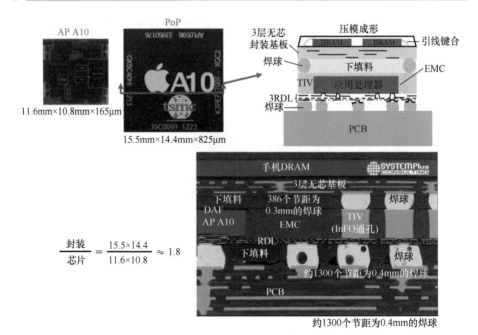

图 5.26　台积电的 InFO-WLP 用于苹果 iPhone 的应用处理器（A10）芯片组（来源：SystemPlus）

5.5　芯片后置或 RDL 先置

5.5.1　芯片后置或 RDL 先置的原因

　　根据参考文献 [19, 20]，芯片先置（芯片面朝上或芯片面朝下）型 FOWLP 的挑战之一同时也是提出芯片后置或 RDL 先置型 FOWLP 的关键原因，是因为 KGD 已经完成嵌入且在 RDL 工艺过程中的生产良率很低。正确做法应该是在芯片到晶圆键合之前对芯片后置（RDL 先置）馈通转接板（FTI）进行全面功能测试。如果在系统测试时发现 FTI 的 RDL 有坏的情况，KGD 必须丢弃。此外，应该注意的是，在 FTI 上对 RDL 进行全功能测试不仅非常昂贵，而且非常困难。

5.5.2　芯片后置或 RDL 先置工艺

　　图 5.27 所示是芯片后置或 FOWLP 的工艺流程。它与 5.3 节和 5.4 节中讨论的芯片先置 FOWLP 工艺完全不同。首先，RDL 先置 FOWLP 要求在馈通转接板（FTI）或玻璃晶圆上制备 RDL，形成晶圆凸点，进行助焊剂、芯片与晶圆键合及清洁，执行下填料的点胶和固化。晶圆凸点、芯片与晶圆键合和底部填充等技术，可参见参考文献 [89, 90]。这些工序中的每一项都是一项重要任务，都需要投入额外的材料、工艺、设备、制造场地空间和人力。因此，与芯片先置

FOWLP 相比，芯片后置（RDL 先置）FOWLP 具有较高的成本，更有可能导致更大的良率损失，只能用在如高端服务器和计算机等等要求很高密度和优异性能的领域。

如图 5.27a~f 所示，RDL 先置的第一步是在裸硅、玻璃晶圆或面板上制备 RDL。首先，在玻璃晶圆或面板上旋涂牺牲层，如图 5.27a 所示。然后，制备铜焊盘和介质层，并在介电层上制作开口，如图 5.27b 所示。接着为 RDL1 电镀铜金属层，如图 5.27c 所示。重复上述所有过程来制造其他 RDL，如图 5.27d、e 所示。然后，制作最后的介质层（钝化层）和微凸点焊盘，如图 5.27f 所示。

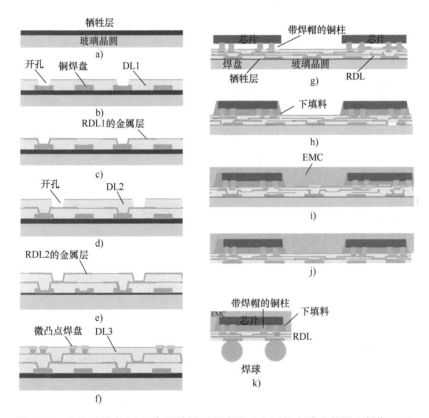

图 5.27 芯片后置或 RDL 先置关键工艺步骤（左图是在玻璃晶圆上制作 RDL，右图是在玻璃晶圆上制作封装）

同时，在器件晶圆上的第一步是形成晶圆凸点，如图 5.28 所示[89]。下一步是测试 KGD，然后将晶圆切割成单独的 KGD。接着拾取 KGD，施加助焊剂，然后将 KGD 正面朝下放置在全厚玻璃或硅晶片（或面板）的微凸点接触焊盘（位于 RDL 顶部）上完成芯片与晶圆键合，如图 5.27g 所示。然后清洁助焊剂残留物，进行填充下填料和固化，如图 5.27h 所示。使用 EMC 对整个重组晶圆载体进行模压成型，如图 5.27i 所示。然后，取下玻璃载体，如图 5.27j 所示。最后，焊球安装在底部 RDL 上，重组后的晶圆被切割成单独的封装，如图 5.27k 所示。

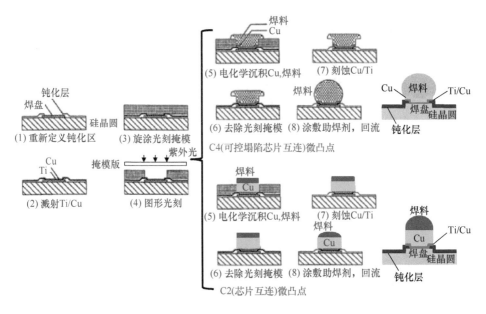

结构	主体材料	热导率（W/m·K）	电阻率/(μΩ·m)	焊盘节距	自对准
C2 凸点	Cu	400	0.0172	更小	更小
C4 凸点	焊料	55~60	0.12~0.14	更大	更大

图 5.28 C4 凸点和 C2 凸点的晶圆凸点关键工艺步骤

5.5.3 芯片后置或 RDL 先置 FOW/PLP 的应用

与台积电为苹果公司的 AP 芯片组采用芯片先置且芯片面朝上的 InFO PoP 相比，三星公司在 AP 芯片组中使用芯片后置或 RDL 先置工艺[91-93]，如图 5.29 所示。可以看出，AP 和移动 DRAM 与芯片后置 FOW/PLP 并排放置（并非 PoP 堆叠放置）。三星公司的并排放置的封装外形尺寸小于苹果公司/台积电公司的 PoP。另一方面，三星公司封装的横向尺寸应该更大一点。对于芯片后置或 RDL 先置的其他潜在应用，可参阅参考文献 [94-104]。

采用芯片后置的 FOWLP 的芯片尺寸可以很大，并且 RDL 的金属线宽和节距可以很小。但是，这项技术非常昂贵，只能用于要求很高密度和优异性能的应用领域。另一方面，对于要求很高密度和优异性能的应用，为什么要坚持使用 FOWLP 技术？因为有很多封装可供选择。

第 5 章 异构集成的扇出晶圆级/板级封装

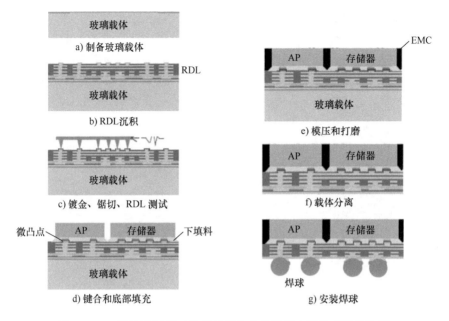

图 5.29 三星公司的用于智能手机的芯片后置应用处理器芯片组

5.6 RDL 制造

RDL[105-108] 是晶圆级封装（WLP）或面板级封装中最不可或缺的部分。RDL 由两层组成，分别是介质层和铜导电层。至少有 8 种方法可以制造 FOW/PLP 的 RDL[108]。本节仅简要介绍其中的 4 种方法。

5.6.1 有机 RDL（聚合物和 ECD 铜 + 刻蚀）

有机 RDL 是扇入 WLP 制作 RDL 的最古老的方法，可参见参考文献 [88]。介质层由聚合物制成，如聚酰亚胺（PI）、苯并环丁烯（BCB）或聚苯并二恶唑（PBO），而导体层由电化学沉积（ECD）铜并通过刻蚀制成。关键工艺步骤描述如下：首先，在整个晶片上旋涂聚合物；接下来旋涂光刻胶；然后用掩模对准机或步进机对光刻胶进行对准曝光；然后刻蚀聚合物，剥离抗蚀剂；使用物理气相沉积（PVD）溅射黏合层/种子层（钛/铜）；然后旋涂光刻胶，用掩模对准机或步进机对光刻胶进行对准曝光；接下来是电镀铜。在剥离抗蚀剂并完成钛/铜刻蚀之后，就形成了第一个 RDL。通过重复上述过程，就可以制备其他 RDL。现在，大多数外包半导体组装和测试（OSAT）都是使用这种方法为芯片先置和芯片后置 FOWLP 制造 RDL。

图 5.30 所示是一个更好、更简单的 RDL 制备工艺过程[46, 77]。可以看出，对于 PI 显影，整个重组晶片旋涂有感光 PI。随后使用步进机（对于高良率）利

143

用光刻技术进行对准、曝光和制备 PI 的通孔。最后，PI 在 200℃下固化 1h，形成 5μm 厚的 PI 层。然后在 175℃下通过 PVD 在整个重组晶片上溅射钛和铜，使用光刻胶和步进机利用光刻技术进行曝光 RDL 的位置。接下来在室温下通过 ECD 在光刻胶开腔中的钛/铜上电镀铜。完成上述步骤后剥离光刻胶并刻蚀掉钛/铜，从而获得 RDL1。

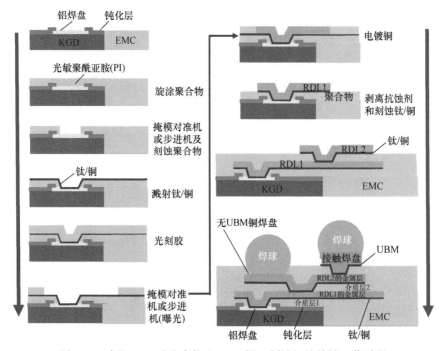

图 5.30　有机 RDL（聚合物和 ECD 铜 + 刻蚀）的关键工艺过程

5.6.2　无机 RDL（PECVD 和镶嵌铜 +CMP）

无机 RDL 是最古老的后端半导体工艺。该工艺使用 SiO_2 或 SiN 作为介质层，使用 ECD 将铜沉积在整个晶圆或面板载体上。然后使用 CMP（化学机械抛光）去除覆盖层的铜和种子层，以制造 RDL 的铜导体层。关键流程步骤如图 5.31 所示。首先，使用等离子增强化学气相沉积（PECVD）在全厚裸硅晶圆（或玻璃面板）上形成一层薄薄的 SiO_2（或 SiN），然后使用旋涂机涂层光刻胶。完成这些步骤之后，使用步进机开腔抗蚀剂并使用反应离子刻蚀（RIE）去除 SiO_2。然后，使用步进机将抗蚀剂开腔得更宽，并使用 RIE 刻蚀更多的 SiO_2。接下来，剥离抗蚀剂，溅射钛/铜，并通过 ECD 在整个晶圆或面板载体上沉积铜。完成上述步骤之后通过 CMP 法去除多余的铜和钛/铜，这样就制造了第一个 RDL1 和 V01（连接 Si 和 RDL1 的通孔），如图 5.32 所示。称为双铜镶嵌法[105, 106]。最后，重复上述所有过程以制造其他 RDL。这种方法可用于芯片先置和芯片后置FOWLP工艺。由PECVD和铜镶嵌+CMP制成的RDL称为无机RDL。RDL的金属线宽和节距可以小于2μm甚至低至亚微米。无机RDL有时应用于具有很高性能应用的芯片后置构件。

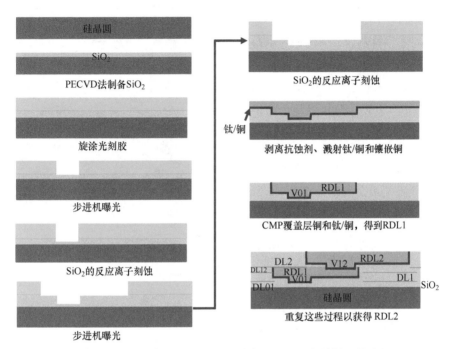

图5.31 无机RDL（PECVD和镶嵌铜+CMP）的关键工艺步骤

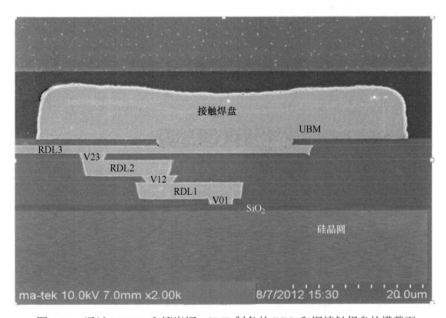

图5.32 通过PECVP和镶嵌铜+CMP制备的RDL和铜接触焊盘的横截面

5.6.3 混合 RDL（先无机 RDL，后有机 RDL）

时至今日，混合 RDL 方法仅适用于芯片后置或 RDL 先置，也就是说对于混合 RDL 方法晶圆凸点和芯片与晶圆键合是必要的。图 5.33 所示是通过混合 RDL 进行芯片后置的关键工艺步骤。可以看出，玻璃载板 1 涂有牺牲层，如图 5.33a 所示。然后通过 PECVD 法制备 SiO_2 电介质层，双铜镶嵌法 +CMP 制备导体层，从而制造接触焊盘和第一个 RDL（RDL1），如图 5.33b 所示。其余的 RDL 采用普通聚合物（或光敏聚合物）和镀铜加刻蚀的方法制造。然后将另一个载板 2 贴在重组晶圆的另一侧，如图 5.33c 所示。该步骤之后是载板 1 的剥离，如图 5.33d 所示。接着是上助焊剂、芯片与晶圆的键合、清洁、底部填充和固化，如图 5.33e 所示。然后，通过使用 EMC 进行模压成型为重组晶圆，如图 5.33f 所示。接下来是剥离载板 2 和安装焊球，如图 5.33g 所示。图 5.34 所示是由 SPIL 公司在参考文献 [109] 中发表的具有混合 RDL 的 FOWLP 的横截面。更多的其他混合 RDL，可参阅参考文献 [110，111]。

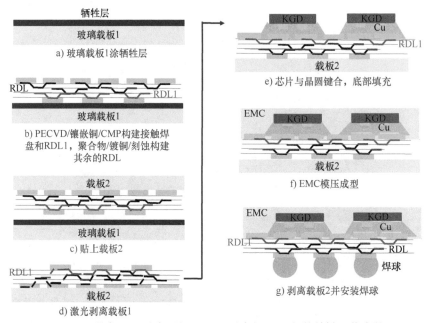

图 5.33 混合 RDL（先无机 RDL，后有机 RDL）的关键工艺步骤

5.6.4 纯 PCB 技术的 RDL（ABF/SAP/LDI 和镀铜 + 刻蚀）

图 5.35 所示是参考文献 [48-50] 中发表的在板级载体上制造 RDL 的工艺流程。首先在重组的 ECM 板上层压一层 Ajinomoto 增强膜（ABF）。然后是激光打孔及进行化学镀铜种子层，依次进行干膜层压、LDI 光刻、干膜显影和 PCB 电镀铜等步骤来制造 RDL1。接着剥离干膜并刻蚀掉种子层。重复上述步骤制取

第 5 章 异构集成的扇出晶圆级/板级封装

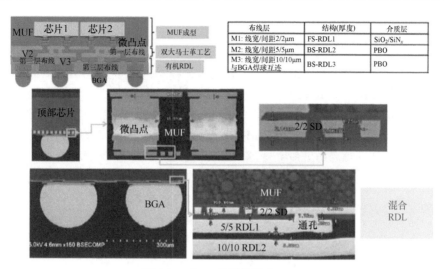

图 5.34 SPIL 公司的混合 RDL

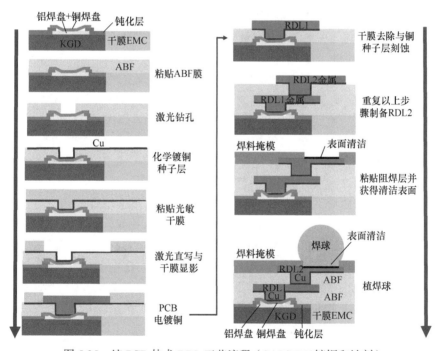

图 5.35 纯 PCB 技术 RDL 工艺流程（SAP/LDI/镀铜和蚀刻）

其他的 RDL。最后的 RDL 可用作接触焊盘。接下来的步骤是层压、光刻和固化阻焊层（定义的阻焊层或定义的非阻焊层），再安装焊球。在这种情况下，电介质层厚度可以小至 10μm，导体层厚度可以小至 5μm。如图 5.12 和图 5.13 中的 RDL 就是用这种方法制作的。一般来说，纯 PCB 技术 RDL 的金属线宽和间距 ≥ 10μm。

5.7 翘曲

翘曲是 FOW/PLP 的重要问题，翘曲问题得到了大量的研究。Lin 等人[112]研究了芯片先置和芯片面朝下晶圆级封装的翘曲，Che 等人[113]研究了模具先置晶圆级封装的翘曲，Che 等人[114]研究了芯片后置或 RDL 先置晶圆级封装的翘曲，Hou 等人[115]研究了板级封装的翘曲，Shen 等人[116]研究了在带有金属盖的封装基板上进行底部填充后并带有焊球凸点的倒装芯片的单个封装的翘曲，Lau 等人[117, 118]研究了芯片先置和芯片面朝上封装的翘曲。

5.7.1 FOW/PLP 中的各种翘曲

根据封装的形式和 RDL 的数量，有几种不同的翘曲会影响 FOW/PLP 工艺。以在 300 mm FOWLP 上具有 3 个 RDL 的芯片先置和芯片面朝上为例，如图 5.17 和图 5.22 所示，在这种情况下，至少有 6 种不同的翘曲会影响 FOWLP 工艺。

第一种翘曲发生在重组晶圆的模压固化（PMC）之后（见图 5.17e）。如果这种翘曲太大，则无法将重组晶圆放置和（或）无法在背磨设备上进行 EMC 背磨以暴露铜接触焊盘。

第二种翘曲发生在 EMC 背面研磨以暴露铜接触焊盘之后（见图 5.17f）。如果这种翘曲太大，则重组晶片不能放置和（或）无法在 RDL 设备上进行诸如步进、光刻、物理气相沉积（PVD）、电化学沉积和刻蚀等操作。

第三种翘曲发生在第一个 RDL 制作完成之后。PVD 的温度约为 200℃，因此 EMC、硅芯片和玻璃载体之间存在热膨胀失配。如果这种翘曲太大，则制作第二个 RDL 时会出现问题。第四种翘曲发生在第二个 RDL 制作完成之后。如果这种翘曲太大，那么在制作第三个 RDL 时就会出现问题。第五种翘曲发生在第三个 RDL 制作完成之后。如果这种翘曲太大，则在安装焊球时会出现问题（如设备让对重组晶片进行固定和 / 或操作以及控制落球的精度，见图 5.17g）。

第六种翘曲发生在焊球安装之后。无铅回流焊温度约为 250℃，因此 EMC、硅芯片和玻璃载体之间存在非常大的热膨胀失配。如果切割后的单个封装的翘曲太大（见图 5.17h），那么在 PCB 组装中就会出现问题（如焊点间距变化、拉伸焊点和元件倾斜）。

5.7.2 允许的最大翘曲

经验法则表明，对于 300mm 的重组晶圆，前五种翘曲的最大允许翘曲为 1mm，但为了较高的良率，首选 0.5mm。单个封装（≤20mm×20mm）的最大允许翘曲为 0.2mm，但为实现高良率，首选 0.1mm。

5.7.3 翘曲的测量与模拟

图 5.36a 所示是 300mm 重组晶圆（见图 5.22）在模压固化（PMC，笑脸形为 609μm，见图 5.17e）后以及立刻进行 EMC 背磨以暴露铜接触焊盘（哭脸形为 811.9μm，见图 5.17f）后进行的投影波纹（shadow moire）技术翘曲测量。值得注意的是，没有 RDL 和焊球的重组晶圆在经过 EMC 背磨暴露铜接触焊盘之后，其翘曲形式从笑脸形变为哭脸形。图 5.36b 所示模拟过程中也发现了类似的趋势[117, 118]。为了减小模压固化后重组晶圆的翘曲，玻璃载体和 EMC 的 CTE 应尽可能接近。另外，为了减小背面研磨后重组晶圆的翘曲，EMC 的 CTE 应该比玻璃载体的 CTE 大（创造更大的笑脸形）。但是，在 PMC 后的重组晶圆的翘曲不可以超过 1mm，否则重组晶片便无法放置在背磨设备上。

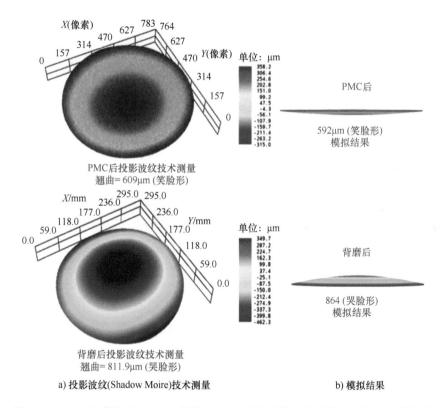

图 5.36　PMC 和背磨后 300mm 晶圆 FOWLP（芯片先置和芯片正面朝上）的翘曲

根据 JEDEC 标准（JESD22-B112A）[119]，激光反射法（共焦位移计量）可以用于测量单个封装的翘曲。如图 5.37a 所示为一个当前封装的典型翘曲测量轮廓显示图。图 5.37b 为两个不同的独立封装样品的翘曲测量值与温度的关系，其中还包括模拟结果。可以看出，两个样品与模拟结果比较吻合。

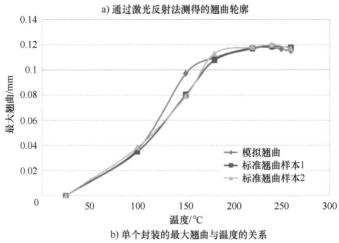

a) 通过激光反射法测得的翘曲轮廓

b) 单个封装的最大翘曲与温度的关系

图 5.37　翘曲的测量与模拟

5.8　临时晶圆与面板载体

首先，FOWLP 本身就是一种非常高吞吐量的 WLP 技术。从理论上讲，FOPLP 可能比 FOWLP 具有更高的吞吐量和更低的成本。然而，为了实现这些目标，需要注意和/或解决 FOPLP 的以下问题[120-123]：

1）大多数 OSATS 和代工厂已经拥有 FOWLP 所需的设备。对于 FOPLP，必须将新资金用于新开发的设备。

2）晶圆的检查是众所周知的过程，必须开发 FOPLP 检查。

3）FOWLP 的良率高于 FOPLP（假设面板尺寸大于晶圆）。

4）需要仔细确定面板相对于晶圆的成本优势（面板载体吞吐量更高，但拾取、放置和 EMC 涂胶时间更长，良率更低）。

5）满载的高良率晶圆生产线可能比部分装载的低良率面板生产线成本低廉。

6）面板设备比晶圆设备需要更长的清洁时间。

7）与 FOWLP 不同，FOPLP 适用于中等芯片尺寸及金属线宽和间距。

8）如果确实面板工艺得到了发展，并且在细线宽和间距方面具有高产量，那么就有可能产生严重的产能过剩。

9）需要进行知识产权、材料背景、设备自动化以及大幅面面板空间稳定性和良率的管理。

10）FOPLP 缺乏面板标准意味着设备供应商无法制造设备。

5.9 异构集成的 FOW/PLP 机会

由于人工智能的驱动，半导体的密度和 I/O 增加，焊盘节距减小。即使是 12 层（6-2-6）有机封装基板也不足以支撑半导体，因此需要 TSV 转接板[124-142]。台积电将这种结构称为 CoWoS（chip-on-wafer-on substrate）[140, 141]。Leti 公司[124, 125]将其称为 SoW（system-on-wafer）。

图 5.38 所示是 SoC 的异构集成示意图[126]，如 CPU（中央处理器）或 GPU（图形处理器）和由 DRAM（动态随机存取存储器）堆栈组成的 HBM 和基本逻辑通过 TSV（硅通孔）和微凸点垂直互连。这些 SoC 和 HBM 通过带有 RDL 的 TSV 转接板上的微凸点并排连接。TSV 转接板（通过 Cu-C4 凸点）连接在封装基板上，然后通过焊球将其连接到 PCB。英伟达的 Pascal 100 GPU 就是一个例子。由于 TSV 转接板非常昂贵[10, 11]，这为 FOWLP 带来了机会。

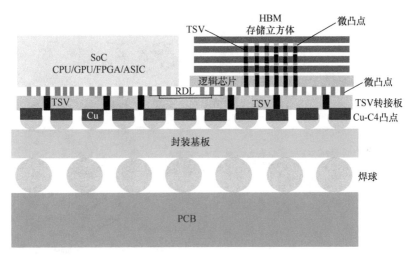

下填料应用于
➤ 在TSV转接板与SoC/ 逻辑芯片 + 存储立方体之间
➤ 在TSV转接板与封装基板之间

图 5.38 台积电的 CoWoS-2

星科金朋[34, 143]公司提出使用扇出倒装芯片（FOFC）-eWLB 制作芯片的 RDL，以实现大部分横向互连，如图 5.39 所示。可以看出，TSV 转接板、晶圆凸点、助焊剂、芯片到晶圆键合、清洗以及下填料填充和固化都被省略了。

日月光半导体公司[35]提出使用 FOWLP 技术（芯片先置和芯片面朝下在临时晶圆载体上，然后通过模压方法成型）使芯片的 RDL 主要执行横向互连，如图 5.40 所示，称其为扇出型晶圆级基板芯片（FOCoS）。FOCoS 取消了

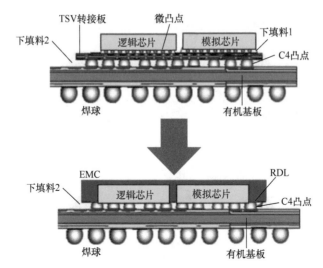

图 5.39　星科金朋公司采用芯片先置工艺的 FOCoS

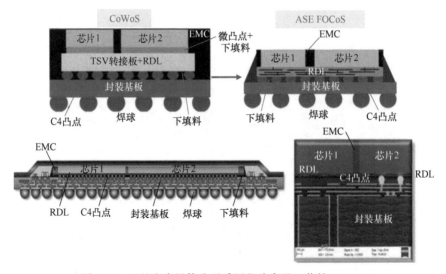

图 5.40　日月光半导体公司采用芯片先置工艺的 FOCoS

TSV 转接板、芯片的晶圆凸点、助焊剂、芯片到晶圆的键合和清洁，以及下填料填充和固化。底部 RDL 使用凸点金属化（UBM）和 C4 凸点连接到封装基板，如图 5.40 所示。最近，台积电[144]将此称为 InFO_oS（基板上直接扇出）。

最近，三星公司[145]提出使用芯片后置或 RDL 先置 FOWLP 来消除 TSV 转接板，如图 5.41 所示。首先，在硅片或玻璃面板上制备 RDL，以及逻辑芯片和 HBM 的晶圆凸点。然后，执行助焊剂、芯片到晶圆或面板键合、清洁、下填料填充芯片和固化。其次是 EMC 压缩成型，背磨硅晶圆和 C4 晶圆凸点，然后将整个模块贴在封装基板上。最后，进行植球和封盖连接。三星公司称其为无硅 RDL 转接板[145]。

第 5 章 异构集成的扇出晶圆级/板级封装

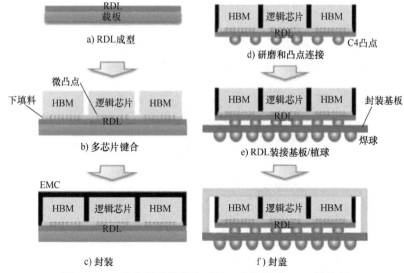

图 5.41 三星公司采用芯片后置工艺的无硅 RDL 转接板

TSMC[146] 提出使用 FOWLP 技术 + 铜柱/焊凸点来消除 HBM 模块中的 TSV。在单个芯片（无凸点的 DRAM）中，通过使用 FOWLP 使 RDL 将所有电路扇出到封装的外围。各个封装的垂直互连通过铜柱和焊凸点，如图 5.42 所示。图 5.43 所示是另一个示例[147]，其中 6 层芯片在没有 TSV 的情况下垂直互连在一起。FOWLP 用于制备 RDL，将电路扇出到封装的外围，并且垂直互连通过铜柱和微凸点。

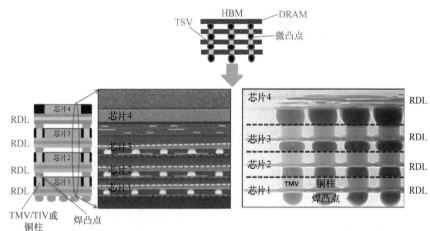

图 5.42 TSMC 的 InFO-WLP + HBM 用铜柱/焊凸点

自从英特尔提出[148, 149]使用 EMIB（嵌入式多芯片互连桥）作为异构集成系统中芯片之间的高密度互连以来，如图 5.44 所示，"桥"一直非常流行。基本上，桥是一片带有 RDL 和接触焊盘但没有 TSV 的虚拟（无器件）硅。如最近 IMEC[150] 提出使用桥 +FOWLP 用于互连逻辑芯片、宽 I/O DRAM 和闪存，如图 5.45 所示，目标是所有器件芯片都不使用 TSV。

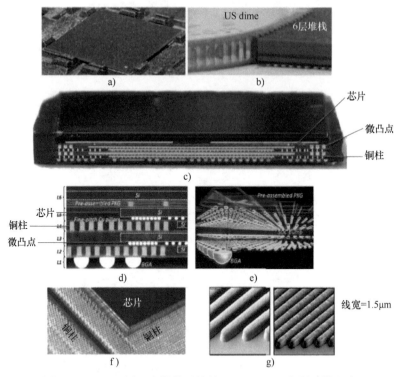

图 5.43　TSMC 用于高性能系统的 InFO-WLP + 铜柱 / 微凸点

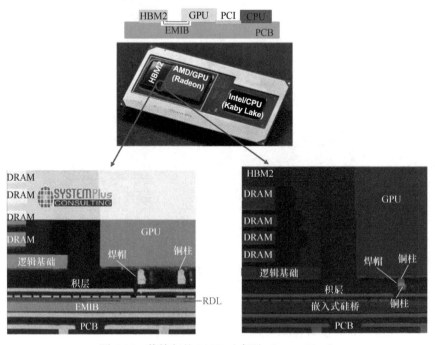

图 5.44　英特尔的 EMIB（来源：SystemPlus）

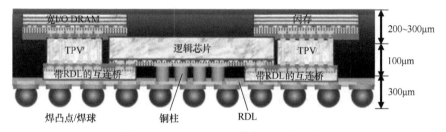

图 5.45　IMEC 用于异构集成系统的桥

台积电[151]证明，用于高性能和紧凑型 5G 毫米波系统集成的 InFO_AiP（封装天线）优于基板上的焊料凸点倒装芯片 AiP，如图 5.46 所示。可以看出，在 28GHz 频率范围内，InFO RDL 传输损耗（0.175dB/mm）比倒装芯片基板走线（0.288dB/mm）低 65%；在 38GHz 频率范围内，InFO RDL 的传输损耗（0.225dB/mm）比倒装芯片基板走线上的传输损耗（0.377dB/mm）低 53%。

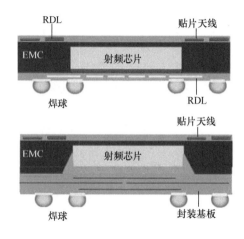

28GHz 和 38GHz 时的 RDL 和基板线路传输损耗

频率	InFO RDL	基板线路
28GHz	0.175dB/mm	0.228dB/mm
38GHz	0.225dB/mm	0.377dB/mm

图 5.46　台积电用于 5G 的 InFO_AiP

5.10　总结和建议

本章介绍了 FOW/PLP 的最新进展和趋势。一些重要的结果和建议总结如下：

1）对于便携式、移动和可穿戴产品的基带、RF/ 模拟、PMIC、AP 和低端 ASIC、CPU 和 GPU 等半导体 IC 的异构集成，芯片先置是一个不错的选择。芯片后置（RDL 先置）是潜在的适用于超级计算机、服务器、网络和电信产品的高端 CPU、GPU、ASIC 和 FPGA（现场可编程门阵列）等半导体 IC 器件的异构集成技术。

2）芯片先置和芯片面朝下是最简单和低成本的形式，一般来说适用于较小的芯片，RDL 的金属线宽和间距 ≥ 10μm。

3）芯片先置和芯片面朝上的工艺步骤比芯片先置和芯片面朝下的工艺步骤稍微复杂一些，因此成本略高，一般来说适用于较大的芯片，RDL 的金属线宽和间距 ≥ 5μm。

4）芯片后置或 RDL 先置的工艺步骤最复杂、成本最高。但是，它适用于非常大的芯片和金属线的宽度和间距 RDL 小于 5μm 至亚微米。因此，这种工艺只能由非常高密度和高性能的应用程序提供。另外，对于高密度和高性能的应用，因为还有很多其他的封装替代方案，因此没必要坚持使用 FOWLP 技术。

5）通过聚合物（光敏或非光敏）和 ECD 铜 + 刻蚀制造的有机 RDL 是 OSAT 甚至代工厂最常用的 FOWLP 方法，可以应用于芯片先置和芯片后置。

6）通过 PECVD 和铜镶嵌 +CMP 制造的无机 RDL 是一种形成芯片后置 FOWLP 的后端半导体方法。从应用处理器芯片组（从 iPhone 7 的 A10 到 iPhone 8 的 A11）的 RDL 的金属线宽和间距（5~10μm）的变化，可以看到使用 PECVD 和铜镶嵌 +CMP 方法为 FOWLP 制造 RDL 的机会非常少（可能仅适用于小众应用）。但是，如果需要无机 RDL，就没必要坚持使用 FOWLP 技术。

7）先由无机 RDL 再由有机 RDL 制造的混合 RDL 是一种用于形成芯片后置 FOWLP 的混合方法。同样，如果需要混合 RDL，就没必要坚持使用 FOWLP 技术。

8）纯 PCB/LDI 技术的 RDL 用于 FOPLP，不需要任何半导体设备，是产量最高、成本最低的技术，但是它适用于尺寸小的芯片（<8mm × 8mm）与金属线宽和间距较大的 RDL（≥ 10μm）。

9）为了使用 FOPLP 提高吞吐量和产量并降低成本，一些（应注意并解决）重要问题已在前文列出。

10）翘曲是 FOW/PLP 的一个关键问题。根据封装的形式和 RDL 的数量，有几种不同的翘曲会影响 FOW/PLP 工艺。经验法则表明，对于 300mm 的重组晶圆，重组晶圆的最大允许翘曲为 1 mm，但为了高良率，0.5mm 是首选。单个封装的最大允许翘曲（≤ 20mm × 20mm）为 0.2 mm，但为实现高良率，首选 0.1mm。

参考文献

[1] Hedler, H., T. Meyer, and B. Vasquez, "Transfer Wafer-Level Packaging", US Patent 6,727,576, Filed on October 31, 2001, Patented on April 27, 2004.
[2] Lau, J.H. 2015. Patent Issues of Fan-Out Wafer/Panel-Level Packaging. *Chip Scale Review* 19: 42–46.
[3] Brunnbauer, M., E. Furgut, G. Beer, T. Meyer, H. Hedler, J. Belonio, E. Nomura, K. Kiuchi, and K. Kobayashi, "An Embedded Device Technology Based on a Molded Reconfigured Wafer", *IEEE/ECTC Proceedings*, May 2006, pp. 547–551.

[4] Brunnbauer, M., E. Furgut, G. Beer, and T. Meyer, "Embedded Wafer Level Ball Grid Array (eWLB)", *IEEE/EPTC Proceedings*, December 2006, pp. 1–5.
[5] Eichelberger, C., and R. Wojnarowski, "High-density interconnect with high volumetric efficiency", US Patent 5,019,946, Filed on September 27, 1988, Patented on May 28, 1991.
[6] Fillion, R., R. Wojnarowski, M. Gdula, H. Cole, E. Wildi, and W. Daum, "Method for fabricating an integrated circuit module", US Patent 5,353,498, Filed on July 9, 1993, Patented on October 11, 1994.
[7] Eichelberger, C., "Single-chip modules, repairable multi-chip modules, and methods of fabrication thereof", US Patent 5,841,193, Filed on May 20, 1996, Patented on November 24, 1998.
[8] Lau, J. H., "Patent Issues of Embedded Fan-Out Wafer/Panel Level Packaging", *Proceedings of CSTIC*, March 2016, pp. 1–7.
[9] Lau, J. H., *Reliability of RoHS compliant 2D & 3D IC Interconnects*, McGraw-Hill, New York, 2011.
[10] Lau, J. H., *Through-Silicon Via (TSV) for 3D Integration*, McGraw-Hill, New York, 2013.
[11] Lau, J. H., *3D IC Integration and Packaging*, McGraw-Hill Book Company, NY, 2015.
[12] Keser, B., C. Amrine, T. Duong, O. Fay, S. Hayes, G. Leal, W. Lytle, D. Mitchell, and R. Wenzel, "The Redistributed Chip Package: A Breakthrough for Advanced Packaging", *Proceedings of IEEE/ECTC*, 2007, pp. 286–291.
[13] Kripesh, V., V. Rao, A. Kumar, G. Sharma, K. Houe, X. Zhang, K. Mong, N. Khan, and J. H. Lau, "Design and Development of a Multi-Die Embedded Micro Wafer Level Package", *IEEE/ECTC Proceedings*, 2008, pp. 1544–1549.
[14] Khong, C., A. Kumar, X. Zhang, S. Gaurav, S. Vempati, V. Kripesh, J. H. Lau, and D. Kwong, "A Novel Method to Predict Die Shift During Compression Molding in Embedded Wafer Level Package", *IEEE/ECTC Proceedings*, 2009, pp. 535–541.
[15] Sharma, G., S. Vempati, A. Kumar, N. Su, Y. Lim, K. Houe, S. Lim, V. Sekhar, R. Rajoo, V. Kripesh, and J. H. Lau, "Embedded Wafer Level Packages with Laterally Placed and Vertically Stacked Thin Dies", *IEEE/ECTC Proceedings*, 2009, pp. 1537–1543. Also, *IEEE Transactions on CPMT*, Vol. 1, No. 5, May 2011, pp. 52–59.
[16] Kumar, A., D. Xia, V. Sekhar, S. Lim, C. Keng, S. Gaurav, S. Vempati, V. Kripesh, J. H. Lau, and D. Kwong, "Wafer Level Embedding Technology for 3D Wafer Level Embedded Package", *IEEE/ECTC Proceedings*, 2009, pp. 1289–1296.
[17] Lim, Y., S. Vempati, N. Su, X. Xiao, J. Zhou, A. Kumar, P. Thaw, S. Gaurav, T. Lim, S. Liu, V. Kripesh, and J. H. Lau, "Demonstration of High Quality and Low Loss Millimeter Wave Passives on Embedded Wafer Level Packaging Platform (EMWLP)", *IEEE/ECTC Proceedings*, 2009, pp. 508–515. Also, *IEEE Transactions on Advanced Packaging*, Vol. 33, 2010, pp. 1061–1071.
[18] Lau, J. H., N. Fan, and M. Li, "Design, Material, Process, and Equipment of Embedded Fan-Out Wafer/Panel-Level Packaging", *Chip Scale Review*, Vol. 20, 2016, pp. 38–44.
[19] Kurita, Y., T. Kimura, K. Shibuya, H. Kobayashi, F. Kawashiro, N. Motohashi, et al., "Fan-out wafer-level packaging with highly flexible design capabilities", *IEEE/ECTC Proceedings*, 2010, pp. 1–6.
[20] Motohashi, N., T. Kimura, K. Mineo, Y. Yamada, T. Nishiyama, K. Shibuya, et al., "System in wafer-level package technology with RDL-first process", *IEEE/ECTC Proceedings*, 2011, pp. 59–64.
[21] Kurita, Y., K. Soejima, K. Kikuchi, M. Takahashi, M. Tago, M. Koike, et al., "A novel "SMAFTI" package for inter-chip wide-band data transfer", *IEEE/ECTC Proceedings*, 2006, pp. 289–297.
[22] Kawano, M., S. Uchiyama, Y. Egawa, N. Takahashi, Y. Kurita, K. Soejima, et al., "A 3D packaging technology for 4 Gbit stacked DRAM with 3 Gbps data transfer", *Proceedings of IEMT*, 2006, pp. 581–584.
[23] Kurita, Y., S. Matsui, N. Takahashi, K. Soejima, M. Komuro, M. Itou, et al., "A 3D stacked memory integrated on a logic device using SMAFTI technology", *IEEE/ECTC Proceedings*, 2007, pp. 821–829.
[24] Kawano, M., N. Takahashi, Y. Kurita, K. Soejima, M. Komuro, S. Matsui, "A 3-D Packaging Technology for Stacked DRAM with 3 Gb/s Data Transfer," *IEEE Transaction on Electron*

[25] Motohashi, N., Y. Kurita, K. Soejima, Y. Tsuchiya, M. Kawano, "SMAFTI Package with Planarized Multilayer Interconnects", *IEEE/ECTC Proceedings*, 2009, pp. 599–606.
[26] Kurita, M., S. Matsui, N. Takahashi, K. Soejima, M. Komuro, M. Itou, et al., "Vertical Integration of Stacked DRAM and High-Speed Logic Device Using SMAFTI Technology," *IEEE Transaction on Advanced Packaging*, 2009, pp. 657–665.
[27] Kurita, Y., N. Motohashi, S. Matsui, K. Soejima, S. Amakawa, K. Masu, et al., "SMAFTI Packaging Technology for New Interconnect Hierarchy," *Proceedings of IITC*, 2009, pp. 220–222.
[28] Yoon, S., J. Caparas, Y. Lin, and P. Marimuthu, "Advanced Low Profile PoP Solution with Embedded Wafer Level PoP (eWLB-PoP) Technology", *IEEE/ECTC Proceedings*, 2012, pp. 1250–1254.
[29] Yu, D., "Wafer-Level System Integration (WLSI) Technologies for 2D and 3D System-in-Package", *SEMIEUROPE* 2014.
[30] Lin, J., J. Hung, N. Liu, Y. Mao, W. Shih, and T. Tung, "Packaged Semiconductor Device with a Molding Compound and a Method of Forming the Same", US Patent 9,000,584, Filed on December 28, 2011, Patented on April 7, 2015.
[31] Tseng, C., Liu, C., Wu, C., and D. Yu, "InFO (Wafer Level Integrated Fan-Out) Technology", *IEEE/ECTC Proceedings*, 2016, pp. 1–6.
[32] Hsieh, C., Wu, C., and D. Yu, "Analysis and Comparison of Thermal Performance of Advanced Packaging Technologies for State-of-the-Art Mobile Applications", *IEEE/ECTC Proceedings*, 2016, pp. 1430–1438.
[33] Lau, J. H., "TSV-Less Interposers", *Chip Scale Review*, Vol. 20, September/October 2016, pp. 28–35.
[34] Yoon, S., P. Tang, R. Emigh, Y. Lin, P. Marimuthu, and R. Pendse, "Fanout Flipchip eWLB (Embedded Wafer Level Ball Grid Array) Technology as 2.5D Packaging Solutions", *IEEE/ECTC Proceedings*, 2013, pp. 1855–1860.
[35] Lin, Y., W. Lai, C. Kao, J. Lou, P. Yang, C. Wang, and C. Hseih, "Wafer Warpage Experiments and Simulation for Fan-out Chip on Substrate", *IEEE/ECTC Proceedings*, 2016, pp. 13–18.
[36] Chen, N., T. Hsieh, J. Jinn, P. Chang, F. Huang, J. Xiao, A. Chou, and B. Lin, "A Novel System in Package with Fan-out WLP for high speed SERDES application", *IEEE/ECTC Proceedings*, 2016, pp. 1495–1501.
[37] Hayashi, N., T. Takahashi, N. Shintani, T. Kondo, H. Marutani, Y. Takehara, K. Higaki, O. Yamagata, Y. Yamaji, Y., Katsumata, and Y. Hiruta, "A Novel Wafer Level Fan-out Package (WFOP^{TM}) Applicable to 50 μm Pad Pitch Interconnects", *IEEE/EPTC Proceeding*, December 2011, pp. 730–733.
[38] Hayashi, N., H. Machida, N. Shintani, N. Masuda, K. Hashimoto, A. Furuno, K. Yoshimitsu, Y. Kikuchi, M. Ooida, A. Katsumata and Y. Hiruta, "A New Embedded Structure Package for Next Generation, WFOP^{TM} (Wide Strip Fan-Out Package)". *Pan Pacific Symposium Conference Proceedings*, February 2014, pp. 1–7.
[39] Hayashi, N., M. Nakashima, H. Demachi, S. Nakamura, T. Chikai, Y. Imaizumi, Y. Ikemoto, F. Taniguchi, M. Ooida, and A. Yoshida, "Advanced Embedded Packaging for Power Devices", *IEEE/ECTC Proceedings,* 2017, pp. 696–703.
[40] Braun, T., K.-F. Becker, S. Voges, T. Thomas, R. Kahle, J. Bauer, R. Aschenbrenner, and K.-D. Lang, "From Wafer Level to Panel Level Mold Embedding", *IEEE/ECTC Proceedings*, 2013, pp. 1235–1242.
[41] Braun, T., K.-F. Becker, S. Voges, J. Bauer, R. Kahle, V. Bader, T. Thomas, R. Aschenbrenner, and K.-D. Lang, "24"×18" Fan-out Panel Level Packing", *IEEE/ECTC Proceedings*, 2014, pp. 940–946.
[42] Braun, T., S. Raatz, S. Voges, R. Kahle, V. Bader, J. Bauer, K. Becker, T. Thomas, R. Aschenbrenner, and K. Lang, "Large Area Compression Molding for Fan-out Panel Level Packing", *IEEE/ECTC Proceedings*, 2015, pp. 1077–1083.
[43] Chang, H., D. Chang, K. Liu, H. Hsu, R. Tai, H. Hunag, Y. Lai, C. Lu, C. Lin, and S. Chu, "Development and Characterization of New Generation Panel Fan-Out (PFO) Packaging Technology", *IEEE/ECTC Proceedings*, 2014, pp. 947–951.
[44] Liu, H., Y. Liu, J. Ji, J. Liao, A. Chen, Y. Chen, N. Kao, and Y. Lai, "Warpage Characterization of Panel Fab-out (P-FO) Package", *IEEE/ECTC Proceedings*, 2014, pp. 1750–1754.

[45] Lau, J. H., M. Li, M. Li, T. Chen, I. Xu, X. Qing, Z. Cheng, et al., "Fan-Out Wafer-Level Packaging for Heterogeneous Integration", *Proceedings of IEEE/ECTC*, May 2018, pp. 2354–2360.

[46] Lau, J. H., M. Li, M. Li, T. Chen, I. Xu, X. Qing, Z. Cheng, N. Fan, E. Kuah, Z. Li, K. Tan, Y. Cheung, E. Ng, P. Lo, K. Wu, J. Hao, S. Koh, R. Jiang, X. Cao, R. Beica, S. Lim, N. Lee, C. Ko, H. Yang, Y. Chen, M. Tao, J. Lo, and R. Lee, "Fan-Out Wafer-Level Packaging for Heterogeneous Integration", *IEEE Transactions on CPMT*, 2018, September 2018, pp. 1544–1560.

[47] Lau, J. H., M. Li, Y. Lei, M. Li, I. Xu, T. Chen, Q. Yong, Z. Cheng, et al., "Reliability of Fan-Out Wafer-Level Heterogeneous Integration", *IMAPS Transactions, Journal of Microelectronics and Electronic Packaging*, Vol. 15, Issue 4, October 2018, pp. 148–162.

[48] Ko, CT, H. Yang, J. H. Lau, M. Li, M. Li, C. Lin, et al., "Chip-First Fan-Out Panel-Level Packaging for Heterogeneous Integration", *IEEE/ECTC Proceedings*, May 2018, pp. 355–363.

[49] Ko, CT, H. Yang, J. H. Lau, M. Li, M. Li, C. Lin, J. W. Lin, T. Chen, I. Xu, C. Chang, J. Pan, H. Wu, Q. Yong, N. Fan, E. Kuah, Z. Li, K. Tan, Y. Cheung, E. Ng, K. Wu, J. Hao, R. Beica, M. Lin, Y. Chen, Z. Cheng, S. Koh, R. Jiang, X. Cao, S. Lim, N. Lee, M. Tao, J. Lo, and R. Lee, "Chip-First Fan-Out Panel-Level Packaging for Heterogeneous Integration", *IEEE Transactions on CPMT*, September 2018, pp. 1561–1572.

[50] Ko, C., H. Yang, J. H. Lau, M. Li, M. Li, et al., "Design, Materials, Process, and Fabrication of Fan-Out Panel-Level Heterogeneous Integration", *IMAPS Transactions, Journal of Microelectronics and Electronic Packaging*, Vol. 15, Issue 4, October 2018, pp. 141–147.

[51] Hsieh, C., C. Tsai, H. Lee, T. Lee, H. Chang, "Fan-out Technologies for WiFi SiP Module Packaging and Electrical Performance Simulation", *IEEE/ECTC Proceedings*, May 2015, pp. 1664–1669.

[52] Lin, Y., C. Kang, L. Chua, W. Choi, and S. Yoon, "Advanced 3D eWLB-PoP (embedded Wafer Level Ball Grid Array - Package on Package) Technology", *IEEE/ECTC Proceedings*, May 2016, pp. 1772–1777.

[53] Lau, J. H., "Fan-Out Wafer-Level Packaging for 3D IC Heterogeneous Integration", *Proceedings of CSTIC*, March 2018, pp. VII_1–6.

[54] Lau, J. H., "Heterogeneous Integration with Fan-Out Wafer-Level Packaging", *Proceedings of IWLPC,* October 2017, pp. 1–25.

[55] Lau, J. H., "3D IC Heterogeneous Integration by FOWLP", *Chip Scale Review*, Vol. 22, January/February 2018, pp. 16–21.

[56] Lim, J., and V. Pandey, "Innovative Integration Solutions for SiP Packages Using Fan-Out Wafer Level eWLB Technology", *IMAPS Proceedings*, October 2017, pp. 263–269.

[57] Kyozuka, M., T. Kiso, H. Toyazaki, K. Tanaka, and T. Koyama, "Development of Thinner POP base Package by Die Embedded and RDL Structure", *IMAPS Proceedings*, October 2017, pp. 715–720.

[58] Cardoso, A., M. Pires, and R. Pinto, "Thermally Enhanced FOWLP Development of a Power-eWLB Demonstrator", *IEEE/ECTC Proceedings*, May 2015, pp. 1682–1688.

[59] Cardoso, A., M. Pires, R. Pinto, E. Fernades, I. Barros, H. Kuisma, and S. Nurmi, "Implementation of Keep-Out-Zones to Protect Sensitive Sensor Areas During Backend Processing in Wafer Level Packaging Technology", *IEEE/ECTC Proceedings*, May 2016, pp. 1160–1166.

[60] Cardoso, A., L. Dias, E. Fernandes, A. Martins, A. Janeiro, P. Cardoso, and H. Barros, "Development of Novel High Density System Integration Solutions in FOWLP—Complex and Thin Wafer-Level SiP and Wafer-Level 3D Packages", *IEEE/ECTC Proceedings*, May 2017, pp. 14–21.

[61] Seler, E., M. Wojnowski, W. Hartner, J. Böck, R. Lachner, R. Weigel, A. Hagelauer, "3D Rectangular Waveguide Integrated in embedded Wafer Level Ball Grid Array (eWLB) Package", *IEEE/ECTC Proceedings*, May 2014, pp. 956–962.

[62] Rodrigo, A., B. Isabel, C. José, C. Paulo, C. José, H. Vítor, O. Eoin, and P. Nelson, "Enabling of Fan-Out WLP for More Demanding Applications by Introduction of Enhanced Dielectric Material for Higher Reliability", *IEEE/ECTC Proceedings*, May 2014, pp. 935–939.

[63] Wojnowski, M., G. Sommer1, K. Pressel, and G. Beer, "3D eWLB—Horizontal and Vertical Interconnects for Integration of Passive Components", *IEEE/ECTC Proceedings*, May 2013, pp. 2121–2125.

[64] Liu, K., R. Frye, M. Hlaing, Y. Lee, H. Kim, G. Kim, S. Park, and B. Ahn, "High-Speed Packages with Imperfect Power and Ground Planes", *IEEE/ECTC Proceedings,* May 2013, pp. 2046–2051.
[65] Pachler, W., K. Pressel, J. Grosinger, G. Beer, W. Bösch, G. Holweg, C. Zilch, M. Meindl, "A Novel 3D Packaging Concept for RF Powered Sensor Grains", *IEEE/ECTC Proceedings*, May 2014, pp. 1183–1188.
[66] Osenbach, J., S. Emerich, L. Golick, S. Cate, M. Chan, S.W. Yoon, Y. Lin, and K. Wong, "Development of Exposed Die Large Body to Die Size Ratio Wafer Level Package Technology", *IEEE/ECTC Proceedings*, May 2014, pp. 952–955.
[67] Fan, X., "Wafer Level Packaging (WLP): Fan-in, Fan-out and Three-Dimensional Integration", *Proceedings of International Conference on Thermal, Mechanical and Multiphysics Simulation and Experiments in Micro-Electronics and Micro-Systems*, April 2010, pp. 1–6.
[68] Wojnowski, M., K. Pressel, and G. Beer, "Novel Embedded Z Line (EZL) Vertical Interconnect Technology for eWLB", *IEEE/ECTC Proceedings*, May 2015, pp. 1071–1076.
[69] Ishibashi, D., S. Sasaki, Y. Ishizuki, S. Iijima, Y. Nakata, Y. Kawano, T. Suzuki, and M. Tani, "Integrated Module Structure of Fan-out Wafer Level Package for Terahertz Antenna", *IEEE/ECTC Proceedings*, May 2015, pp. 1084–1089.
[70] Chen, S., S. Wang, J. Hunt, W. Chen, L. Liang, G. Kao, and A. Peng, "A Comparative study of a Fan Out Packaged Product: Chip First and Chip Last", *IEEE/ECTC Proceedings*, May 2016, pp. 1483–1488.
[71] Spinella, L., J. Im, and P. S. Ho, "Reliability Assessment of Fan-out Packages Using High Resolution Moiré Interferometry and Synchrotron X-ray Microdiffraction", *IEEE/ECTC Proceedings*, May 2016, pp. 2016–2021.
[72] Braun, T., K.-F. Becker, S. Raatz, M. Minkus, V. Bader, J. Bauer, R. Aschenbrenner, R. Kahle, L. Georgi, S. Voges, M. Wohrmann, and K.-D. Lang, "Foldable Fan-out Wafer Level Packaging", *IEEE/ECTC Proceedings*, May 2016, pp. 19–24.
[73] Takahashii, H., H. Nomai, N. Suzuki, Y. Nomura, A. Kasahara, N. Takano, and T. Nonaka, "Large Panel Level Fan Out Package Built Up Study with Film Type Encapsulation Material", *IEEE/ECTC Proceedings*, May 2016, pp. 134–139.
[74] Lau, J.H, The Roles of DNP (Distance to Neutral Point) on Solder Joint Reliability of Area Array Assemblies. *Journal of Soldering & Surface Mount Technology*, 1997, 9 (2): 58–60.
[75] Lau, J.H., and R.S.W. Lee. 1999. *Chip Scale Package*. New York: McGraw-Hill Book Company.
[76] Lau, J. H., M. Li, N. Fan, E. Kuah, Z. Li, K. Tan, T. Chen, et al., "Fan-out wafer-level packaging (FOWLP) of large chip with multiple redistribution-layers (RDLs)", *Proceedings of IMAPS Symposium*, 2017, pp. 576–583.
[77] Lau, J. H., M. Li, N. Fan, E. Kuah, Z. Li, K. Tan, T. Chen, et al., "Fan-out wafer-level packaging (FOWLP) of large chip with multiple redistribution-layers (RDLs)", *IMAPS Transactions Journal of Microelectronics and Electronic Packaging*, Oct, 2017, pp. 123–131.
[78] Lau, J. H., M. Li, Q. Li, I. Xu, T. Chen, Z. Li, K. Tan, X. Qing, C. Zhang, K. Wee. R. Beica, C. Ko, S. Lim, N. Fan, E. Kuah, K. Wu, Y. Cheung, E. Ng, X. Cao, J. Ran, H. Yang, Y. Chen, N. Lee, M. Tao, J. Lo, and R. Lee, "Design, Materials, Process, and Fabrication of Fan-Out Wafer-Level Packaging", *IEEE Transactions on CPMT*, June 2018, pp. 991–1002.
[79] Lau, J. H., M. Li, D. Tian, N. Fan, E. Kuah, K. Wu, M. Li, et al., "Warpage and Thermal Characterization of Fan-out Wafer-Level ackaging", *IEEE/ECTC Proceedings*, May 2017, pp. 595–602.
[80] Lau, J. H., M. Li, D. Tian, N. Fan, E. Kuah, K. Wu, M. Li, J. Hao, Y. Cheung, Z. Li, K. Tan, R. Beica, T. Taylor, CT Lo, H. Yang, Y. Chen, S. Lim, NC Lee, J. Ran, X. Cao, S. Koh, and Q. Young, "Warpage and Thermal Characterization of Fan-Out Wafer-Level Packaging", *IEEE Transactions on CPMT*, October 2017, pp. 1729–1738.
[81] Li, M., Q. Li, J. H. Lau, N. Fan, E. Kuah, K. Wu, et al., "Characterizations of fan-out wafer-level packaging", *Proceedings of IMAPS Symposium*, Oct. 2017, pp. 557–562.
[82] Lim, S., Y. Liu, J. H. Lau, M. Li "Challenges of ball-attach process using Flux for fan-out wafer/panel level (FOWL/PLP) packaging", *Proceedings of IWLPC*, Oct. 2017, pp. S10_P3_1–7.
[83] Kuah, E., W. Chan, J. Hao, N. Fan, M. Li, J. H. Lau, K. Wu, et al., "Dispensing challenges of large format packaging and some of its possible solutions", *IEEE/EPTC Proceedings*, December 2017, pp. S27_1–6.

[84] Hua, X., H. Xu, Z. Li, D. Chen, K. Tan, J. H. Lau, M. Li, et al., "Development of chip-first and die-up fan-out wafer-level packaging", *IEEE/EPTC Proceeding*, December 2017, pp. S23_1–6.
[85] Rogers, B., C. Scanlan, and T. Olson, "Implementation of a Fully Molded Fan-Out Packaging Technology", *Proceeding of IWLPC*, October 2013, pp. S10_P1_1–6.
[86] Bishop, C., B. Rogers, C. Scanlan, and T. Olson, "Adaptive Patterning Design Methodologies", *IEEE/ECTC Proceedings*, May 2016, pp. 7–12.
[87] Lau, J. H., M. Li, Y. Lei, M. Li, Q. Yong, Z. Cheng, T. Chen, I. Xu, et al., "Reliability of FOWLP with Large Chips and Multiple RDLs", *IEEE/ECTC Proceedings*, May 2018, pp. 1568–1576.
[88] Lau, J. H., "Recent Advances and Trends in Advanced Packaging", *Chip Scale Review*, Vol. 21, May/June 2017, pp. 46–54.
[89] Lau, J. H., "Recent Advances and New Trends in Flip Chip Technology", *ASME Transactions, Journal of Electronic Packaging*, September 2016, Vol. 138, Issue 3, pp. 1–23.
[90] Lau, J. H., *Fan-Out Wafer-Level Packaging*, Springer Book Company, 2018.
[91] Hwang, T., D. Oh, E. Song, K. Kim, J. Kim, and S. Lee, "Study of Advanced Fan-Out Packages for Mobile Applications", *IEEE/ECTC Proceedings*, May 2018, pp. 343–348.
[92] Suk, K., S. Lee, J. Kim, S. Lee, H. Kim, S. Lee, P. Kim, D. Kim, D. Oh, and J. Byun, "Low Cost Si-less RDL Interposer Package for High Performance Computing Applications", *IEEE/ECTC Proceedings*, May 2018, pp. 64–69.
[93] You, S., S. Jeon, D. Oh, K. Kim, J. Kim, S. Cha, and G. Kim, "Advanced Fan-Out Package SI/PI/Thermal Performance Analysis of Novel RDL Packages", *IEEE/ECTC Proceedings*, May 2018, pp. 1295–1301.
[94] Huemoeller, R., and C. Zwenger. 2015. *Silicon Wafer Integrated Fan-Out Technology*, 34–37. Chip Scale Review: Mar/Apr.
[95] Lin, Y., S. Wu, W. Shen, S. Huang, T. Kuo, A. Lin, T. Chang, H. Chang, S. Lee, C. Lee, J. Su, X. Liu, Q. Wu, and K. Chen, "An RDL-First Fan-out Wafer Level Package for Heterogeneous Integration Applications", *IEEE/ECTC Proceedings*, May 2018, pp. 349–354.
[96] Che, F. X., D. Ho, M. Ding, and D. MinWoo, "Study on Process Induced Wafer Level Warpage of Fan-Out Wafer Level Packaging", *IEEE/ECTC Proceedings*, 2016, pp. 1879–1885.
[97] Rao, V., C. Chong, D. Ho, D. Zhi, C. Choong, S. Lim, D. Ismael, and Y. Liang, "Development of High Density Fan Out Wafer Level Package (HD FOWLP) With Multi-layer Fine Pitch RDL for Mobile Applications", *IEEE/ECTC Proceedings*, May 2016, pp. 1522–1529.
[98] Li, H., A. Chen, S. Peng, G. Pan, and S. Chen, "Warpage Tuning Study for Multi-chip Last Fan Out Wafer Level Package", *IEEE/ECTC Proceedings*, May 2017, pp. 1384–1391.
[99] Chen, Z., F. Che, M. Ding, D. Ho, T. Chai, V. Rao, "Drop Impact Reliability Test and Failure Analysis for Large Size High Density FOWLP Package on Package", *IEEE/ECTC Proceedings*, May 2017, pp. 1196–1203.
[100] Ki, W., W. Lee, I. Lee, I. Mok, W. Do, M. Kolbehdari, A. Copia, S. Jayaraman, C. Zwenger, and K. Lee, "Chip Stackable, Ultra-thin, High-flexibility 3D FOWLP (3D SWIFT® Technology) for Hetero-integrated Advanced 3D WL-SiP", *IEEE/ECTC Proceedings*, May 2018, pp. 580–586.
[101] Cheng, W., C. Yang, J. Lin, W. Chen, T. Wang, and Y. Lee, "Evaluation of Chip-Last Fan-Out Panel Level Packaging with G2.5 LCD Facility Using FlexUPTM and Mechanical De-bonding Technologies", *IEEE/ECTC Proceedings*, May 2018, pp. 386–391.
[102] Shih, M., R. Chen, P. Chen, Y. Lee, K. Chen, I. Hu, T. Chen, L. Tsai, E. Chen, E. Tsai, D. Tarng, C. Hung, "Comparative Study on Mechanical and Thermal Performance of eWLB, M-Series™ and Fan-Out Chip Last Packages", *IEEE/ECTC Proceedings*, May 2018, pp. 1670–1676.
[103] Lee, C., J. Su, X. Liu, Q. Wu, J. Lin, P. Lin, C. Ko, Y. Chen, W. Shen, T. Kou, S. Huang, Y. Lin, K. Chen, and A. Lin, "Optimization of Laser Release Process for Throughput Enhancement of Fan-Out Wafer-Level Packaging", *IEEE/ECTC Proceedings*, May 2018, pp. 1818–1823.
[104] Zhang, H., X. Liu, S. Rickard, R. Puligadda, and T. Flaim, "Novel Temporary Adhesive Materials for RDL-First Fan-Out Wafer-Level Packaging", *IEEE/ECTC Proceedings*, May 2018, pp. 1925–1930.
[105] Lau, J. H., P. Tzeng, C. Lee, C. Zhan, M. Li, J. Cline, et al., "Redistribution layers (RDLs) for 2.5D/3D IC integration", *Proceedings of IMAPS Symposium*, 2013, pp. 434–441.

[106] Lau, J. H., P. Tzeng, C. Lee, C. Zhan, M. Li, J. Cline, et al., "Redistribution Layers (RDLs) for 2.5D/3D IC Integration", *IMAPS Transactions, Journal of Microelectronic Packaging*, Vol. 11, No. 1, First Quarter 2014, pp. 16–24.

[107] Garrou, P., and C. Huffman, "RDL: an integral part of today's advanced packaging technologies", *Solid State Technology*, May 2011, pp. 18–20.

[108] Lau, J. H., "8 Ways to Make RDLs for FOW/PLP", *Chip Scale Review*, Vol. 22, May/Jun, 2018, pp. 11–19.

[109] Ma, M., S. Chen, P. I. Wu, A. Huang, C. H. Lu, A. Chen, et al., "The development and the integration of the 5 μm to 1 μm half pitches wafer level Cu redistribution layers", *IEEE/ECTC Proceedings*, 2016, pp. 1509–1614.

[110] Kim, Y., J. Bae, M. Chang, A. Jo, J. Kim, S. Park, et al., "SLIM^{TM}, high-density wafer-level fan-out package development with sub-micron RDL", *IEEE/ECTC Proceedings*, 2017, pp. 18–13.

[111] Hiner, D., M. Kolbehdari, M. Kelly, Y. Kim, W. Do, J. Bae, "SLIM^{TM} advanced fan-out packaging for high-performance multi-die solutions", *IEEE/ECTC Proceedings*, 2017, pp. 575–580.

[112] Lin, Y., W. Lai, C. Kao, J. Lou, P. Yang, C. Wang, and C. Hseih, "Wafer warpage experiments and simulation for fan-out chip on substrate", *IEEE/ECTC Proceedings*, May 2016, pp. 13–18.

[113] Che, F., D. Ho, M. Z. Ding, and X. Zhang, "Modeling and design solutions to overcome warpage challenge for Fan-out wafer level packaging (FO-WLP) technology", *IEEE/EPTC Proceedings*, Dec. 2015, pp. 2–4.

[114] Che, F., D. Ho, M. Z. Ding, and D. R. MinWoo, "Study on process induced wafer level warpage of fan-out wafer level packaging", *IEEE/ECTC Proceedings*, May 2016, pp. 1879–1885.

[115] Hou, F., T. Lin, L. Cao, F. Liu, J. Li, X. Fan, and G. Zhang, "Experimental verification and optimization analysis of warpage for panel-level fan-out package", *IEEE Transactions on Components, Packaging, and Manufacturing Technology*, Vol. 7, No. 10, Oct. 2017, pp. 1721–1728.

[116] Shen, Y., L. Zhang, W. Zhu, J. Zhou, and X. Fan, "Finite-element analysis and experimental test for a capped-die flip chip package design", *IEEE Transactions on Components, Packaging, and Manufacturing Technology*, Vol. 6, No. 9, Sept. 2016, pp. 1308–1316.

[117] Lau, J. H., M. Li, Y. Lei, M. Li, I. Xu, T. Chen, S. Chen, et al., "Warpage Measurements and Characterizations of Fan-Out Wafer-Level Packaging with Large Chips and Multiple Redistributed Layers", *IEEE/ECTC Proceedings*, May 2018, pp. 594–600.

[118] Lau, J. H., M. Li, Y. Lei, M. Li, I. Xu, T. Chen, S. Chen, Q. Yong, J. Madhukumar, K. Wu, F. Fan, E. Kuah, Z. Li, K. Tan, W. Bao, A. Lim, R. Beica, C. Ko, and X. Cao, "Warpage Measurements and Characterizations of Fan-Out Wafer-Level Packaging with Large Chips and Multiple Redistributed Layers", *IEEE Transactions on Components, Packaging, and Manufacturing Technology*, Vol. 8, No. 10, Oct. 2018, pp. 1729–1737.

[119] JEDEC Standard JESD22-B112A, *Package Warpage Measurement of Surface-Mount Integrated Circuits at Elevated Temperature*, October 2009.

[120] Lau, J. H., Extracted from the 2017 IEEE/ECTC panel session: "Panel Fan-out Manufacturing: Why, When, and How?".

[121] Lau, J. H., M. Li, Q. Li, I. Xu, T. Chen, Z. Li, et al., "Design, Materials, Process, and Fabrication of Fan-Out Wafer-Level Packaging", *IEEE Transaction on CPMT*, June 2018, pp. 991–1002.

[122] Ko, C. T., H. Yang, J. H. Lau, M. Li, M. Li, C. Lin, et al., "Design, materials, process, and fabrication of fan-out panel-level heterogeneous integration", *IMAPS Transaction on Journal of Microelectronics and Electronic Packaging*, Oct. 2018, pp. 141–147.

[123] Lau, J. H., M. Li, Y. Lei, M. Li, I. Xu, T. Chen, et al., "Reliability of fan-out wafer-level heterogeneous integration", *IMAPS Transaction on Journal of Microelectronics and Electronic Packaging*, Oct. 2018, pp. 148–162.

[124] Souriau, J., O. Lignier, M. Charrier, and G. Poupon, "Wafer Level Processing Of 3D System in Package for RF and Data Applications", *IEEE/ECTC Proceedings*, 2005, pp. 356–361.

[125] Henry, D., D. Belhachemi, J-C. Souriau, C. Brunet-Manquat, C. Puget, G. Ponthenier, J. Vallejo, C. Lecouvey, and N. Sillon, "Low Electrical Resistance Silicon Through Vias: Technology and Characterization", *IEEE/ECTC Proceedings*, 2006, pp. 1360–1366.

[126] Hou, S., W. Chen, C. Hu, C. Chiu, K. Ting, T. Lin, W. Wei, W. Chiou, V. Lin, V. Chang, C. Wang, C. Wu, and D. Yu, "Wafer-Level Integration of an Advanced Logic-Memory Sys-

tem Through the Second-Generation CoWoS Technology", *IEEE Transactions on Electron Devices*, October 2017, pp. 4071–4077.
[127] Selvanayagam, C., J. H. Lau, X. Zhang, S. Seah, K. Vaidyanathan, and T. Chai, "Nonlinear Thermal Stress/Strain Analysis of Copper Fill TSV (Through Silicon Via) and Their Flip-Chip Microbumps", *IEEE/ECTC Proceedings*, May 27–30, 2008, pp. 1073–1081.
[128] Selvanayagam, C., J. H. Lau, X. Zhang, S. Seah, K. Vaidyanathan, and T. Chai, "Nonlinear Thermal Stress/Strain Analyses of Copper Filled TSV (Through Silicon Via) and Their Flip-Chip Microbumps", *IEEE Transactions on Advanced Packaging*, Vol. 32, No. 4, November 2009, pp. 720–728.
[129] Lau, J. H., and G. Tang, "Thermal Management of 3D IC Integration with TSV (Through Silicon Via)", *IEEE/ECTC Proceedings*, May 2009, pp. 635–640.
[130] Lau, J. H., Y. S. Chan, and R. S. W. Lee, "3D IC Integration with TSV Interposers for High-Performance Applications", *Chip Scale Review*, Vol. 14, No. 5, September/October, 2010, pp. 26–29.
[131] Lau, J. H., "TSV Manufacturing Yield and Hidden Costs for 3D IC Integration", *IEEE/ECTC Proceedings,* May 2010, pp. 1031–1041.
[132] Zhang, X., T. Chai, J. H. Lau, C. Selvanayagam, K. Biswas, S. Liu, D. Pinjala, et al., "Development of Through Silicon Via (TSV) Interposer Technology for Large Die (21 × 21 mm) Fine-pitch Cu/low-k FCBGA Package", *IEEE Proceedings of ECTC*, May 2009, pp. 305–312.
[133] Chai, T.C., X. Zhang, J.H. Lau, C.S. Selvanayagam, D. Pinjala, et al. Development of Large Die Fine-Pitch Cu/low-*k* FCBGA Package with through Silicon via (TSV) Interposer. *IEEE Transactions on CPMT* , 2011, 1 (5): 660–672.
[134] Lau, J. H., et al, "Apparatus Having Thermal-Enhanced and Cost-Effective 3D IC Integration Structure with Through Silicon Via Interposer". US Patent No: 8,604,603, Date of Patent: December 10, 2013.
[135] Chien, H.C., J.H. Lau, Y. Chao, R. Tain, M. Dai, S.T. Wu, W. Lo, and M.J. Kao. Thermal Performance of 3D IC Integration with Through-Silicon Via (TSV). *IMAPS Transactions, Journal of Microelectronic Packaging* , 2012, 9: 97–103.
[136] Chaware, R., K. Nagarajan, and S. Ramalingam, "Assembly and reliability challenges in 3D integration of 28 nm FPGA die on a large high-density 65 nm passive interposer", *IEEE/ECTC Proceedings*, May 2012, pp. 279–283.
[137] Banijamali, B., S. Ramalingam, K. Nagarajan, and R. Chaware, "Advanced reliability study of TSV interposers and interconnects for the 28 nm technology FPGA", *IEEE/ECTC Proceedings*, May 2011, pp. 285–290.
[138] Banijamali, B., S. Ramalingam, H. Liu, and M. Kim, "Outstanding and innovative reliability study of 3D TSV interposer and fine-pitch solder micro-bumps", *IEEE/ECTC Proceedings*, May 2012, pp. 309–314.
[139] Xie, J., H. Shi, Y. Li, Z. Li, A. Rahman, K. Chandrasekar, et al., "Enabling the 2.5D integration", *Proceedings of IMAPS International Symposium on Microelectronics*, October 2012, pp. 254–267.
[140] Banijamali, B., C. Chiu, C. Hsieh, T. Lin, C. Hu, S. Hou, et al., "Reliability Evaluation of a CoWoS-Enabled 3D IC Package", *IEEE/ECTC Proceedings*, May 2013, pp. 35–40.
[141] Chuang, Y., C. Yuan, J. Chen, C. Chen, C. Yang, W. Changchien, C. Liu, and F. Lee, "Unified Methodology for Heterogeneous Integration with CoWoS Technology", *IEEE/ECTC Proceedings*, May 2013, pp. 852–859.
[142] Lau, J. H., C. Lee, C. Zhan, S. Wu, Y. Chao, M. Dai, R. Tain, H. Chien, et al., "Low-Cost Through-Silicon Hole Interposers for 3D IC Integration", *IEEE Transactions on CPMT*, Vol. 4, No. 9, September 2014, pp. 1407–1419.
[143] Pendse, R. D., Semiconductor Device and Method of Forming Extended Semiconductor Device with Fan-Out Interconnect Structure to Reduce Complexity of Substrate, Filed in the US Patent Office on December 23, 2011, US 2013/0161833.
[144] Yu, D., "Advanced system integration technology trends", *Sip Global Summit* , SEMICON Taiwan, China, September 6, 2018.
[145] Hong, J., K. Choi, D. Oh, S Park, S. Shao, H. Wang, Y. Niu, and V. Pham, "Design Guideline of 2.5D Package with Emphasis on Warpage Control and Thermal Management", *IEEE/ECTC Proceedings*, May 2018, pp. 682–692.

[146] Jeng, S.-P., S.-M. Chen, F.-C. Hsu, P.-Y. Lin, J.-H. Wang, T.-J. Fang, P. Kavle, and Y.-J. Lin, "High Density 3D Fanout Package for Heterogeneous Integration", *IEEE/VLSI Circuits Proceedings*, August 2017, pp. T114–T115.

[147] Hsu, F., J. Lin, S. Chen, P. Lin, J. Fang, J. Wang, and S. Jeng, "3D Heterogeneous Integration with Multiple Stacking Fan-Out Package", *IEEE/ECTC Proceedings*, May 2018, pp. 337–342.

[148] Chiu, C., Z. Qian, and M. Manusharow, "Bridge interconnect with air gap in package assembly", *US Patent No. 8,872,349*, Filed on September 11, 2012, Patented on October 28, 2014.

[149] Mahajan, R., R. Sankman, N. Patel, D. Kim, K. Aygun, Z. Qian, et al., "Embedded Multi-Die Interconnect Bridge (EMIB)—A High-Density, High-Bandwidth Packaging Interconnect", *IEEE/ECTC Proceedings*, May 2016, pp. 557–565.

[150] Podpod, A., J. Slabbekoorn, A. Phommahaxay, F. Duval, A. Salahouedlhadj, M. Gonzalez, K. Rebibis, R.A. Miller, G. Beyer, and E. Beyne, "A Novel Fan-Out Concept for Ultra-High Chip-to-Chip Interconnect Density with 20-μm Pitch", *IEEE/ECTC Proceedings*, May 2018, pp. 370–378.

[151] Wang, C. -T., T. -C. Tang, C. -W. Lin, C. -W. Hsu, J. -S. Hsieh, C. -H. Tsai, K. -C. Wu, H. -P. Pu, and D. Yu, "InFO_AiP Technology for High Performance and Compact 5G Millimeter Wave System Integration", *IEEE/ECTC Proceedings*, May 2018, pp. 202–207.

第 6 章

基于扇出型 RDL 基板的异构集成

6.1 引言

如第 4 章所述，TSV（硅通孔）转接板非常昂贵[1-10]，并且硅桥已多次被提出取代 TSV 转接板应用于异构集成。最近，采用晶圆/面板级扇出封装技术[11-20]，在基板上制造 RDL（再分布层），以替代 TSV 转接板实现异构集成的方法已经引起了广泛关注。在 ECTC 2013 期间，星科金朋（STATSChipPAC）有限公司提出使用扇出型倒装芯片，通过 RDL 实现大部分的横向互连。在 ECTC 2016 期间，日月光半导体公司（ASE）和联发科技（Media Tek）股份有限公司在其晶圆级扇出封装技术（FOWLP）中使用类似的技术制造了 RDL，表明在这种情况下，TSV 转接板、晶圆级微凸点、助焊剂涂覆、芯片到晶圆键合、助焊剂清洗及下填料填充与固化，都可以省去。在 ECTC 2018 期间，三星公司（Samsung）使用芯片后置（或 RDL 先置）工艺在硅转接板上制作 RDL。2018 台积电公司（TSMC）也提到了这项技术，称其为基板上集成扇出（InFO_oS）。

6.2 星科金朋公司的 FOFC-eWLB 技术

2011 年 12 月 23 日，星科金朋公司申请了一项名为"一种通过扇出式互连结构降低基板复杂度的扩展半导体器件及其制备方法"的专利[21]。在 ECTC 2013 期间，星科金朋公司提出并发表了[22]通过嵌入式晶圆级球栅阵列封装的扇出型倒装芯片（FOFC-eWLB）中的 RDL，实现了芯片间大部分的横向互连，如图 6.1 所示。可以看出，TSV 转接板、晶圆微凸点、助焊剂、芯片到晶圆键合、助焊剂清洗及下填料填充与固化都可以被省去，无须使用。

异构集成技术

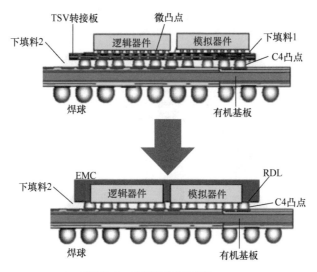

> 无微凸点、下填料1和硅转接板
> 通过扇出技术制备RDL

图 6.1 星科金朋公司通过扇出型基板（FOFC-eWLB）实现异构集成

6.3 日月光半导体公司的 FOCoS 技术

2016 年，日月光半导体公司[23]提出通过晶圆级扇出封装（FOWLP）技术（芯片先置，芯片面朝下贴装在临时晶圆载体上，并通过压缩成型实现塑封包覆）制备 RDL，实现了芯片间大部分的横向互连，如图 6.2 所示。该技术被称为晶圆级基板上芯片扇出技术（FOCoS）。在这种结构中，没有硅转接板、芯片表面的晶圆微凸点、助焊剂、芯片到晶圆键合、助焊剂清洗、下填料填充与固化。通过凸点下金属层（UBM）和可控塌陷芯片互连凸点（C4 凸点）技术实现底部 RDL 与封装基板的互连，如图 6.2 所示。

6.3.1 关键工艺步骤

日月光半导体公司的 FOCoS 技术组装工艺步骤如图 6.3 所示。可以看出，共有晶圆重构与模压、RDL 制备、倒装焊接 3 道主要工序。首先，使用芯片先置、芯片面朝下的晶圆级扇出形式（见 5.3 节）。具体来说，首先将芯片面朝下并排贴装在覆有双面热解胶膜的重构晶圆载体上。然后，使用环氧模塑料（EMC）压缩成型并塑封后固化。接着移除双面热解胶膜并开始 RDL 制备工艺（见 5.6.1 节）。RDL 由 3 层金属布线层和 4 层聚合物介质层构成。第三道主要工序是使用 C4 凸点技术对芯片进行晶圆级的凸点制备，并将重构晶圆切割为独立的封装体。随后是芯片倒装和下填料填充与固化。最后，贴装散热片并基板植球。图 6.4 给出了日月光半导体公司的 FOCoS 样品横截面的扫描电镜图像。

第 6 章 基于扇出型 RDL 基板的异构集成

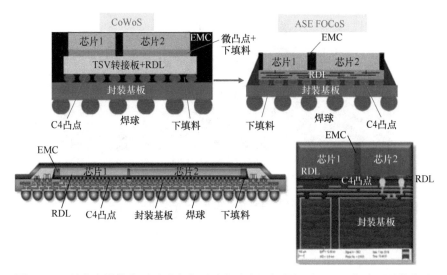

图 6.2 日月光半导体公司通过扇出型（芯片先置）基板（FOCoS）实现异构集成

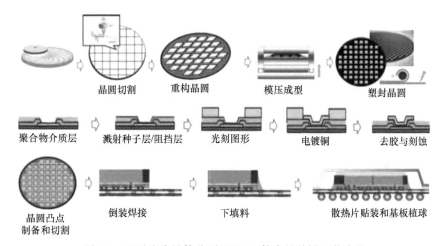

图 6.3 日月光半导体公司 FOCoS 技术的关键工艺步骤

6.3.2 FOCoS 的可靠性

在异构集成中，为了实现最优的电气性能，芯片间距必须尽可能小。然而，如果芯片彼此之间贴装得太近，它们之间"过短的" RDL 可能带来可靠性问题。日月光半导体公司通过热循环试验、高温存储试验和热冲击试验等试验发现，设计不当的 RDL 可能产生断裂[24]。

图 6.5 给出了热循环试验中的热-机械行为示意。可以看出，在低温下，封装体呈"哭脸"状态，RDL1 受到压应力。另一方面，当它处于高温时，封装体呈"笑脸"状态，RDL1 受到拉应力。经过一定的温度循环后，铜导线断裂，如图 6.6 所示。

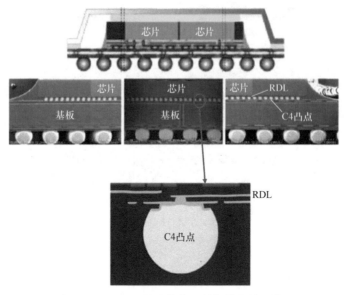

图 6.4　FOCoS 样品横截面的扫描电镜图像

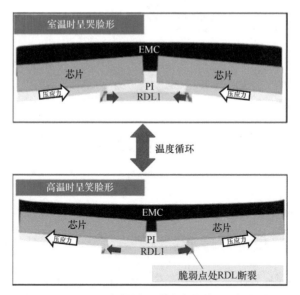

图 6.5　FOCoS 在热循环试验中的热 - 机械行为

对图 6.7 所示的 4 种不同的 RDL 设计方案进行了有限元建模、制造及实验测量[24]。图 6.7a、b 设计方案在芯片区域下方的 RDL 上都有角度弯曲，图 6.7c、d 设计方案在芯片区域下方的 RDL 上无角度弯曲。图 6.7a ~ c 设计方案增加了 RDL 布线厚度以增强 RDL 的强度。另一方面，图 6.7d 设计方案在 RDL 端点焊盘附近具有非直角走线布局。

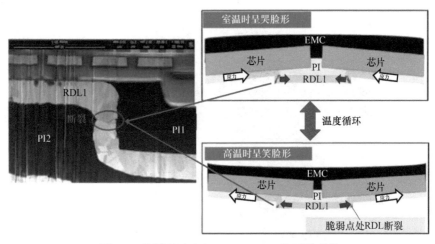

图 6.6　热循环试验中 FOCoS RDL 的开裂现象

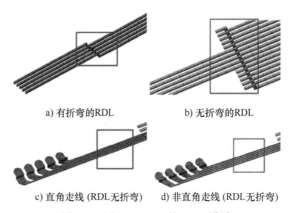

图 6.7　用于 FOCoS 的 RDL 设计

仿真结果表明，图 6.7c、d 设计的应力值小于图 6.7a、b 设计的应力值。热循环试验结果也印证了超过 1000 次循环后，图 6.7c、d 设计方案没有产生失效，而图 6.7a、b 设计方案在 500 次循环内产生了失效。因此，应采用图 6.7c、d 的 RDL 设计。

6.4　联发科公司通过 FOWLP 技术实现的 RDL

2016 年，联发科公司[25]提出了一种类似的无 TSV（硅通孔）转接板 RDL 结构。该结构通过晶圆级扇出技术制备，如图 6.8 和图 6.9 所示。联发科公司没有使用 C4 凸点，而是使用微凸点（铜柱 + 焊帽）将底层 RDL 连接到 6-2-6 封装基板上（见图 6.8）。

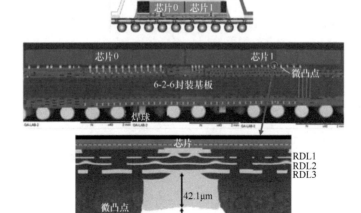

图 6.8　无 TSV 转接板：联发科公司通过晶圆级扇出（FOWLP）技术制备的 RDL

回流后的凸点高度约为 96.4μm。图 6.9a 给出了一个扇出芯片（DIE0/1）的水平横截面图像。可以看出，RDL1 和 RDL2 的金属平均厚度为 2.04μm，RDL3 的金属厚度为 4.23μm。图 6.9b 给出了芯片 0 和芯片 1 之间的垂直横截面图像。可以看出，金属布线的平均宽度和间距分别为 2.09μm 和 2.00μm。

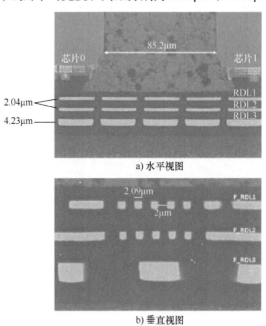

图 6.9　联发科公司的无 TSV 转接板横截面图像

6.5 台积电公司的 InFO_oS 技术

图 6.10 给出了台积电公司的基板上集成扇出技术（InFO_oS）示意图。RDL 由台积电的芯片先置、芯片面朝上的集成扇出技术制备。InFO_oS 技术适用于高性能应用，但性能不及 CoWoS 或 CoWoS-2 技术。从第 6.2 节到第 6.5 节，芯片先置的晶圆级扇出封装技术已替代了 TSV 转接板。

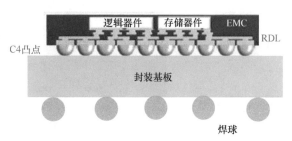

图 6.10　台积电用于异构集成的无 TSV 转接板（InFO_oS）

6.6 三星公司的无硅 RDL 转接板技术

最近，三星公司提出使用芯片后置或 RDL 先置的晶圆/板级扇出封装，如图 6.11 所示，替代 TSV 转接板在高性能计算异构集成器件中的应用。三星公司将称此结构为无硅 RDL 转接板[26]。

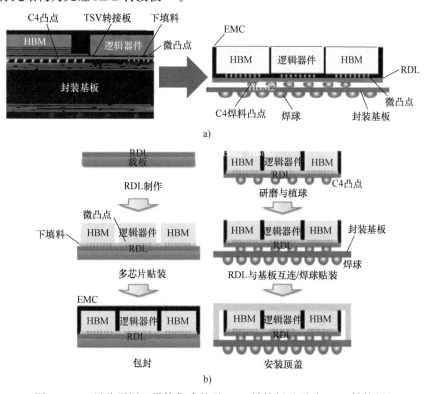

图 6.11　三星公司用于异构集成的无 TSV 转接板（无硅 RDL 转接板）

6.6.1 关键工艺步骤

首先，无论晶圆还是面板形式，RDL 都制备在裸玻璃片表面。同时，在逻辑和 HBM 芯片上制备晶圆凸点。然后，完成上助焊剂、芯片到晶圆或芯片到面板键合、助焊剂清洁、下填料填充与固化。这些步骤之后是 EMC 压缩成型。然后，对 EMC、芯片、HBM 立方体和 C4 晶圆凸点进行背面研磨。完成这些步骤后，可以将整个模块贴装到封装基板上。最后，完成植球和盖板安装。图 6.12 为参考文献 [26] 报道的试验件。可以看出，封装（RDL 转接层）基板尺寸为 55mm × 55mm，为包括黏结层、信号层和接地层在内的 5 层 RDL。

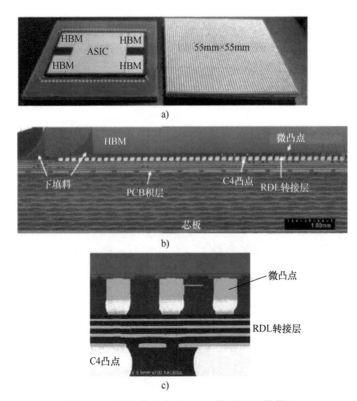

图 6.12 试验件和无硅 RDL 转接板的横截面

6.6.2 无硅 RDL 转接板的可靠性

图 6.13 为通过无硅 RDL 转接板实现的异构集成。图 6.14 显示了常规的 CoWoS 技术。由于 TSV 转接板与有机封装基板之间的热膨胀失配大于在 EMC/Si 芯片和有机封装基板之间的热膨胀失配，无硅 RDL 转接板中的 C4 焊点受到的应力小于 CoWoS 中的 C4 焊点受到的应力。实际上根据三星公司的建模结果，应力减少了 34%。此外，应力最大值的位置不同。对于无硅 RDL 转接

板，最大应力产生在芯片角附近，而对于常规 CoWoS，最大应力产生在 TSV 转接板角附近[26]。

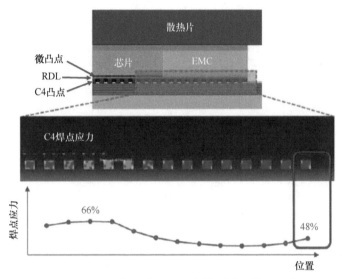

图 6.13　三星公司无硅 RDL 转接板的建模和结果

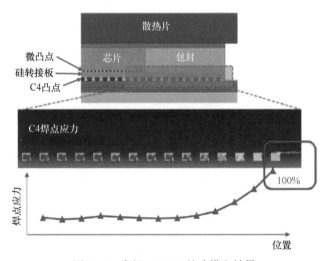

图 6.14　常规 CoWoS 的建模和结果

6.7　总结和建议

一些重要结论和建议总结如下：

1）无 TSV 转接板（如扇出基板）是非常有效的制备异构集成 RDL 的方法。相比在硅转接板上实现的异构集成，它不仅减少了工艺步骤，而且有希望降低成本。

2）采用芯片先置的扇出型 RDL 基板工艺（如日月光集团的 FOCoS）制备的异构集成器件（SoC 和存储器）已实现小批量产。

3）采用芯片后置的扇出型 RDL 基板实现的异构集成，正在进入批量生产。

4）采用芯片先置工艺的扇出型基板实现的异构集成，其尺寸通常小于采用芯片后置工艺的异构集成。

5）采用芯片先置工艺的扇出型基板实现的异构集成，其金属布线线宽和间距通常大于采用芯片后置工艺的异构集成。

参 考 文 献

[1] Selvanayagam, C., J. H. Lau, X. Zhang, S. Seah, K. Vaidyanathan, and T. Chai, "Nonlinear Thermal Stress/Strain Analyses of Copper Filled TSV (Through Silicon Via) and Their Flip-Chip Microbumps", *IEEE Transactions on Advanced Packaging*, Vol. 32, No. 4, November 2009, pp. 720–728.

[2] Chai, T. C., X. Zhang, J. H. Lau, C. S. Selvanayagam, D. Pinjala, et al., "Development of Large Die Fine-Pitch Cu/low-*k* FCBGA Package with through Silicon Via (TSV) Interposer", *IEEE Transactions on CPMT*, Vol. 1, No. 5, May 2011, pp. 660–672.

[3] Chien, H. C., J. H. Lau, Y. Chao, R. Tain, M. Dai, S. T. Wu, W. Lo, and M. J. Kao, "Thermal Performance of 3D IC Integration with Through-Silicon Via (TSV)", *IMAPS Transactions, Journal of Microelectronic Packaging*, Vol. 9, 2012, pp. 97–103.

[4] Chaware, R., K. Nagarajan, and S. Ramalingam, "Assembly and Reliability Challenges in 3D Integration of 28 nm FPGA Die on a Large High-Density 65 nm Passive Interposer", *IEEE/ECTC Proceedings*, May 2012, pp. 279–283.

[5] Banijamali, B., S. Ramalingam, K. Nagarajan, and R. Chaware, "Advanced Reliability Study of TSV Interposers and Interconnects for the 28 nm Technology FPGA", *IEEE/ECTC Proceedings*, May 2011, pp. 285–290.

[6] Banijamali, B., S. Ramalingam, H. Liu, and M. Kim, "Outstanding and Innovative Reliability Study of 3D TSV Interposer and Fine-Pitch Solder Micro-Bumps", *IEEE/ECTC Proceedings*, May 2012, pp. 309–314.

[7] Banijamali, B., C. Chiu, C. Hsieh, T. Lin, C. Hu, S. Hou, et al., "Reliability Evaluation of a CoWoS-Enabled 3D IC Package", *IEEE/ECTC Proceedings*, May 2013, pp. 35–40.

[8] Lau, J. H., *3D IC Integration and Packaging*. New York: McGraw-Hill, 2016.

[9] Lau, J. H., *Through-Silicon Via (TSV) for 3D Integration*. New York: McGraw-Hill, 2013.

[10] Lau, J. H., *Reliability of RoHS Compliant 2D & 3D IC Interconnects*. New York: McGraw-Hill, 2011.

[11] Lau, J. H., *Fan-Out Wafer-Level Packaging*. New York: Springer, 2018.

[12] Ko, C. T., H. Yang, J. H. Lau, M. Li, M. Li, C. Lin, J. W. Lin, T. Chen, I. Xu, C. Chang, J. Pan, H. Wu, Q. Yong, N. Fan, E. Kuah, Z. Li, K. Tan, Y. Cheung, E. Ng, K. Wu, J. Hao, R. Beica, M. Lin, Y. Chen, Z. Cheng, S. Koh, R. Jiang, X. Cao, S. Lim, N. Lee, M. Tao, J. Lo, and R. Lee, "Chip-First Fan-Out Panel-Level Packaging for Heterogeneous Integration", *IEEE Transactions on CPMT*, September 2018, pp. 1561–1572.

[13] Lau, J. H., M. Li, M. Li, T. Chen, I. Xu, X. Qing, Z. Cheng, N. Fan, E. Kuah, Z. Li, K. Tan, Y. Cheung, E. Ng, P. Lo, K. Wu, J. Hao, S. Koh, R. Jiang, X. Cao, R. Beica, S. Lim, N. Lee, C. Ko, H. Yang, Y. Chen, M. Tao, J. Lo, and R. Lee, "Fan-Out Wafer-Level Packaging for Heterogeneous Integration", *IEEE Transactions on CPMT*, 2018, September 2018, pp. 1544–1560.

[14] Lau, J. H., M. Li, Y. Lei, M. Li, I. Xu, T. Chen, Q. Yong, Z. Cheng, et al., "Reliability of Fan-Out Wafer-Level Heterogeneous Integration", *IMAPS Transactions, Journal of Microelectronics and Electronic Packaging*, Vol. 15, Issue: 4, October 2018, pp. 148–162.

[15] Ko, C. T., H. Yang, J. H. Lau, M. Li, M. Li, I. Xu, et al., "Design, Materials, Process, and Fabrication of Fan-Out Panel-Level Heterogeneous Integration", *IMAPS Transactions, Journal*

of *Microelectronics and Electronic Packaging*, Vol. 15, Issue: 4, October 2018, pp. 141–147.
[16] Yoon, S., J. Caparas, Y. Lin, and P. Marimuthu, "Advanced Low Profile PoP Solution with Embedded Wafer Level PoP (eWLB-PoP) Technology", *IEEE/ECTC Proceedings*, 2012, pp. 1250–1254.
[17] Lau, J. H., C. Lee, C. Zhan, S. Wu, Y. Chao, M. Dai, R. Tain, H. Chien, et al., "Low-Cost Through-Silicon Hole Interposers for 3D IC Integration", *IEEE Transactions on CPMT*, Vol. 4, No. 9, September 2014, pp. 1407–1419.
[18] Lau, J. H., M. Li, N. Fan, E. Kuah, Z. Li, K. Tan, T. Chen, et al., "Fan-Out Wafer-Level Packaging (FOWLP) of Large Chip with Multiple Redistribution-Layers (RDLs)", *IMAPS Transactions Journal of Microelectronics and Electronic Packaging*, October 2017, pp. 123–131.
[19] Lau, J. H., M. Li, Q. Li, I. Xu, T. Chen, Z. Li, K. Tan, X. Qing, C. Zhang, K. Wee. R. Beica, C. Ko, S. Lim, N. Fan, E. Kuah, K. Wu, Y. Cheung, E. Ng, X. Cao, J. Ran, H. Yang, Y. Chen, N. Lee, M. Tao, J. Lo, and R. Lee, "Design, Materials, Process, and Fabrication of Fan-Out Wafer-Level Packaging", *IEEE Transactions on CPMT*, June 2018, pp. 991–1002.
[20] Lin, Y., W. Lai, C. Kao, J. Lou, P. Yang, C. Wang, and C. Hseih, "Wafer Warpage Experiments and Simulation for Fan-Out Chip on Substrate", *IEEE/ECTC Proceedings*, May 2016, pp. 13–18.
[21] Pendse, R., "*Semiconductor Device and Method of Forming Extended Semiconductor Device with Fan-Out Interconnect Structure to Reduce Complexity of Substrate*", Filed on December 23, 2011, US 2013/0161833 A1, pub. Date: June 27, 2013.
[22] Yoon, S. W., P. Tang, R. Emigh, Y. Lin, P. C. Marimuthu, and R. Pendse, "Fan-Out Flip-Chip eWLB (Embedded Wafer-Level Ball Grid Array) Technology as 2.5D Packaging Solutions", *Proceedings of IEEE/ECTC*, May 2013, pp. 1855–1860.
[23] Lin, Y., W. Lai, C. Kao, J. Lou, P. Yang, C. Wang, et al., "Wafer Warpage Experiments and Simulation for Fan-Out Chip-on-Substrate", *Proceedings of IEEE/ECTC*, May 2016, pp. 13–18.
[24] Lee, Y., W. Lai, I. Hu, M. Shih, C. Kao, D. Tarng, and C. Hung, "Fan-Out Chip on Substrate Device Interconnection Reliability Analysis", *Proceedings of IEEE/ECTC*, May 2017, pp. 22–27.
[25] Chen, N. C., T. Hsieh, J. Jinn, P. Chang, F. Huang, J. Xiao, A. Chou, and B. Lin, "A Novel System in Package with Fan-Out WLP for High Speed SERDES Application", *IEEE/ECTC Proceedings*, China, May 2016, pp. 1496–1501.
[26] Suk, K., S. Lee, J. Kim, S. Lee, H. Kim, S. Lee, et al., "Low-Cost Si-Less RDL Interposer Package for High-Performance Computing Applications", *Proceedings of IEEE/ECTC*, May 2018, pp. 64–69.

第 7 章

PoP 异构集成

7.1 引言

PoP（堆叠封装）技术[1-21]是异构集成技术的一大类别。它可以将多个芯片组合在一个 PoP 结构中。在近 10 年中，PoP 技术已经应用在移动设备的处理器及存储器中。

7.2 引线键合 PoP

图 7.1a 为一个引线键合 PoP 封装示意图。它包括底部封装体以及一个堆叠于底部封装体上的顶部封装体。其中顶部封装体内有两个芯片，芯片交叉堆叠并以引线键合形式连接在一个封装基板上。底部封装体包含一个芯片，芯片同样以引线键合形式连接在另一个封装基板上。所有芯片都进行注塑处理，所有基板都带有焊球。

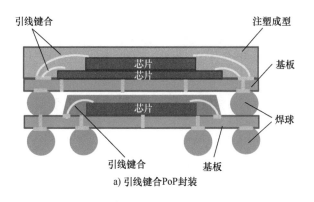

a) 引线键合PoP封装

图 7.1　PoP 封装

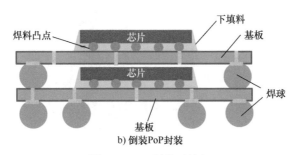

图 7.1 PoP 封装（续）

7.3 倒装 PoP

图 7.1b 为一个倒装 PoP 封装示意图。它包括两个堆叠在一起的相同封装体，封装体中的芯片以焊球倒装的形式连接在封装基板上。所有的倒装芯片底部都填充有下填料，所有的基板都带有焊球。

7.4 引线键合与倒装混合 PoP 封装

图 7.2 所示为一个引线键合与倒装混合 PoP 封装体的横截面图。可以看出，顶部封装是引线键合芯片，底部封装是倒装芯片。图 7.3 展示了一个应用于 iPhone 5S 的混合 PoP 封装实例。

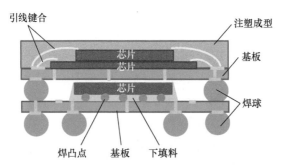

图 7.2 引线键合与倒装混合 PoP 封装体的横截面图

7.5 iPhone 5S 中的 PoP

iPhone 5S 是一款高端智能手机。图 7.3 展示了 iPhone 5S 中 AP（应用处理器）芯片组的 PoP 封装。从封装的横截面示意图可以看出，顶部封装体包含尔必达（Elpida）公司（现已被美光科技有限公司收购）的 1GB 内存 LPDDR3 移动 RAM 芯片（尺寸约为 11mm × 7.8mm），这些芯片交叉堆叠并以引线键合的形式连接在一个无芯 FBGA（细节距球栅阵列）封装基板上，然后进行注塑成型。FBGA 基板上有 3 排（456 个）焊球，如图 7.4 所示。底部封装包含一个 64 位 A7 处理器芯片（尺寸约为 10mm×10mm），该芯片倒装于积层封装基板上，共有 38×34=1292 个焊球。图 7.5 为支撑 A7 处理器的封装基板的横截面示意图。可以看出，此封装基板 2-2-2（基板顶面的两层积层、两层芯层以及基板底面的两层积层）结构简单，基板上带有两个 0201 和一个 0402 的嵌入式无源器件。A7 处理器芯片中 C4（可控塌陷芯片连接）凸点的焊盘节距约为 200μm，球栅阵列焊球的焊盘节距约为 400μm。

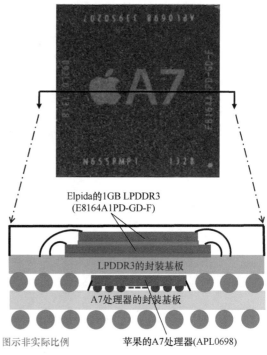

图 7.3　iPhone 5s 中的 PoP 封装（包含移动 DRAM 和 A7 应用处理器）顶视图和横截面图

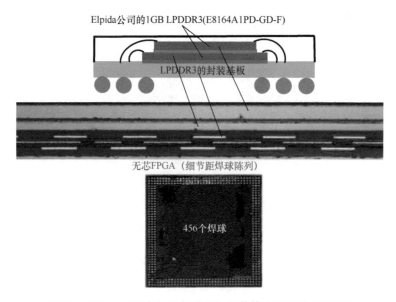

图 7.4　iPhone 5S 中 PoP 封装顶部封装体中的移动 DRAM

第 7 章　PoP 异构集成

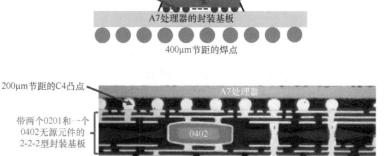

图 7.5　A7 应用处理器倒装封装于 PoP 结构底部封装体

7.6　安靠科技 / 高通 / 新科公司的 PoP

安靠科技公司率先开发了采用 Cu 柱锡帽焊点的高键合力热压技术[4]，这种 TC-NCP（热压 - 绝缘膏）技术采用 NCP（绝缘膏）作为下填料，在键合前就将膏体涂覆于封装基板上。如图 7.6 所示，三星 Galaxy 智能手机中的 PoP 封装结构中的底部封装体就采用了此技术，用于组装高通的骁龙处理器芯片。NCP 下填料可以采用旋涂、针管点胶或真空辅助的方式进行涂覆。MCeP（模芯嵌入式封装）基板由新科公司加工，如图 7.7 所示。可以看到，此基板采用铜芯焊球支撑处理器。基板组装过程如图 7.8 所示。首先，对 MCeP 基板进行预烘焙处理，去除挥发性化合物和水分。随后，预涂覆 NCP，并采用合适的加热曲线和压力进行热压键合。

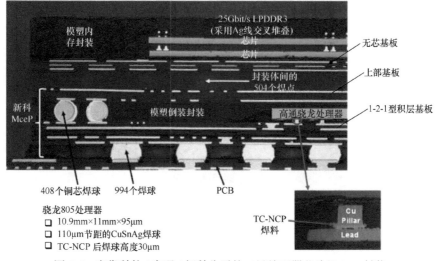

图 7.6　安靠科技 / 高通 / 新科公司的三星处理器芯片组 PoP 封装

179

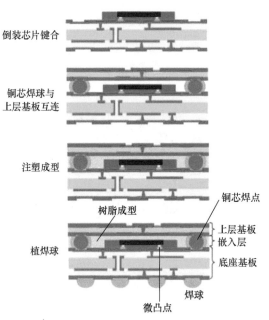

图 7.7 新科公司的 MCeP 封装

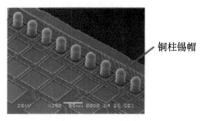

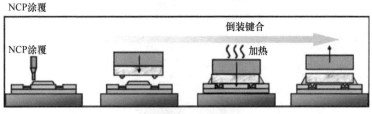

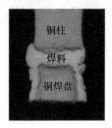

图 7.8 安靠公司 TC-NCP 工艺的关键工序

7.7 苹果公司的焊点倒装 PoP 封装

图 7.9 显示了一部 iPhone 6 Plus 手机的横截面图。可以看到 A9 处理器被封装于 PoP 结构的底部封装体中,以倒装方式采用回流焊工艺将处理器的 C4 焊点与一个带下填料的 2-2-2 型有机积层封装基板进行键合。LPDDR4(4 代低功耗双倍数据率)内存被封装于上部基板中。

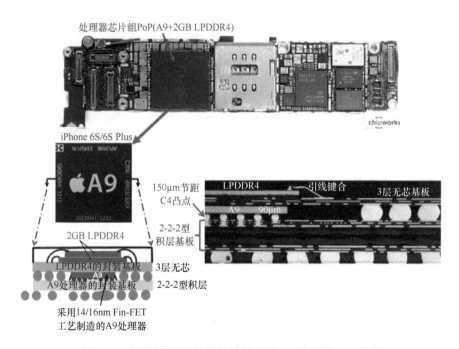

图 7.9　采用倒装芯片技术的苹果公司 A9 处理器 PoP 封装

7.8 星科金朋公司的处理器 PoP 封装

为了得到比图 7.3 ~ 图 7.9 所示 PoP 封装更小尺寸的封装,星科金朋公司开发了 3D eWLB(嵌入式晶圆级球栅阵列)技术,其横截面 SEM(扫描电子显微镜)图像如图 7.10 所示,其中包括一个 eWLB(如替换图 7.3 ~ 图 7.9 所示的焊点倒装芯片封装体)底部封装体和一个顶部内存封装体。由图 7.10 可知,eWLB 厚度仅 450μm,eWLB 里封装有一个应用处理器,顶部封装厚度为 520μm,内存芯片采用引线键合的形式封装于其中,电互连从 PCB 开始,沿焊球、RDL(再布线层)到处理器,再通过焊球、RDL 到存储芯片。

异构集成技术

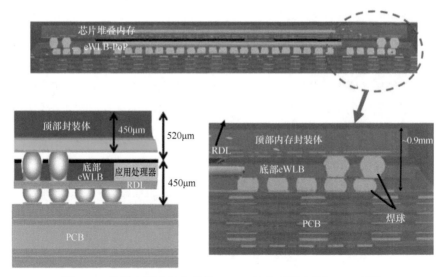

图 7.10　星科金朋有限公司带 eWLB 的处理器芯片组 PoP 封装

7.9　英飞凌公司的 eWLB 上 3D eWLB 封装

图 7.11 展示了英飞凌公司的 3D eWLB 上 eWLB 封装[10, 11]。eWLB 之间的垂直互连通过焊点和 TEV（封装通孔）实现。TEV 制备首先采用激光打孔技

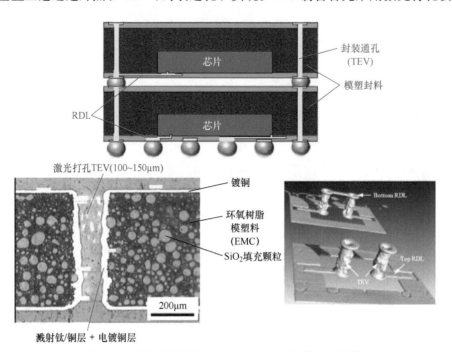

图 7.11　英飞凌公司的 eWLB 上 eWLB 技术 PoP 封装

术加工出贯穿模塑化合物的通孔。随后采用溅射工艺制备钛黏附层。然后在黏附层上溅射薄铜层作为化学沉积铜种子层。最后化学沉积铜电极。制备完成的通孔横截面图及 SEM 图如图 7.11 所示。可以看出 TEV 没有被完全填充。图中 TEV 的直径是 150μm，电镀铜厚度为 15～30μm。

7.10 台积电 / 苹果公司的处理器 PoP 封装

7.10.1 台积电 / 苹果公司的 A10 处理器 PoP 封装

2016 年 9 月，台积电采用该公司 InFO-PoP（集成扇出 PoP）技术的处理器芯片组投入大批量生产，在世界范围内首次实现了采用 InFO-PoP 技术芯片的量产[12]。图 7.12 展示了台积电制造的采用 InFO-PoP 技术封装的处理器（A10）芯片组的横截面示意图和 SEM 图。可以看出，A10 处理器的尺寸为 11.6mm×10.8mm×150μm，封装尺寸为 15.5mm×14.4mm×825μm，封装/芯片面积比为 1.8，RDL 层数为 3 层，TIV（穿 InFO 通孔）采用电镀铜填充，手机 DRAM 引线键合在一个 3 层无芯封装基板上，DRAM 和引线采用注塑成型工艺，顶部封装体和底部封装体之间采用 386 个焊球（0.3mm 节距）进行互连，顶部和底部封装体之间有下填料，底部封装体上有约 1300 个焊球（直径 200μm，节距 0.4mm），PoP 封装体和 PCB 之间有抗跌落的下填料。

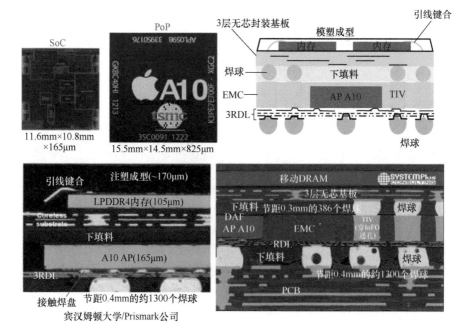

图 7.12 台积电 / 苹果公司采用 InFO-WLP 技术的 A10 处理器 PoP 封装

台积电 PoP 关键工艺如图 7.13 所示。首先在玻璃晶圆上旋涂一层非常薄的 LTHC 光热转换层。随后，采用电镀工艺制备 RDL、接触焊盘和铜柱。采用 DAF（芯片黏贴薄膜）将 A10 处理器的 KGD（已知合格芯片）黏贴于载体上，铜接触焊盘面朝上置于载板上。然后进行 EMC 压缩成型及固化。接下来背面研磨 EMC 至铜接触焊盘和铜柱露出。随后去除晶圆载体，加工 RDL，植焊球。最后划切再分布晶圆，得到单独封装体。

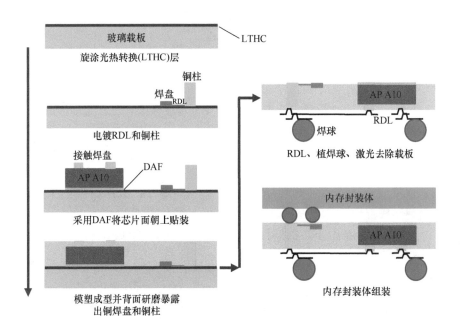

图 7.13　台积电 PoP 关键工艺

7.10.2　台积电 / 苹果公司的 A11 处理器 PoP 封装

图 7.14 展示了台积电制造的采用 InFO-WLP（集成扇出晶圆级封装）技术并于 2017 年 9 月投入量产的处理器（A11）芯片组的示意图和 SEM 图。基本上，A10 和 A11 处理器的 PoP 封装形式相差不大。但是台积电在制造 A11 的过程中采用 10nm 制程工艺替换了 A10 中的 16nm 制程，因此 A11 芯片的尺寸（10mm×8.7mm×150μm）更小，其芯片面积较 A10 减小约 30%。A11 芯片组的 PoP 封装体尺寸为 13.9mm×14.8mm，相比 A10 封装面积仅减小约 8%，变化不大。封装 / 芯片面积比由 A10 芯片组的 1.8 上升至 A11 的 2.3，而且 RDL 层数由 3 层升至 4 层。4 层 RDL 的线宽和间距为 10μm。

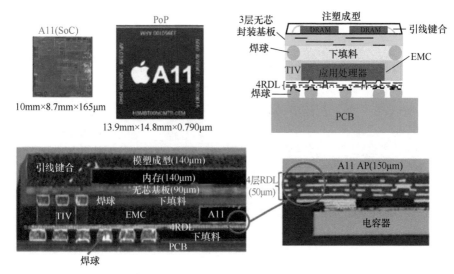

图 7.14　台积电/苹果公司采用 InFO-WLP 技术的 A11 处理器 PoP 封装

7.10.3　台积电/苹果公司的 A12 处理器 PoP 封装

图 7.15 展示了台积电制造的采用 InFO-WLP 技术封装并于 2018 年 10 月投入量产的处理器（A12）芯片组的示意图和 SEM 图。基本上，所有处理器（A10、A11 和 A12）的 PoP 封装形式都相差不大。但是，由于台积电在制造 A12 的过程中采用 7nm 制程工艺，因此即便芯片添加了更多功能，如人工智能等，A12 芯片仍略小于 A11。为了取得更好的电学性能，焊点倒装型集成无源器件芯片被嵌入扇出封装体的底部，如图 7.15 所示。4 层 RDL 的最小金属线宽和间距为 8μm。

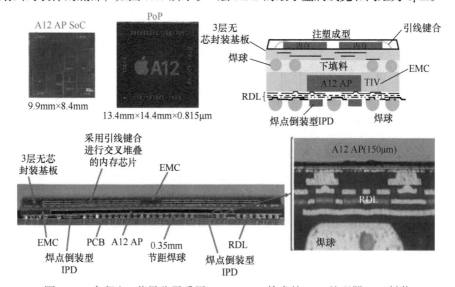

图 7.15　台积电/苹果公司采用 InFO-WLP 技术的 A12 处理器 PoP 封装

7.11 三星智能手表 PoP 封装

图 7.16 展示了 2018 年夏季发布的三星智能手表的 PoP 封装，其中上部封装体包含 2 个 DRAM、2 个 NAND 闪存和 1 个 NAND 控制器的内存 ePoP（嵌入式 PoP）。内存 ePoP 结构中的芯片均采用引线键合的形式固定在一个 3 层无芯封装基板上，如图 7.17 所示。上部封装体的尺寸为 8mm×9.5mm×1mm。采用 FOPLP（面板级扇出型封装）技术将处理器和 PMIC（电源管理集成电路）芯片并排封装于底部封装体，如图 7.18 所示。处理器和 PMIC 芯片的大小约为 3mm×3mm。关键工艺如图 7.19 所示[15]。首先在 PCB 上加工空腔。随后，将芯片放置于空腔中并覆盖 EMC 层。之后将其固定在载板上，制备 RDL 并植焊球。

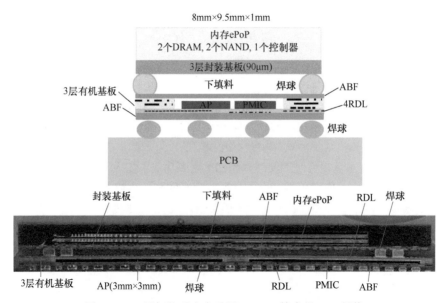

图 7.16　三星智能手表中采用 FOPLP 技术的 PoP 封装

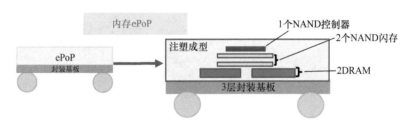

图 7.17　三星公司的内存 ePoP

第 7 章 PoP 异构集成

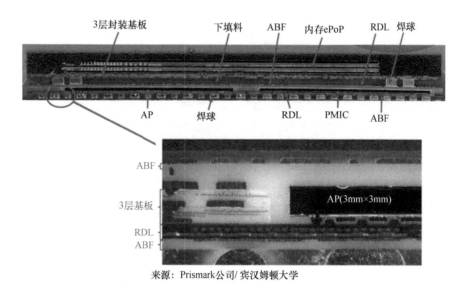

图 7.18 三星智能手表的 PoP 封装横截面图

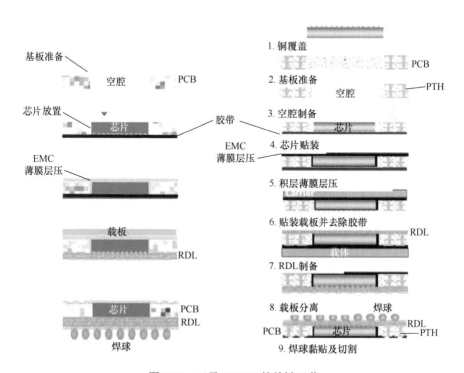

图 7.19 三星 FOPLP 的关键工艺

7.12 总结和建议

一些重要的结论和建议总结如下：

1）PoP 是一种异构集成封装技术。

2）PoP 技术已经被应用于封装平板电脑、智能手机及智能手表等电子产品的应用芯片组（包括应用处理器及内存）。

3）PoP 结构的关键是上部封装体及底部封装体之间的互连。如安靠公司使用的 TMV 结构[22]、星科金朋有限公司采用的焊球结构[6]、英飞凌公司使用的 TEV 结构[10, 11]、台积电使用的 TIV 结构[12] 以及 Invensas 公司使用的 BVA（键合通孔阵列）结构[23]。

4）台积电的 InFO 技术采用扇出晶圆级封装，用于嵌入苹果公司的平板电脑及 iPhone 手机的应用芯片组 PoP 封装。

5）三星公司的 FOPLP 技术采用扇出面板级封装，用于嵌入三星智能手表的应用芯片组 PoP 封装。

参考文献

[1] Dreiza, M., A. Yoshida, K. Ishibashi, and T. Maeda, "High Density PoP (Package-on-Package) and Package Stacking Development", *IEEE/ECTC Proceedings*, May 2007, pp. 1397–1402.

[2] Carson, F., and S. Lee, "Controlling Top Package Warpage for POP Applications", *IEEE/ECTC Proceedings*, May 2007, pp. 737–742.

[3] Vijayaragavan, N., F. Carson, and A. Mistry, "Package on Package Warpage—Impact on Surface Mount Yields and Board Level Reliability", *IEEE/ECTC Proceedings*, May 2008, pp. 389–386.

[4] Lee, M., Yoo, M., Cho, J., Lee, S., Kim, J., Lee, C., Kang, D., Zwenger, C., and Lanzone, R., "Study of Interconnection Process for Fine Pitch Flip Chip", *IEEE/ECTC Proceedings*, May 25–28, 2009, pp. 720–723.

[5] Eslampour, H., S. Lee, S. Park, T. Lee, I. Yoon, and Y. Kim, "Comparison of Advanced PoP Package Configurations", *IEEE/ECTC Proceedings*, May 2010, pp. 1946–1950.

[6] Yoon, S., J. Caparas, Y. Lin, and P. Marimuthu, "Advanced Low Profile PoP Solution with Embedded Wafer Level PoP (eWLB-PoP) Technology", *Proceedings of IEEE/ECTC*, May 2012, pp. 1250–1254.

[7] Eslampour, H., M. Joshi, K. Kang, H. Bae, and Y. Kim, "fcCuBE Technology: A Pathway to Advanced Si-node and Fine Pitch Flip Chip", *IEEE/ECTC Proceedings*, May 2012, pp. 904–909.

[8] Eslampour, H., Y. Kim, S. Park, and T. Lee, "Low Cost Cu Column fcPoP Technology", *IEEE/ECTC Proceedings*, May 2012, pp. 871–876.

[9] Eslampour, H., M. Joshi, S. Park, H. Shin, and J. Chung, "Advancements in Package-on-Package (PoP) Technology, Delivering Performance, Form Factor & Cost Benefits in Next Generation Smartphone Processors", *Proceedings of IEEE/ECTC*, Las Vegas, NV, May 2013, pp. 1823–1828.

[10] Wojnowski, M., G. Sommer, K. Pressel, and G. Beer, "3D eWLB—Horizontal and Vertical Interconnects for Integration of Passive Components", *IEEE/ECTC Proceedings*, May 2013, pp. 2121–2125.

[11] Wunderle, B., J. Heilmann, S. G. Kumar, O. Hoelck, H. Walter, O. Wittler, G. Engelmann, et al., "Accelerated Reliability Testing and Modeling of Cu-Plated Through Encapsulant Vias (TEVs) for 3D-Integration", *Proceeding of IEEE/ECTC*, May 2013, pp. 372–382.

[12] Tseng, C., Liu, C., Wu, C., and D. Yu, "InFO (Wafer Level Integrated Fan-Out) Technology", *IEEE/ECTC Proceedings*, 2016, pp. 1–6.
[13] Dhandapani, K., J. Zheng, B. Roggeman, and M. Hsu, "Improving Solder Joint Reliability for PoP Packages in Current Mobile Ecosystem", *IEEE/ECTC Proceedings*, May 2018, pp. 1639–1644.
[14] Lee, J., C. Chen, L. Brown, E. Mehretu, T. Obrien, and F. Lu, "A Dynamic Bending Method for PoP Package Board Level Reliability Validation", *IEEE/ECTC Proceedings*, May 2018, pp. 2217–2223.
[15] Hwang, T., D. Oh, E. Song, K. Kim, J. Kim, and S. Lee, "Study of Advanced Fan-Out Packages for Mobile Applications", *IEEE/ECTC Proceedings*, May 2018, pp. 343–348.
[16] Ki, W., W. Lee, I. Lee, I. Mok, W. Do, M. Kolbehdari, A. Copia, S. Jayaraman, C. Zwenger and K. Lee, "Chip Stackable, Ultra-thin, High-flexibility 3D FOWLP (3D SWIFT® Technology) for Hetero-integrated Advanced 3D WL-SiP", *IEEE/ECTC Proceedings*, May 2018, pp. 580–586.
[17] Wei, H., and B. Han, "Stacking Yield Prediction of Package-on-Package Considering the Statistical Distributions of Top/Bottom Package Warpages and Solder Ball Heights", *IEEE/ECTC Proceedings*, May 2018, pp. 693–702.
[18] You, S., S. Jeon, D. Oh, K Kim, J. Kim, S. Cha, and G. Kim, "Advanced Fan-Out Package SI/PI/Thermal Performance Analysis of Novel RDL Packages", *IEEE/ECTC Proceedings*, May 2018, pp. 1295–1301.
[19] Wu, B., and B. Han, "Effects of Underfill on Thermo-Mechanical Behavior of Fan-out Wafer Level Package Used in PoP: An Experimental Study by Advancements of Real-time Moiré Interferometry", *IEEE/ECTC Proceedings*, May 2018, pp. 1609–1616.
[20] Lujan, A., "Comparison of Package-on-Package Technologies Utilizing Flip Chip and Fan-Out Wafer Level Packaging", *IEEE/ECTC Proceedings*, May 2018, pp. 2083–2088.
[21] Lim, H., J. Yang, and R. Fuentes, "Practical Design Method to Reduce Crosstalk for Silicon Wafer Integrated Fan-out Technology (SWIFT®) Packages", *IEEE/ECTC Proceedings*, May 2018, pp. 2205–2211.
[22] Kim, J., K. Lee, D. Park, T. Hwang, K. Kim, D. Kang, J. Kim, C. Lee, C. Scanlan, C. Berry, C. Zwenger, L. Smith, M. Dreiza, and R. Darveaux, "Application of Through Mold Via (TMV) as PoP base package", *IEEE/ECTC Proceedings*, May 2008, pp. 1089–1092.
[23] Katkar, R., R. Co, and W. Zohni, "Manufacturing Readiness of BVA Technology for Ultra-High Bandwidth Package-on-Package", *IEEE/ECTC Proceedings*, May 2014, pp. 1389–1395.

第 8 章

内存堆叠的异构集成

8.1 引言

第一篇通过芯片贴装材料和引线键合完成 3D 堆叠内存芯片的论文,发表于 1994 年的 IEEE 多芯片组件会议[1]上,如图 8.1 所示。自此,通过引线键合方式堆叠的内存芯片(尤其是 NAND 闪存芯片)一直处于大批量生产状态。如图 8.2 和图 8.3 所示的三星智能手机、平板计算机和固态驱动器。可以看到,有 16 个 48 层的 V-NAND 3D 闪存内存芯片通过引线键合技术堆叠,每个芯片的厚度只有 40μm,而这些都是内存栈的同构集成。

图 8.4 展示了采用芯片贴装材料和引线键合方法的内存堆栈异构集成(内存 + ASIC),最近这种结构的应用也是层出不穷。本章第一部分将展示两个堆叠在一个 ASIC 上的内存芯片的异构集成。

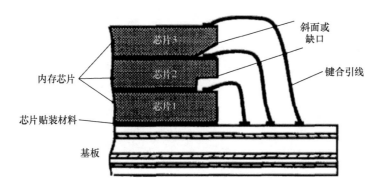

图 8.1 采用芯片贴装和引线键合方法的内存芯片堆叠

第 8 章 内存堆叠的异构集成

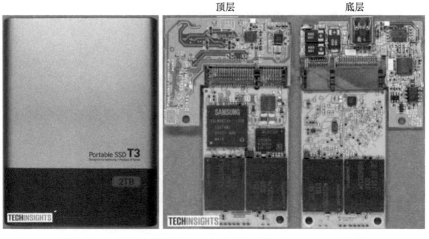

图 8.2 三星固态硬盘（SSD）中的 48 层 V-NAND 3D 闪存

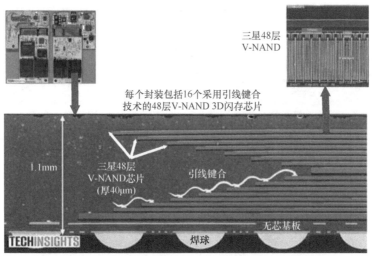

图 8.3 堆叠 16 层的三星 48 层 V-NAND 芯片

为了增加其带宽、集成密度和提升性能，内存芯片将通过微凸点和 TSV（硅通孔）进行堆叠。三星公司在 2014 年 8 月量产了业界首款基于 TSV 的 64 GB 4 代双倍数据速率（DDR4）DRAM 堆栈模块，如图 8.5 所示。每个堆栈有 4 个 DRAM，每个 DRAM 裸片有 78 个 TSV，并且此 64GB 的 DDR4 DRAM 是在一个 PCB 上。该模块的运行速度是使用引线键合封装技术模块的 2 倍，同时只有一半功耗。此模块可用于环保型服务器的应用。2015 年 11 月 26 日，三星公司开始生产 128GB 带寄存器的双线内存模块（RDIMM），这是内存堆栈的同构集成。

异构集成技术

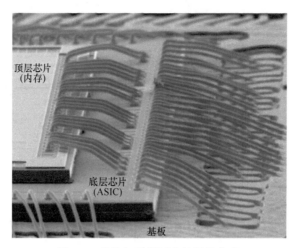

图 8.4　芯片与引线键合的异构集成

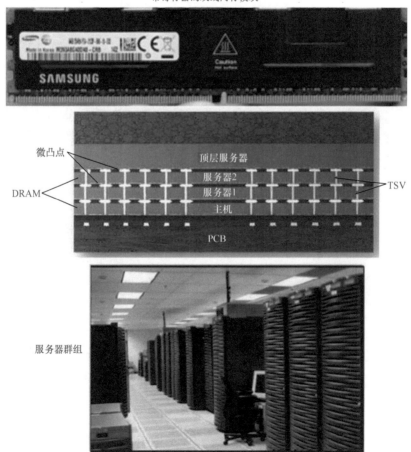

图 8.5　三星公司为服务器群组量产的业界首款基于 TSV 的 DDR4 DRAM

图 8.6 展示了 AMD 公司在 2015 年下半年推出的 Radeon R9 Fury X 图形处理器（GPU），该 GPU 用台积电 28nm 工艺制造，并以 4 个海力士公司制造的 128 GB/s 的高带宽内存（HBM）方块为基础。每个 HBM 都由 4 个带有 C2 凸点的 DRAM 和一个带有 TSV 的基础逻辑芯片组件构成。这是内存芯片和逻辑芯片的异构集成。每个 DRAM 芯片都有超过 1000 个 TSV 以及在每个引脚 1Gbit/s 速度下达 2Gbit 的数据密度。GPU 和 HBM 方块位于由联华电子公司（UMC）采用 64nm 工艺制造的 TSV 的转接板（28mm×35 mm）顶部。TSV 转接板（带有 C4 凸点）在由日本揖斐电株式会社（IBIDEN）制造的 4-2-4 型有机封装基板上的最终组装是由 ASE 完成。

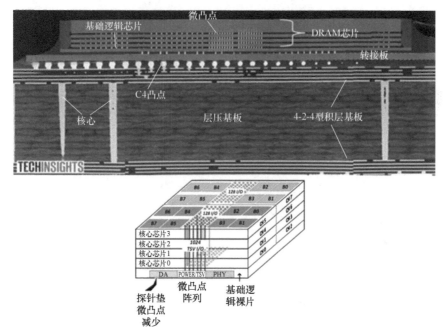

图 8.6　海力士 HBM 制造的 AMD 显卡，由非导电膜（NCF）DRAM
芯片逐一热压键合（TCB）制成

图 8.7 展示了英伟达公司的 Pascal 100 GPU，该型号 GPU 于 2016 年下半年上市，使用了台积电的 16nm 工艺制造，并由 4 个三星公司制造的 HBM2（16GB）为基础，带宽为 256GB/s。每个 HBM2 都由 4 个带有 C2 凸点的 DRAM 和一个带有 TSV 的基础逻辑裸片组成。每个 DRAM 芯片都具有超过 1000 个 TSV 和在每个引脚 2Gbit/s 速度下达 8Gbit 的数据密度。GPU 和 HBM2 都位于由台积电采用 64nm 工艺制造的 TSV 转接板（1200mm^2）顶部。TSV 转接板被附着到一个 5-2-5 型有机封装基板上，同时还带有 C4 凸点。这也是内存芯片和逻辑芯片的异构集成。

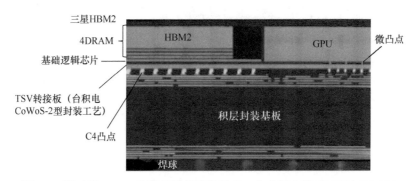

图 8.7　以三星 HBM2 为基础的英伟达显卡，由非导电膜（NCF）DRAM 芯片逐一热压键合（TCB）制成

三星公司和海力士公司都使用参考文献 [2] 中微凸点 DRAM 的高键合力 TCB 和带有 NCF 的逻辑芯片来制造 3D 异构集成的堆栈。这个 3D 内存块一次堆叠一个芯片，如图 8.8 所示，每个芯片需要约 10s 的时间来使下填料膜凝胶化，焊料熔化，再使下填料膜固化，最后焊料凝固。这样的话还必须关注生产率问题！

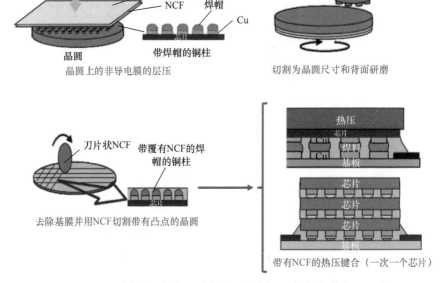

图 8.8　NCF 在铜柱 / 焊帽凸点晶圆上的层压、切割和带有 NCF 的倒装芯片的逐一热压键合

为了解决这个问题，Toray[3, 4] 提出了一种集体键合的方法，如图 8.9 所示。可以看出，带有 NCF 的微凸点芯片在 80℃温度的平台上进行了预键合处理（键合力为 30N，温度 150℃，时间 <1s）。后续的键合在温度为 80℃的平台上进行，第一步（3s）：键合力为 50N，温度 220～260℃；第二步（7s）：键合力 70N，

温度280℃。因此，用传统方法堆叠4个芯片需要40s，而集体键合的方法只需不到14s。该集体键合方法的一些横截面图像如图8.9所示，它在优化的条件下实现了理想的连接。

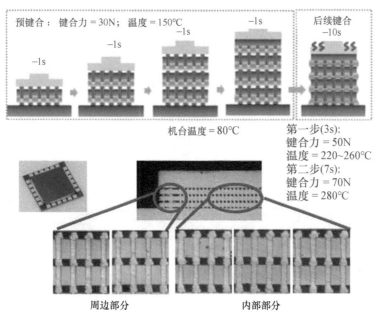

图8.9　由Toray提出的带有NCF倒装芯片的、高键合力的集体热压键合

本章第二部分将介绍内存芯片和逻辑芯片的低温键合异构集成。在低于200℃的温度下进行对准和键合后，所有焊料将会与芯片/晶圆上的UBM（凸点下金属层）发生反应，并反应成为比焊料熔点高出几百度的IMC（金属间化合物）。这样的特性尤其适用于3D IC芯片堆叠。如在低温键合前两个芯片后，所有互连线都将转化为具有更高重熔温度的IMC。当第三个芯片低温键合在这两个芯片之上时，这两个芯片之间的IMC就不会回流也不会移动。

8.2　铜低k芯片上堆叠裸片（存储器）的引线键合

8.2.1　测试装置

图8.10展示了两个测试装置。两个装置中，底层芯片都是65nm的铜低k裸片，此芯片被贴装在有机衬底上。铜低k裸片支持：①测试装置1（TV1）为两个夹有一层引线嵌入式薄膜（WEF）的子芯片，如图8.10a所示；②测试装置2（TV2）为两个有夹有一层芯片贴装薄膜（DAF）的子芯片，如图8.10b所示。两个测试装置中，DAF都被用于底层子裸片和铜低k裸片之间，芯片贴装胶都被用于铜低k裸片和基板之间。

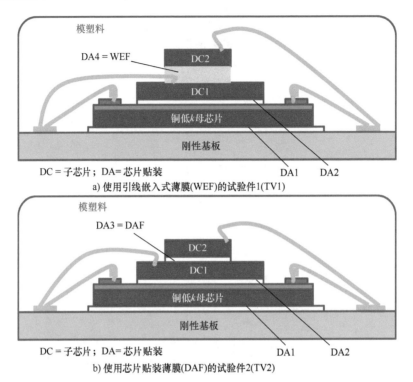

图 8.10　65nm 铜低 k 裸片堆叠封装示意图（两个子芯片摆放在互相都合适的角度）

这个 65nm 的铜低 k 裸片有 6 层金属层，共 408 个输入输出引脚（I/O），尺寸为 7mm × 7mm，厚度为 0.3mm。子芯片的尺寸为 3mm × 5mm × 0.075mm。以上使用的有机基板是含有塑料球栅阵列（PBGA）的两层金属层。最终的封装尺寸是 17mm × 17mm × 1.1mm。基板和模具的成分都是绿色材料，符合环境保护法。

8.2.2　铜低 k 焊盘处的应力

铜低 k 材料较差的机械性会使封装过程复杂化，包括切割过程中的低 k 脱落、引线键合后的键合点成坑（bond-pad cratering）、可靠性测试中的钝化层分层[5-7]。切割低 k 晶圆时常见的失效模式包括金属/中间电介质层（ILD）分层、金属/ILD 脱落、金属层变色[8-15]。在铜低 k 封装项目中，新加坡微电子研究所开发了聚合物封装切割方法（PEDL）[8, 16]，其目的在于解决铜低 k 封装切割中的难题，同时降低传统切割方法施加在铜低 k 芯片上的角应力，提高封装的可靠性。

局部最大的剪应力集中点位于硅裸片最外围的角上。这是因为不同材料的热膨胀系数（TCE）不同，同时芯片角上又存在应力奇点。裸片边角附近的剪应力是导致低 k 层和氟硅酸盐玻璃（FSG）层分层的主要因素之一[16]。结果表明，在硅裸片的边缘，从45°切一个35μm的倒角可以把最大应力减少34%或更多[16]。

如图8.10b所示封装配置，材料的相关参数见表8.1，其中使用了PEDL技术来减小低k层中的应力，PEDL技术即斜面切削和苯并环丁烯（BCB）涂覆，如图8.11所示。图8.12为有限元分析（FEA）结果，其中给出了整个结构在25~175℃温度变化的边界条件[17]。图8.12a展示了有限元模型，图8.12b是芯片贴装2附近的低k层中的剪应力分布，图8.12c是低k角处的剪应力分布。可以看到，在4种情况下，最大的剪应力几乎都出现在了同样的位置。

表8.1 有限元分析中使用的材料参数

材料	热膨胀系数/（ppm/℃）	弹性模量/GPa	泊松比
苯并环丁烯（BCB）	52	3	0.35
低k（BD）	10	8	0.3
FTEOS	2	105.44	0.3
TEOS	0.57	66	0.18
硅	2.7	131	0.28
铜	17	110	0.34
模塑料	8.9/40.4（T_g=147.8℃）	20.7（25℃）0.296（250℃）	0.3
芯片贴装胶	50/100（T_g=150℃）	5.9（25℃）0.56（260℃）	0.3
DAF	80/170（T_g=128℃）	1.66（25℃）0.56（260℃）	0.3
基板	15	29	0.3
阻焊层	52	4	0.4

图8.13展示了低k层、芯片贴装层、EMC中的剪应力，这3个案例研究的条件为有无斜面切削、有无BCB涂覆，即直线切削、无BCB涂覆；斜面切削、无BCB涂覆；非斜面切削、有BCB涂覆；斜面切削、有BCB涂覆。由于低k层很容易分层，应密切监测低k角处和芯片贴装2附近这两个关键位置的剪应力。可以看出，斜面切削有效地减小了低k角处的剪应力。当基准样本（非斜面切削、无BCB涂覆）被替换为斜面切削、无BCB涂覆时，低k角处的剪应力从100.90MPa降低至23.75MPa。另一方面，BCB涂覆可以有效减小芯片贴装2附近的低k层中的剪应力。与基准样本相比，斜面切削、有BCB涂覆的案例中，芯片贴装2附近的低k层中的剪应力从32.30MPa降低至11.90MPa。

总的来说，如图8.13所示，斜面切削和BCB涂覆都应该采用以减小整个低k层中的剪应力。在封装中实施PEDL技术不仅可以减小芯片角上的剪应力，还能减小贴装在第二块堆叠芯片（如子芯片1）下方的低k层中的剪应力。

异构集成技术

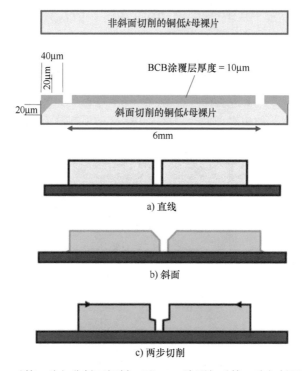

图 8.11 （第一片）非斜面切削、无 BCB 涂覆与（第二片）斜面切削、有 BCB 涂覆的堆叠裸片示意图

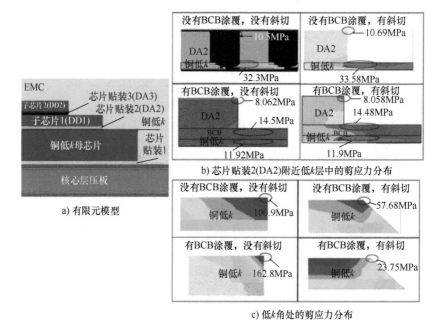

图 8.12 65nm 铜低 k 裸片堆叠封装有限元模型及剪应力分布

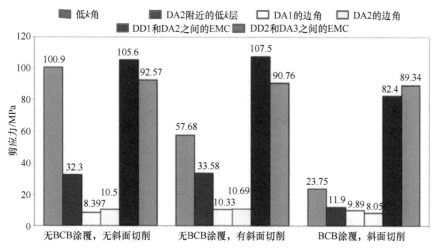

图 8.13 不同情况下（有无 BCB 涂覆和斜面切削）低 k 层中低 k 角和芯片贴装 2 附近的剪应力

8.2.3 组装和处理

所有 3D 异构集成都要求较薄的晶圆和芯片。在此研究中，铜低 k 焊盘的晶圆被减薄至 300μm，铝焊盘晶圆（子芯片上的）被减薄至 75μm。所有的晶圆都先通过机械粗磨减薄，接着再经过机械精磨。为了去除减薄后晶圆背面的微裂痕，还需要另一项工艺——干抛光。所有铜低 k 焊盘晶圆的内部厚度的总厚度偏差（TTV）为 2.21μm。铝焊盘晶圆的 TTV 为 2.28μm。

8.2.4 切割方法的测评

芯片切割分离是每一块集成电路芯片都要经历的主要过程。随着 3D 异构集成的发展和低 k 晶圆越来越薄的趋势，芯片分离切割的过程变得很重要。本研究从芯片强度和崩边结果两方面比较了直线切削（见图 8.11a）、斜面切削（见图 8.11b）和两步切削（见图 8.11c）的晶圆。三种晶圆切割方法的参数见表 8.2。

表 8.2 三种晶圆切割方法的参数

切割方法	参变量	刀片种类	刀片速度/（r/min）	进给速度/（mm/s）	刀片高度/mm
直线切削		27HCAA	30000	15	0.06
斜面切削	刀片 1	2050D-TI	30000	30	0.115
	刀片 2	HDDCC-B	30000	30	0.06
两步切削	刀片 1	27HCCC	30000	20	0.1
	刀片 2	27HAAA	30000	20	0.06

抗弯试验已被指出是一种确定芯片强度的可靠测量方法[18]。本研究使用三点式抗弯试验来确定芯片强度，结果见图 8.14 和表 8.3。每个案例中的样本容量都是 33。可以看出，斜面切削得到的芯片强度比另外两种切割方式的更强；相比直线切削，斜面切削可以将芯片强度增强 130%；斜面切削引起的切屑比直线切削和两步切削少；相比直线切削，斜面切削会使正面崩边减少 10%；相比两步切削，斜面切削会使正面崩边减少 15%。

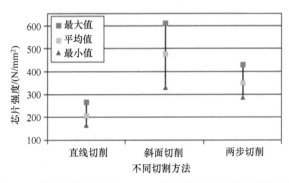

图 8.14　不同切割方法下 300μm 厚的铜低 k 芯片的芯片强度

表 8.3　不同切割方法下 300μm 厚的铜低 k 芯片的芯片强度

	背面崩边 /μm	正面崩边 /μm
直线切削	9.2μm	18.2μm
斜面切削	9μm	16.5μm
两步切削	9μm	19.5μm

8.2.5　芯片贴装工艺

薄芯片非常柔软，为了防止芯片弯曲、保证均一的胶层厚度，芯片在拾取和放置时需要全面的保护，这对封装的可靠性来说非常重要。利用薄膜型材料更容易控制薄芯片贴装层。所以，DAF 被用于 TV2 中的两个顶部子芯片，而 WEF 被用于 TV1 中的两个顶部子芯片（见图 8.10）。TV1 和 TV2 中，子芯片 1 和铜低 k 母芯片之间的芯片贴装也都是 DAF。这里往往采用特殊的芯片贴装胶将铜低 k 母芯片贴装至基板上，这是因为基板的表面不平滑，需要有较好流动性的材料才能实现良好、无空洞和分层的覆盖。

表 8.4a 展示了 DAF 的键合参数设置。可以发现，空洞在第四种工艺参数下消失了。因此，最终 DAF 的键合参数由此确定。DAF 工艺过程中的主要难点是薄膜排空隙。典型的 DAF 温度在 120~150℃ [19]。本研究最终的 DAF 温度是 150℃。一旦 DAF 不再有空隙或被完全固化，使用传递模塑工艺来去除被困在 DAF 内部的空洞的效果就会下降。

表 8.4 DAF 的键合参数设置与 WEF 的键合后固化参数设置

	序号	工艺参数	键合后固化的超声波扫描图像（一步键合后固化：160℃下固化 60min）
a）DAF 的键合参数设置	1	键合强度：1.0kgf 键合温度：150℃ 保持时间：1s	
	2	键合强度：1.0kgf 键合温度：150℃ 保持时间：2s	
	3	键合强度：2.0kgf 键合温度：150℃ 保持时间：2s	
	4	键合强度：2.5kgf 键合温度：150℃ 保持时间：2s	

（续）

键合后固化的参数	不同键合后固化和芯片剪切测试后的光学图像	观察结果
b）WEF的键合后固化参数设置	一步键合后固化：160℃下固化 60min	裸片剪切测试后出现亮斑
	两步键合后固化：120℃下固化 60min，然后 140℃下固化 60min	裸片剪切测试后，薄膜上的颜色很均匀

WEF 是一种芯片贴装材料技术，它允许同一尺寸、通过引线键合的芯片直接顶部堆叠。WEF 在中心和外围引线键合焊盘上都可以使用。最终的 WEF 键合参数确定如下：键合强度 =0.3kgf；键合温度 =150℃；150℃下的保持时间为 1s。WEF 工艺中，两步键合后固化和一步键合后固化的比较见表 8.4。根据芯片剪切样品的光学图像，两步键合后固化的数据比一步键合后固化的数据更好。因此，本研究在 WEF 中使用了两步键合后固化。

图 8.15 为采用最终的 WEF 工艺参数后，TV1 中两个子芯片间引线的 X 光图像（CT 重建后）。WEF 将引线变湿并将引线完全封装在内部。没有观察到引线变形。本例中 WEF 有两层。顶层 WEF 贴附在顶部芯片背部，其芯片键合条件下的流动性很有限。底层 WEF 面对底部芯片的引线键合，为了避免引线变形、完全填充底部芯片的间隙，其芯片键合条件下的黏性很低。

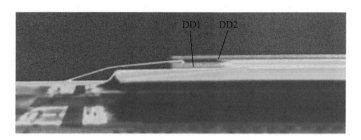

图 8.15　TV1 中两个子芯片间引线的 X 光图像

8.2.6　引线键合工艺

引线键合是 3D 异构集成的关键互连方法之一。最先进的引线键合机具有针对复杂引线环、细长引线和高键合精度的软件和功能，用于堆叠不同尺寸的芯片、键合薄悬垂芯片、极低高度引线环的正反向键合、低 k 介质焊盘上的键合，以及在焊盘结构下对有源电路进行键合。

图 8.16a 展示了 TV1 中第一个子芯片上的超低引线环高度（50μm）的引线键合[17]。它采用一种新的前向运动方法，引线环剖面专门为 50μm 的引线环高度设计，如 8.16b 所示。引线使用的是 0.8mil（$1mil=25.4×10^{-6}m$，20μm）的 NL4 金导线。引线键合机使用 ASM Eagle 60，引线间距为 53μm，钝化层开孔为 43μm×100μm。关键的引线键合参数见表 8.5。新的前向运动为堆叠芯片封装提供以下优势：①提供了较低的引线环高度和比传统正向键合更少的颈部损伤；②提供了比反向键合更少的焊盘损伤、更高的产量和更好的模具去除效果；③由于在一次键合上的低形变，它还提供了比反向键合更细的节距。

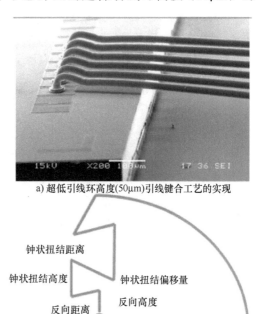

a) 超低引线环高度(50μm)引线键合工艺的实现

b) 超低引线环运动剖面：钟状引线环

图 8.16　超低引线环键合工艺

表 8.5　关键的引线键合参数

键合参数	设置	
	一次键合	二次键合
键合时间 /ms	10	8
键合功耗 /（Dac）	36	30
键合力 /gf	8	27
EFO 电流 /mA	3800	
EFO 时间 /ms	223	
键合温度 /℃	170	

引线键合后，样品进行了拉线试验和球剪试验，结果发现所有的球剪读数在预期的剪切失效模式下都高于最低要求（8gf），所有的测试样品在球剪测试中都没有观察到剥落，引线拉力读数在预期的引线拉力失效模式（颈断裂）下高于最低要求（3gf），如图 8.17 所示。最终，如图 8.18 所示，制造了一些引线键合的试验件 TV1、TV2。

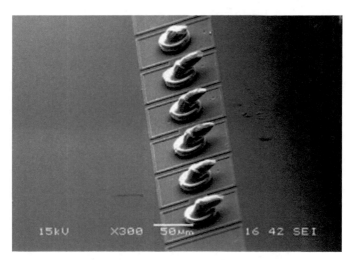

图 8.17 引线拉力试验后颈断裂的失效模式

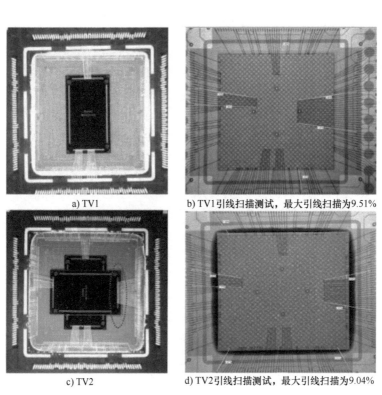

a) TV1

b) TV1引线扫描测试,最大引线扫描为9.51%

c) TV2

d) TV2引线扫描测试,最大引线扫描为9.04%

图 8.18 引线键合试验件

8.2.7 成型工艺

3D 芯片堆叠所增加的引线密度和长度导致成型工艺相比于传统单个芯片更加困难。不同层的引线键合环承受不同大小的拖曳力，导致引线扫描不同，增加了引线短路的可能性。因此，为了制造可靠的 3D 芯片堆叠封装，需要精心选择模塑料（MC）。

在 MC 的材料选择上，对四种 MC 进行了黏附力测试（拉拔测试）和湿敏度（MSL）测试（MSL 2 和 MSL 3），见表 8.6。样品制备过程如下：将伪低 k 芯片附着在有机基板上；用四种 MC 材料进行封装，将模制基板切割成 17mm×17mm 用于 MSL 测试，切割成 8mm×8mm 用于拉力测试。在 260℃ 的峰值温度下，样品经过 MSL 3 次回流后进行拉拔测试（黏附力测试）。表 8.6 第四列显示，4 个 MC 的拉力值没有显著差异[17]。

表 8.6 选择 MC

模塑料序号	MSL 3（失效的样品数/总样品数）	MSL 2（失效的样品数/总样品数）	拉力测试中的拉力值/kgf	翘曲/μm
MC1	0/7	1/6	2.33	377
MC2	0/7	0/7	2	504
MC3	0/7	5/7	2.33	692
MC4	0/7	3/7	1.33	620

在 MSL 测试中，样品分别进行 MSL 3 和 MSL 2 测试，并在 260℃ 的峰值温度下进行 3 次回流，采用 CSAM 和 Thru-Scan 两种分析方法检验四种 MC 的封装性能，采用联合电子器件工程委员会（JEDEC）推荐的 10% 分层失效准则。表 8.6 第二列显示，四种 MC 都通过了 MSL 3，没有出现空洞和分层。MSL 2 中，MC2 的 7 个样本均通过 MSL 2，未出现失效（见表 8.6 第三列）。然而，MC1 的 6 个样本中有 1 个在 MSL 2 中失败，MC3 的 7 个样品中有 5 个在 MSL 2 中出现失效，MC4 的 7 个样品中有 3 个在 MSL 2 不合格。因此，选择 MC2 用于后续的组装构建。

MC 材料选定后，在 TV1 和 TV2 上进行成型。TV1、TV2 的引线扫描测试结果如图 8.18 所示。所有的最大引线扫描小于 10%，在可接受的范围内。

8.2.8 可靠性测试和结果

最终的 TV1 和 TV2 须根据 JEDEC 器件标准进行可靠性测试，见表 8.7。采用 CSAM 和 Thru-Scan 两种分析方法检查分层的发生。在热循环（TC）1000 次和高温储存（HTS）1000h 测试后，所有样品均未观察到分层现象，如图 8.19 所示。

表 8.7　65nm 铜低 k 堆叠芯片 FBGA 封装在热循环和高温储存中的可靠性测试结果

可靠性测试	组别	样本容量	电测试 失败	CSAM 分层
热循环 （-40～125℃）	TV1	22	0	0
	TV2	21	0	0
高温存储（150℃）	TV1	22	0	0
	TV2	21	2	0

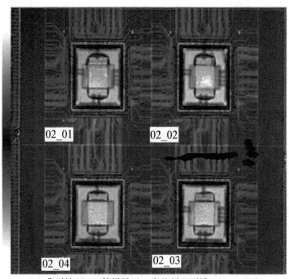

a) 典型的CSAM结果显示，在热循环测试(-40~125℃)
中经过1000次循环后没有分层

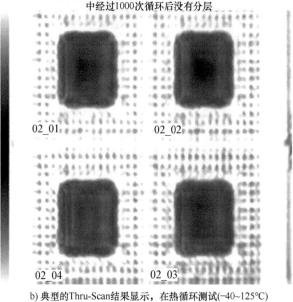

b) 典型的Thru-Scan结果显示，在热循环测试(-40~125℃)
中经过1000次循环后没有分层

图 8.19　TV1、TV2 的引线扫描测试结果

在可靠性 TC 和 HTS 测试中，还要进行电气测试以监测沿着菊花链的电阻。在 TC 测试（-40～125℃）1000 次以上和 1000h HTS 测试（150℃）中，除了 HTS 测试中的两个 TV2 样品，所有的样品都没有发生电气失效（见表 8.7）。两个样品在菊花链上显示为开路。失效分析是通过对 MC 进行剥离，然后对引线键合进行拉力测试和剪切测试。两种试验均表明，引线键合强度较低。在最坏的情况下，球键合从焊盘上脱落，表明发生了典型的 Kirkendall 空洞失效，如图 8.20 所示。发现失败的引线键合发生在顶部芯片的右侧（即子芯片 2）。需要注意的是，在 TV2 样品中，子芯片 2 和子芯片 1 之间有 1mm 的悬垂（见图 8.18b）。然而，TV1 没有任何悬垂问题。虽然 TV2 样品的所有球剪读数和引线拉力读数都超过了最低要求，但可能是由于 1mm 的悬垂造成了零时刻引线键合强度弱。

a) 剥离 MC 后 TV2 样品的失败引线键合处

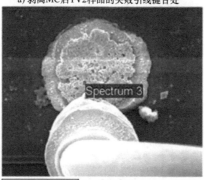

b) 失败的引线键合俯视图

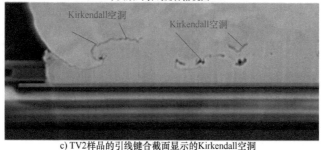

c) TV2 样品的引线键合截面显示的 Kirkendall 空洞

图 8.20　失效分析

8.2.9 总结和建议

一些重要的结果和建议总结如下[17]：

1）目前，世界上 70% 以上的半导体芯片采用引线键合。2015 年以来，铜（包括 PdCu）线材的使用量超过金线材。在过去的几年里，银引线引起了一些关注。

2）采用 PEDL 技术通过降低应力来减少铜低 k 层的分层。铜低 k 层的应力降低了 23.8%。

3）基于切割测试，在崩边结果和芯片强度方面，斜面切削优于单步切削，其中正面崩边减少了 15%，芯片强度提高了 2.3 倍。

4）建立了一种超低引线环（50μm）引线键合工艺。

5）建立了引线嵌入膜工艺，它允许特殊的芯片贴装薄膜（DAF）穿透于金线之间，从而可以去除伪芯片（或间隔芯片）。

6）65nm 铜低 k 堆叠芯片异构集成封装（即 TV1）已经研制完成，并成功通过了 JEDEC 器件级测试，如 1000 次循环（-40 ~ 125℃）的 TC 测试和 1000h 的 HTS 测试（150℃）。

8.3 存储芯片和逻辑芯片的低温键合

本节将介绍使用键合温度低于 200℃ 的 InAg、InCu、InSn、InNi、InSnCu 等低熔焊料进行存储芯片和逻辑芯片的堆叠。

8.3.1 低温键合的工作过程

低温键合的基本原理如图 8.21 所示。在本书中，低温键合[20-54]与诸如瞬

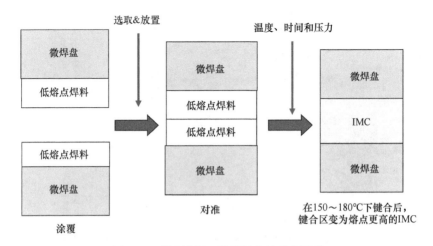

图 8.21　瞬态液相、低温键合的基本原理

态液相（TLP）键合、固液互扩散（SLID）键合和 IMC 键合等有已有清晰定义的键合过程相同。将低温焊料（如 InAg、InCu、InSn、InNi、InAu、InSnAu、InCuNi 和 InSnCu）涂覆在专门设计和制造的凸点下金属化材料（UBM，如 Cu、Cu/Au、Ti/Au、Ni/Au、Ti/Cu、Cu/Ti/Au）上。在低于 200℃ 的温度下对准键合后，所有的键合材料将与芯片 / 晶圆上的 UBM 发生反应并成为熔点比焊料高几百度的 IMC。

8.3.2 低温 SiO_2/Ti/Au/Sn/In/Au 到 SiO_2/Ti/Au 键合

低熔点焊料，如 InAg、InCu、InSn、InNi、InAu、InSnAu、InCuNi 和 InSnCu，其键合温度低于 200℃ [20-54]，降低了由键合芯片的热膨胀失配，即其中一些带有铜填充的硅通孔（TSV）引起的对微结构的损坏（翘曲）。此外，焊点形成的坚固的 IMC 互连，重熔温度高，可以实现稳定的 3D 集成电路芯片堆叠。

需要强调的是，使用基于铟（In）的焊料时（熔点 ≈156℃），选择正确的焊料和 UBM 以形成具有高重熔温度的 IMC 极为重要。由于掺杂其他元素的 In 原子易于形成三元共晶相，其熔点甚至低于 In 二元共晶点 [22]，因此有必要通过设计合适的层组来避免形成三元共晶相。

8.3.3 焊料设计

在 3D 集成中使用 In 与 Ni、Cu 或 Au 进行低温键合 [20-23] 的技术已经有些基础。通常会在 In 层和其他金属层之间将插入一个阻挡层（如 Ti）来防止过度扩散 [24]。

在本节中，选择 In 基焊料层作为低温焊料，并与薄 Au 层进行键合，键合后形成以 AuIn 为基的 IMC。由于 Au 扩散到 In 层中的速度很快，即使在室温下也能形成 AuIn 的 IMC，因此在 In 层和 Au 层之间需要插入一层薄薄的 Sn，以最大限度地减少键合前的相互扩散；也就是说，Sn 与 Au 之间发生反应并在键合之前暂时形成扩散阻挡层，从而能够保留 In 层。另外，该阻挡层不可以在键合期间和键合后中断 In 层和 Au 层之间的界面反应，以保证能够完全形成 IMC。

8.3.4 试验件

用于 3D 集成电路芯片堆叠的试验件如图 8.22 ~ 图 8.24 及表 8.8 所示。硅基测试芯片（8mm × 8mm）上有约 1700 个焊点，每个焊点又由 100μm 直径的 Ti/Au（0.1μm/1μm）UBM 和 80μm 直径的带有 Au（0.05μm）的 Sn/In（0.5μm/2μm）焊料组成。芯片的另一面只有 120μm × 120μm 的 Ti/Au（0.1μm/1μm）UBM。

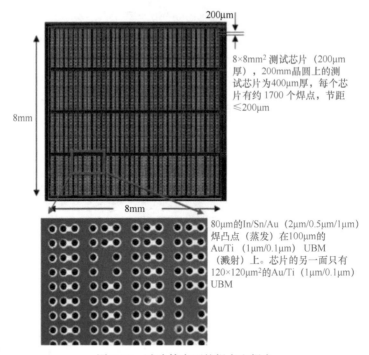

图 8.22　试验件表面的焊点和焊盘

试验件是在 200mm、带有 SiO$_2$ 涂层的硅片上制造的。在晶圆的正面,将 0.1μm 的 Ti 和 1μm 的 Au 溅射到 SiO$_2$ 上形成 120μm × 120μm 的 UBM 图形。然后在晶圆的背面依次进行背面磨至 200μm 厚;涂覆 SiO$_2$ 作为钝化层;沉积 0.1μm 的 Ti 和 1μm 的 Au(在溅射室中);湿法刻蚀,形成 100μm 直径的(Ti/Au) UBM 焊盘图形;用 20μm 的干膜层压;打开焊凸点;在蒸发室中在 Au(UBM)上沉积 Sn(0.5μm),然后在 Sn 上沉积 In(2μm);剥离(剥离工艺)干膜。图 8.23 展示了蒸发复合涂层表面形态的俯视图。可以看出,表面粗糙,呈粒状

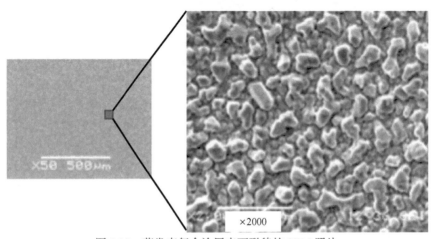

图 8.23　蒸发态复合涂层表面形貌的 SEM 照片

结构，晶粒度为 2~3μm。为了黏合具有这种粗糙特征的表面，需要熔融焊料层和高压来填充表面之间的间隙。

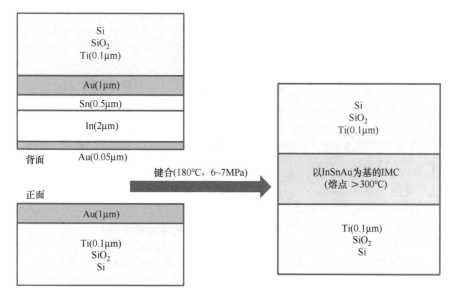

图 8.24 带有 UBM+ 焊凸点的芯片和仅带有 UBM 的芯片（左）、键合后的 IMC 示意图（右）

表 8.8 试验件尺寸

类型	值
晶圆尺寸 /mm	200
芯片尺寸 /mm	8
堆叠芯片厚度 /mm	200
基础芯片厚度 /mm	400
焊凸点直径 /μm	80
焊凸点间距 /mm	≤ 200
焊点数目	~ 1700
UBM（Ti/Au）/μm	0.1/1
铟焊锡凸点 /μm	2.5
总凸点高度 /μm	3.6

8.3.5 采用 InSnAu 低温键合的 3D 集成电路芯片堆叠

图 8.25 显示了采用低温键合的 3D 集成电路芯片堆叠的组装过程。第一层（ASIC）是带 120μm × 120μm UBM 的 200μm 大小晶圆（400μm 厚）。第二层芯片（存储器）带有无铅 InSnAu 焊凸点（背面），在 180℃和 6~7MPa 下，经过 45s

键合在晶圆表面。键合后，无铅焊料互连成为 InSnAu IMC，重熔温度比 In 的熔点高几百度。

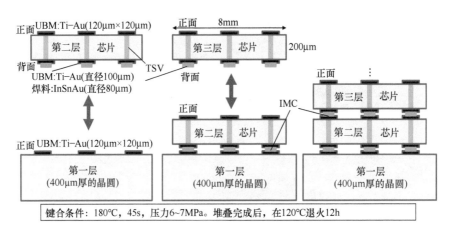

图 8.25　3D 异构集成低温键合

然后在相同条件下将带有焊凸点的第三个芯片（存储器）的背面贴合到第二个芯片的正面。堆叠芯片后，接下来在 120℃下退火 12h。图 8.26 显示了经过目前的低温键合工艺组装的三层芯片堆叠。从横截面看到黏合状态非常好，没有可见的空隙。通过这样的实验设计，得到了目前最佳的键合条件[26]。

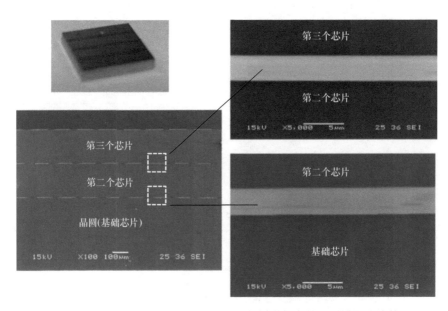

图 8.26　在每个互连层上带有 1700 个焊料微焊点的三层堆叠芯片的装配单元和 SEM 横截面图

8.3.6 InSnAu IMC 的 SEM、TEM、XDR 和 DSC

由于 IMC 层相对较薄，因此使用能提供局部化学分析的 TEM（透射电子显微镜）技术。图 8.27 显示了靠近基材硅的区域以及 Au/Ti 和焊料中间层之间的界面反应区域的 TEM 图像。每层之间的界面是无空隙且均匀的，并且完全转换为 IMC。所选点的元素组成结果（at.%）如图 8.27 所示。沿主反应区确定了两种 IMC，即 InAu 和 InSnAu。该结果与 XRD（X 射线衍射）结果一致，证实了这两种 IMC 在键合界面中的存在。图 8.28 显示了 DSC（差示扫描量热法）曲线，表明重熔温度高于 365℃。SEM 图像显示发生了均匀的黏合，没有任何缺陷，如图 8.29 所示。

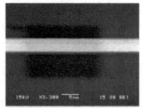
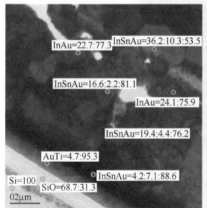
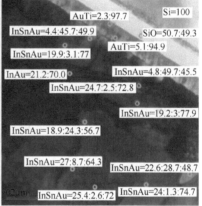

图 8.27　键合界面上的 SEM（上）和 TEM（下）图像

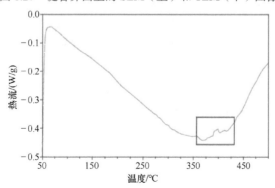

图 8.28　DSC 曲线显示熔点约为 360℃

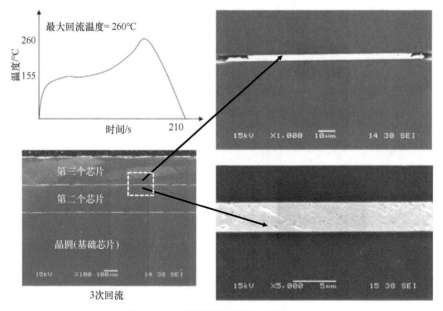

图 8.29　3 次回流的最高温度为 260℃

8.3.7　InSnAu IMC 的杨氏模量和硬度

这里采用纳米压痕法来确定基于 AuSnIn 的 IMC 连接的弹性模量和硬度。由于样品横截面的 IMC 层尺寸对于光学显微镜来说太薄了，因此需要在芯片的剪切测试后沿断裂表面进行测量。可以发现，InSnAu 的杨氏模量约为 81GPa，略高于 Au（78 GPa）[55]，远低于 Cu_6Sn_5（112.6 GPa）和 Cu_3Sn（132.7 GPa）[56]。同时测得的硬度约为 1.5GPa。

8.3.8　InSnAu IMC 的 3 次回流

目前至少需要两种不同的低温组装方法，一种是用于封装中的芯片级互连，另一种是板级互连。对于板级互连来说，由于它是最后进行的组装，因此只要焊点可靠，可用低熔点焊料将部件连接到板上即可。但是对于封装中的芯片级互连，由于必须使用无铅焊料完成，因此在安装到 PCB 上之前，必须对封装进行系列鉴定检验。以下为其中的一些鉴定检验：

1）温度循环测试（-40℃ ↔125℃循环 1000 次），其中封装应经过潮湿环境暴露和至少 3 次无铅 SMT 回流循环进行预处理。

2）加偏置的 HAST（高加速应力测试）（130℃ /85% 相对湿度，1.8V 持续 100h）。

3）高压加热（蒸汽）测试（121℃，100% 相对湿度，在 2 个大气压条件下持续 168h）。

一般来说，芯片级互连对于质量和可靠性的要求往往会比板级互连更严格；也就是说，适用于板级互连的低温焊料可能并不适用于芯片级互连。图 8.29 显示了符合条件的封装的典型回流曲线，同时还显示了 IMC 互连在不同放大倍数下的横截面。可以看出，在 3 次回流之后，3D 堆叠 IC 芯片的 InSnAu IMC 互连没有明显的变化。

8.3.9 InSnAu IMC 的剪切强度

由于 InSnAu IMC 互连非常坚固并且抗剪切强度高，许多 200μm 芯片在剪切（水平方向推动）测试时被损坏了，因此为了确定 InSnAu IMC 互连的剪切强度，往往使用 400μm 芯片进行测试。图 8.30 显示了 InSnAu IMC 互连在剪切测试下的典型失效模式。可以看到，断裂是沿着 IMC 接头以及 TiAu UBM 和 InSnAu IMC 接头之间的界面发生的。

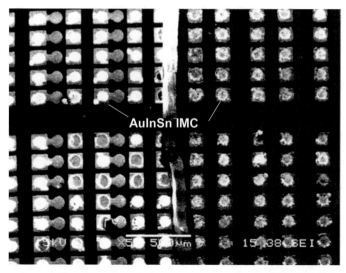

图 8.30　沿 IMC 接头及 UBM 与 IMC 接头之间的界面出现的剪切断口

图 8.31 则显示了 InSnAu IMC 互连在出厂时和经过 3 次回流后的剪切强度测试结果。可以看出，由于 3 次回流，第一个 IMC 接头（在基础芯片和第二个芯片之间）和第二个 IMC 接头（在第二个芯片和第三个芯片之间）的剪切强度降低了大约 15%；虽然 IMC 接头的剪切强度降低了，但仍远高于要求的 20MPa。对于这两种情况，第一个 IMC 接头的剪切强度略（约 8%）高于第二个 IMC 接头。这可能是由于剪切测试设置引起的，即首次剪切（水平推动）第三个芯片并保持底部的两个芯片不动以获得第二个 IMC 强度，之后推动第二个芯片并保持基础芯片不动来获得第一个 IMC 强度。由于支架也不是 100% 刚性，因此某些推力会被缓冲。同时如果支架越大，推力被缓冲的程度也会越大。

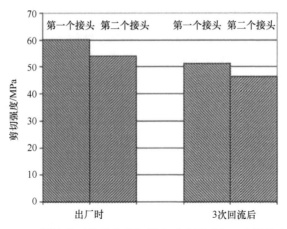

图 8.31　3 层堆叠下的黏合剪切强度（出厂时和 3 次回流之后）

8.3.10　InSnAu IMC 的电阻

InSnAu IMC 互连的电阻可以通过 Kelvin 的四点测量方法来测量，如图 8.32 所示。两个直径为 200μm 的凸点彼此相距很远（800μm）。图 8.33 则显示了测量结果，可以看出 InSnAu IMC 接头的电阻约为 0.12Ω，即使在 260℃下经过 3 次回流后也没有减少。

8.3.11　InSnAu IMC 不稳定分析

DSC 法已经表明，InSnAu IMC 互连的重熔温度大于 365℃。那么 3D 堆叠集成芯片的 IMC 互连失效时的温度到底会是多少呢？由于 DSC 设备中的铂坩

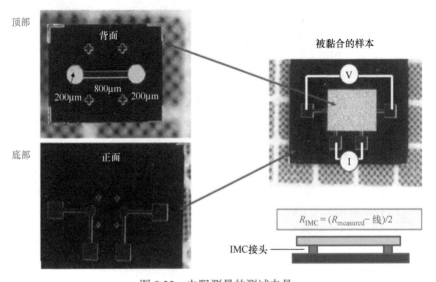

图 8.32　电阻测量的测试夹具

坩尺寸有限制，无法放入整个 3D 堆叠集成芯片进行测量。取而代之的是一种间接方法，即采用 TMA（热机械分析仪）来测量整个 3D 堆叠集成芯片的膨胀与温度的关系，如图 8.34 所示。在 InSnAu IMC 互连开始重熔之前，曲线应该随着温度升高而升高（非常接近直线）。然而，当 InSnAu IMC 互连开始重熔时，膨胀曲线开始变得不稳定，直到出现下降，这意味着 3D 堆叠集成芯片完全失效了。因此，根据在 TMA 测试的 3 个样品，3D 堆叠 IC 芯片中的 IMC 会在 380～400℃左右开始重熔（测试结果会高于 365℃，这是因为整个 3D 堆叠集成芯片在测量中已吸收了更多的热量），并在 450～480℃左右失效。

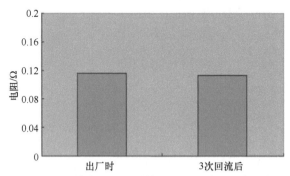

图 8.33　IMC 接头电阻测量结果（出厂时和 3 次回流后）

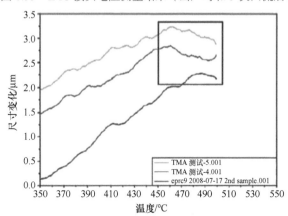

图 8.34　TMA 的测试结果（重熔发生在 400℃左右）

8.3.12　总结和建议

一些重要的结果和建议总结如下[26, 27]：

1) 采用低温焊料键合的内存芯片堆叠的优势是堆叠较高的芯片时，较低的芯片的键合不会重熔，同时它们也不会被移动。

2) 低温焊料体系（InSnAu）专为芯片堆叠的 3D 异构集成而设计，黏合条件为 180℃和 6MPa，持续 45s。黏合之后，整个组装再在 120℃条件下退火 12h。

3）SEM、XDR 和 TEM 图像显示已实现均匀键合，没有任何缺陷，所有键合区域均已转化为 InSnAu 和 InAu IMC。此外，DSC 结果表明 IMC 的重熔温度大于 365℃。

4）3 次回流后（最高温度为 260℃），3D 堆叠 IC 芯片的 InSnAu IMC 没有发生可见的变化，IMC 的接触电阻也没有变化。

5）对于 3D 堆叠 IC 芯片来说，由于芯片可以薄至 20μm，这使得普通的表征方法非常困难甚至完全不可能适用，因此迫切需要新的方法和设备。

参考文献

[1] Tuckerman, D. B., Bauer, L. O., Brathwaite, N. E., Demmin, J., Flatow, K., Hsu, R., Kim, P., Lin, C. M., Lin, K., Nguyen, S., and V. Thipphavong, "Laminated Memory: A New Three-Dimensional Packaging Technology for MCMs", *IEEE Multi-Chip Module Conference*, Santa Cruz, CA, March 15–17, 1994, pp. 58–63.

[2] Lau, J. H., "Recent Advances and New Trends in Flip Chip Technology", *ASME Transactions, Journal of Electronic Packaging*, September 2016, Vol. 138, Issue 3, pp. 1–23.

[3] Matsumura, K., Tomikawa, M., Sakabe, Y., and Shiba, Y., "New Non Conductive Film for High Productivity Process", *IEEE CPMT Symposium Japan (ICSJ)*, Kyoto, Japan, Nov. 9–11, 2015, pp. 19–20.

[4] Asahi, N., Miyamoto, Y., Nimura, M., Mizutani, Y., and Arai, Y., "High Productivity Thermal Compression Bonding for 3D-IC", *IEEE International 3D Systems Integration Conference (3DIC)*, Sendai, Japan, August 31–September 2, 2015, pp. TS8.3.1–TS8.3.5.

[5] van Driel, W. D., G. Wisse, A. Y. L. Chang, J. H. J. Jansen, X. Fan, G. Q. Zhang, and L. J. Ernst, "Influence of Material Combinations on Delamination Failures in a Cavity-Down TBGA Package", *IEEE Transactions of Components and Packaging Technology*, Vol. 27, No. 4, December 2004, pp. 651–658.

[6] Harman, G. G., and C. E. Johnson, "Wire Bonding to Advanced Copper, Low-k Integrated Circuits, the Metal/Dielectric Stacks, and Materials Considerations", *IEEE Transactions of Components and Packaging Technology*, Vol. 25, No. 4, December 2002, pp. 677–683.

[7] Zhang, J., and J. Huneke, "Stress Analysis of Spacer Paste Replacing Dummy Die in a Stacked CSP Package", *Proceedings of IEEE/ICEPT*, 2003, pp. 82–85.

[8] Yoon, S. W., D. Wirtasa, S. Lim, V. Ganesh, A. Viswanath, V. Kripesh, and M. K. Iyer, "PEDL Technology for Copper/Low-k Dielectrics Interconnect", *IEEE Proceedings of EPTC*, Decemeber 2005, pp. 711–715.

[9] Hartfield, C. D., E. T. Ogawa, Y.-J. Park, T.-C. Chiu, and H. Guo, "Interface Reliability Assessments for Copper/Low-k Products," *IEEE Transactions of Device Materials Reliability*, Vol. 4, No. 2, June 2004, pp. 129–141.

[10] Alers, G. B., K. Jow, R. Shaviv, G. Kooi, and G. W. Ray, "Interlevel Dielectric Failures in Copper/Low-k Structures", *IEEE Transactions of Device Materials Reliability*, Vol. 4, No. 2, June 2004, pp. 148–152.

[11] Chiu, C. C., H. H. Chang, C. C. Lee, C. C. Hsia, and K. N. Chiang, "Reliability of Interfacial Adhesion in a multilevel copper/low-k interconnect structure", *Microelectronic Reliability*, Vol. 47, Nos. 9–11, 2007, pp. 1506–1511.

[12] Wang, Z., J. H. Wang, S. Lee, S. Y. Yao, R. Han, and Y. Q. Su, "300-mm Low-k Wafer Dicing Saw Development", *IEEE Transactions of Electronic Packaging & Manufacturing*, Vol. 30, No. 4, October 2007, pp. 313–319.

[13] Chungpaiboonpatana, S., and F. G. Shi, "Packaging of Copper/Low-k IC Devices: A Novel Direct Fine-Pitch Gold Wire-Bond Ball Interconnects onto Copper/Low-k Terminal Pads", *IEEE Transactions of Advanced Packaging*, Vol. 27, No. 3, August 2004, pp. 476–489.

[14] Wang, T. H., Y.-S. Lai, and M.-J. Wang, "Underfill Selection for Reducing Cu/Low-k Delamination Risk of Flip-Chip Assembly", *IEEE Proceedings of EPTC*, December 2006, pp. 233–236.

[15] Zhai, C. J., U. Ozkan, A. Dubey, S. Sidharth, R. Blish II, and R. Master, "Investigation of

Cu/Low-*k* Film Delamination in Flip Chip Packages", *IEEE/ECTC Proceedings*, June 2006, pp. 709–718.

[16] Ong, J. M. G., A. A. O. Tay, X. Zhang, V. Kripesh, Y. K. Lim, D. Yeo, K. C. Chan, J. B. Tan, L. C. Hsia, and D. K. Sohn, "Optimization of the Thermo-Mechanical Reliability of a 65-nm Cu/Low-*k* Large-Die Flip Chip Package", *IEEE Transactions of Components and Packaging Technology*, Vol. 32, No. 4, December 2009, pp. 838–848.

[17] Zhang, X., J. H. Lau, C. Premachandran, S. Chong, L. Wai, V. Lee, T. Chai, V. Kripesh, V. Sekhar, D. Pinjala, and F. Che, "Development of a Cu/Low-k Stack Die Fine Pitch Ball Grid Array (FBGA) Package for System in Package Applications", *IEEE Transactions on CPMT*, Vol. 1, No. 3, March 2011, pp. 299–309.

[18] Chen, S., C. Z. Tsai, E. Wu, I. G. Shih, and Y. N. Chen, "Study on the Effects of Wafer Thinning and Dicing on Chip Strength", *IEEE Transactions of Advanced Packaging*, Vol. 29, No. 1, February 2006, pp. 149–158.

[19] Toh, C., M. Gaurav, H. Tan, and P. Ong, "Die Attached Adhesives for 3D Same-Size Dies Stacked Packages", *IEEE/ECTC Proceedings*, May 2008, pp. 1538–1543.

[20] Fukushima, T., Y. Yamada, H. Kikuchi, T. Tanaka, and M. Koyanagi, "Self-Assembly Process for Chip-to-Wafer Three-Dimensional Integration", *IEEE/ECTC Proceedings*, Reno, NV, 2007, pp. 836–841.

[21] Sakuma, K., P. Andry, C. Tsang, K. Sueoka, Y. Oyama, C. Patcl, B. Dang, S. Wright, B. Webb, E. Sprogis, R. Polastre, R. Horton, and J. Knickerbocker, "Characterization of Stacked Die using Die-to-Wafer Integration for High Yield and Throughput", *IEEE/ECTC Proceedings*, Lake Beuna Vista, FL, May 2008, pp. 18–23.

[22] Wakiyama, S., H. Ozaki, Y. Nabe, T. Kume, T. Ezaki, and T. Ogawa, "Novel Low-Temperature CoC Interconnection Technology for Multichip LSI (MCL)", *IEEE/ECTC Proceedings*, Reno, NV, 2007, pp. 610–615.

[23] Liu, Y. M., and T. H. Chuang, "Interfacial Reactions Between Liquid Indium and Au-Deposited Substrates", *Journal of Electronic Materials*, Vol. 29, No. 4, 2000, pp. 405–410.

[24] Zhang, W., A. Matin, E. Beyne, and W. Ruythooren, "Optimizing Au and In Micro-Bumping for 3D Chip Stacking", *IEEE/ECTC Proceedings*, Lake Beuna Vista, FL, May 2008, pp. 1984–1989.

[25] Liu, X., H. Liu, I. Ohnuma, R. Kainuma, K. Ishida, S. Itabashi, K. Kameda, and K. Yamaguchi, "Experimental determination and thermodynamic calculation of the phase equilibria in the Cu-In-Sn system", *Journal of Electronic Materials*, Vol. 30, No. 9, 2001, pp. 1093–1103.

[26] Choi, W., C. Premachandran, C. Ong, L. Xie, E. Liao, A. Khairyanto, B. Ratmin, K. Chen, P. Thaw, and J. H. Lau, "Development of Novel Intermetallic Joints Using Thin Film Indium Based Solder by Low Temperature Bonding Technology for 3D IC Stacking", *IEEE/ECTC Proceedings*, San Diego, CA, May 2009, pp. 333–338.

[27] Choi, W., D. Yu, C. Lee, L. Yan, A. Yu, S. Yoon, J. H. Lau, M. Cho, Y. Jo, and H. Lee, "Development of Low Temperature Bonding Using In-based Solders", *IEEE/ECTC Proceedings*, Orlando, FL, May 2008, pp. 1294–1299.

[28] Sommadossi, S., and A. F. Guillermet, "Interface Reaction Systematic in the Cu/In-48Sn/Cu System bonded by diffusion soldering", *Intermetallics*, Vol. 15, 2007, pp. 912–918.

[29] Shigetou, A., T. Itoh, K. Sawada, and T. Suga, "Bumpless Interconnect of 6-mm Pitch Cu Electrodes at Room Temperature", *IEEE/ECTC Proceedings*, Lake Buena Vista, FL, May 27–30, 2008, pp. 1405–1409.

[30] Made, R., C. L. Gan, L. Yan, A. Yu, S. U. Yoon, J. H. Lau, and C. Lee, "Study of Low Temperature Thermocompression Bonding in Ag-In Solder for Packaging Applications", *Journal of Electronic Materials*, Vol. 38, 2009, pp. 365–371.

[31] Yan, L.-L., C.-K. Lee, D.-Q. Yu, A.-B. Yu, W.-K. Choi, J. H. Lau, and S.-U. Yoon, "A Hermetic Seal Using Composite Thin Solder In/Sn as Intermediate Layer and Its Interdiffusion Reaction with Cu", *Journal of Electronic Materials*, Vol. 38, 2009, pp. 200–208.

[32] Yan, L.-L., V. Lee, D. Yu, W. K. Choi, A. Yu, A.-U. Yoon, and J. H. Lau, "A Hermetic Chip to Chip Bonding at Low Temperature with Cu/In/Sn/Cu Joint", *IEEE/ECTC Proceedings*, Orlando, FL, May 2008, pp. 1844–1848.

[33] Yu, A., C. Lee, L. Yan, R. Made, C. Gan, Q. Zhang, S. Yoon, and J. H. Lau, "Development of Wafer Level Packaged Scanning Micromirrors", *Proc. Photon. West*, Vol. 6887, 2008, pp. 1–9.

[34] Lee, C., A. Yu, L. Yan, H. Wang, J. Han, Q. Zhang, and J. H. Lau, "Characterization of Intermediate In/Ag Layers of Low Temperature Fluxless Solder Based Wafer Bonding for

MEMS Packaging", *Journal of Sensors and Actuators* (in press).
[35] Yu, D.-Q., C. Lee, L. L. Yan, W. K. Choi, A. Yu, and J. H. Lau, "The Role of Ni Buffer Layer on High Yield Low Temperature Hermetic Wafer Bonding Using In/Sn/Cu Metallization", *Applied Physics Letters* (in press).
[36] Yu, D. Q., L. L. Yan, C. Lee, W. K. Choi, S. U. Yoon, and J. H. Lau, "Study on High Yield Wafer to Wafer Bonding Using In/Sn and Cu Metallization", *Proceedings of the Eurosensors Conference*, Dresden, Germany, 2008, pp. 1242–1245.
[37] Yu, D., L. Yan, C. Lee, W. Choi, M. Thew, C. Foo, and J. H. Lau, "Wafer Level Hermetic Bonding Using Sn/In and Cu/Ti/Au Metallization", *IEEE Proceeding of Electronics Packaging and Technology Conference*, Singapore, December 2008, pp. 1–6.
[38] Chen, K., C. Premachandran, K. Choi, C. Ong, X. Ling, A. Khairyanto, B. Ratmin, P. Myo, and J. H. Lau, "C2W Bonding Method for MEMS Applications", *IEEE Proceedings of Electronics Packaging Technology Conference*, Singapore, December 2008, pp. 1283–1288.
[39] Premachandran, C. S., J. H. Lau, X. Ling, A. Khairyanto, K. Chen, and Myo Ei Pa, "A Novel, Wafer-Level Stacking Method for Low-Chip Yield and Non-Uniform, Chip-Size Wafers for MEMS and 3D SIP Applications", *IEEE/ECTC Proceedings*, Orlando, FL, May 27–30, 2008, pp. 314–318.
[40] Simic, V., and Z. Marinkovic, "Room Temperature Interactions in Ag-Metals Thin Film Couples", *Thin Solid Films*, Vol. 61, 1979, pp. 149–160.
[41] Lin, J.-C., "Solid-Liquid Interdiffusion Bonding Between In-Coated Silver Thick Films", *Thin Solid Films*, Vol. 61, 1979, pp. 212–221.
[42] Roy, R., and S. K. Sen, "The Kinetics of Formation of Intermetallics in Ag/In Thin Film Couples", *Thin Solid Films*, Vol. 61, 1979, pp. 303–318.
[43] Chuang, R. W., and C. C. Lee, "Silver-Indium Joints Produced at Low Temperature for High-Temperature Devices", *IEEE Transactions on Components and Packaging Technologies*, Vol. 25, 2002, pp. 453–458.
[44] Chuang, R. W., and C. C. Lee, "High-Temperature Non-Eutectic Indium-Tin Joints Fabricated by a Fluxless Process", *Thin Solid Films*, Vol. 414, 2002, pp. 175–179.
[45] Lee, C. C., and R. W. Chuang, "Fluxless Non-Eutectic Joints Fabricated Using Gold-Tin Multilayer Composite", *IEEE Transactions on Components and Packaging Technologies*, Vol. 26, 2003, pp. 416–422.
[46] Humpston, G., and D. M. Jacobson, *Principles of Soldering*, ASM International, Materials Park, MD, 2004.
[47] Chuang, T., H. Lin, and C. Tsao, "Intermetallic Compounds Formed During Diffusion Soldering of Au/Cu/Al$_2$O$_3$ and Cu/Ti/Si with Sn/In Interlayer", *Journal of Electronic Materials*, Vol. 35, 2006, pp. 1566–1570.
[48] Lee, C., and S. Choe, "Fluxless In-Sn Bonding Process at 140 °C", *Materials Science and Engineering*, Vol. A333, 2002, pp. 45–50.
[49] Lee, C. "Wafer Bonding by Low-Temperature Soldering," *Sensors & Actuators*, Vol. 85, 2000, pp. 330–334.
[50] Vianco, P. T., "Intermetallic Compound Layer Formation Between Copper and Hot-Dipped 100 In, 50In–50Sn, 100Sn, and 63Sn–37Pb Coatings", *Journal of Electronic Materials*, Vol. 23, 1994, pp. 583–594.
[51] Frear, D. R., "Intermetallic Growth and Mechanical-Behavior of Low and High Melting Temperature Solder Alloys", *Metallurgical and Materials Transactions*, Vol. 25, 1994, pp. 1509–1523.
[52] Morris, J. W., "Microstructure and Mechanical Property of Sn-In and Sn-Bi Solders", *Journal of the Minerals Metals & Materials Society (JOM)*, Vol. 45, 1993, pp. 25–28.
[53] Mei, Z., "Superplastic Creep of Low Melting-Point Solder Joints," *Journal of Electronic Materials*, Vol. 21, 1992, pp. 401–408.
[54] Chuang, T. H., "Phase Identification and Growth Kinetics of the Intermetallic Compounds Formed During In-49Sn/Cu Soldering Reactions", *Journal of Electronic Materials*, Vol. 31, 2002, pp. 640–645.
[55] http://www.webelements.com/WebElements: The periodic table on the website.
[56] Kazumasa, T., "Micro Cu Bump Interconnection on 3D Chip Stacking Technology", *Japanese Journal of Applied Physics*, Vol. 43, No. 4B, 2004, pp. 2264–2270.

第 9 章

芯片到芯片堆叠的异构集成

9.1 引言

正如英特尔公司[1]不久之前指出的,解决亿万级计算的内存带宽挑战的关键是拥有 3D 芯片到芯片和面对面堆叠 MCP(多芯片封装)技术,如图 9.1 所示。顶层芯片是存储器,底层芯片是逻辑芯片或 CPU(中央处理器)。本章将介绍芯片到芯片和面对面异构集成的两个示例,一种是在底层芯片中使用 TSV 来导通信号、电源和接地;另一种是没有 TSV 结构,但会在较大的芯片上使用焊凸点。

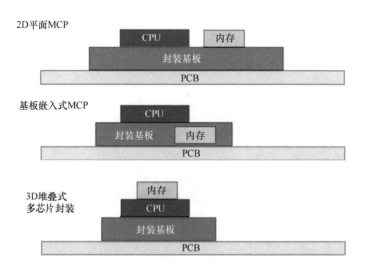

图 9.1 英特尔的封装技术解决了亿万级计算内存带宽的挑战

9.2 带有 TSV 的芯片到芯片的异构集成

图 9.2 展示了封装结构的示意图，包含顶层硅芯片（可以是存储器）和底层硅芯片（可以是 CPU）。在顶层芯片上制作铜柱与焊帽；而在底层芯片上有填充铜的 TSV，还有在 TSV 顶部采用 ENIG（沉镍浸金）方法制成的 UBM 焊盘。通过连接微焊点和 ENIG 焊盘，能够将顶层芯片和底层芯片堆叠在一起。本研究的重点是展示 TSV 铜填充的制造工艺，并开发芯片到芯片的组装工艺。

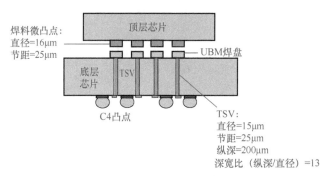

图 9.2 芯片到芯片与 TSV 的异构集成示意图

9.2.1 底层芯片的 TSV 和 UBM 焊盘设计

图 9.3 展示了底层芯片的横截面图，有 TSV 和 UBM 焊盘的信息。TSV 的设计直径为 15μm，TSV 上方是 UBM 焊盘。在本节中，ENIG 即为 UBM 层。对于间距极小的 UBM 焊盘，普通的 UBM 金属薄膜（如 TiCuNiAu 或 AlNiV-Cu）很难用传统的光刻和湿法刻蚀工艺进行刻蚀，因为湿法刻蚀工艺造成的底部切口会损坏较小的 UBM 焊盘。而 ENIG 的形成不需要任何高度真空或光刻设备在键合焊盘顶部形成金属叠层，因此它是一种能够用于高密度和高引脚数积层式集成电路封装的简单且低成本的解决方案。

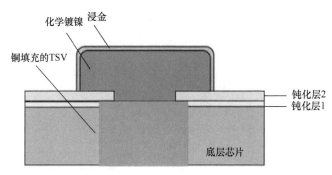

图 9.3 底层芯片上的 UBM 焊盘（ENIG）横截面示意图

9.2.2 顶层芯片的焊料微凸点设计

图9.4展示了顶层芯片的横截面图,其中包含焊料微凸点的详细结构。该焊料微凸点由带有锡焊帽的柱形铜凸点组成。柱形铜凸点和锡焊帽的总厚度为10μm,采用电镀工艺制造。为了确保有足够的锡焊料以实现可靠的连接,同时避免侧壁润湿,锡帽层厚度设计为$3\sim4\mu m$,柱形铜凸点厚度为$6\sim7\mu m$[2, 3]。

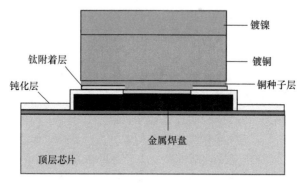

图9.4 顶层芯片上的柱形铜凸点和锡焊帽的横截面示意图

9.2.3 TSV制造

高深宽比的TSV采用BOSCH工艺[4]在基于电感耦合等离子体的深反应离子刻蚀(DRIE)系统中刻蚀。DRIE工艺中有两个工艺周期交替运行,即硅刻蚀周期和侧壁钝化周期。硅刻蚀周期中硅被各向同性地刻蚀,而钝化周期中TSV的侧壁受到聚合物薄层的保护。TSV制造和铜填充的工艺流程如图9.5所示。

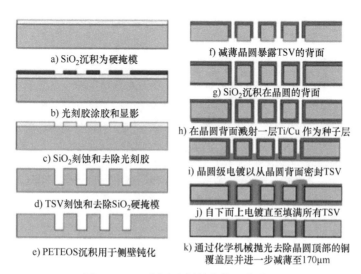

图9.5 TSV制造和铜填充的工艺流程

如图 9.5a 所示，首先在晶圆上沉积一层 SiO_2 作为硬掩模，然后旋涂一层薄的光刻胶并进行显影，如图 9.5b 所示。图 9.5c 为刻蚀 SiO_2 层并剥离光刻胶，图 9.5d 为 TSV 被刻蚀至所需的深度，并且在刻蚀后去除 SiO_2 硬掩模。之后一层等离子体增强的原硅酸四乙酯（PETEOS）沉积在晶圆的正面以进行侧壁钝化，如图 9.5e 所示。对于无器件晶圆，也可以通过热氧化实现侧壁钝化，这样可以提供更好的侧壁均匀性和平滑度。然后减薄晶圆暴露 TSV 的背面，如图 9.5f 所示。为了完全钝化 TSV 的侧壁，在 Plasma Therm 790 系列等离子增强化学气相沉积（PECVD）/反应离子刻蚀（RIE）系统中将另一层 SiO_2 沉积在晶圆的背面，如图 9.5g 所示。

然后在 Balzer LLS 801 溅射系统中将厚度为 $0.3\mu m$ 的 Ti 黏附层和厚度为 $2\mu m$ 的铜种子层溅射到晶圆的背面，如图 9.5h 所示，厚的黏附层和种子层可以确保覆盖 TSV 的侧壁。在图 9.5g、9.5h 两个步骤中，晶圆都被夹持在机器中，不需要真空吸盘。如果这两个步骤使用带真空吸盘的工具进行处理，如住友精密工业株式会社的低温 PECVD 沉积系统和昆山探戈半导体设备有限公司的物理气相沉积系统，则需要临时载体[5]。图 9.5i 表示从晶圆背面用铜电镀工艺密封 TSV。在这个过程中，由于通孔的直径只有 $15\mu m$，因此很容易用高电镀电流（本研究中为 1A）从底部密封通孔。之后进行自下而上的电镀，直到所有的 TSV 都被填满，如图 9.5j 所示。为了减小 TSV 中的空隙，电镀电流随后降低到 0.1A。

这种自下而上的电镀步骤可以在没有支撑晶圆的情况下进行处理。因此，自下而上电镀后不需要移除支撑晶圆，避免了高难度的移除导致晶圆断裂，如参考文献 [6] 的测试。自下而上电镀后，通过化学机械抛光（CMP）去除晶圆顶部的铜覆盖层。为了去除 TSV 中的空隙，晶圆进一步减薄至 $30\mu m$，如图 9.5k 所示。在 CMP 之后可以对晶圆进行铜退火处理，如果需要还可以进行另一个 CMP 步骤以去除退火后的铜凸起[7]，本研究中跳过了这些步骤。

9.2.4 底层芯片 ENIG UBM 焊盘的制造

在 TSV 的顶部制造 ENIG UBM 焊盘，图 9.6 显示了该工艺流程。首先在晶圆顶部沉积一层电介质，如图 9.6a 所示，然后旋涂一层薄的光刻胶并进行显影，如图 9.6b 所示，之后刻蚀介电层暴露 TSV 并去除光刻胶，如图 9.6c 所示，最后通过化学镀镍和磷、沉金形成 UBM 焊盘，如图 9.6d 所示。

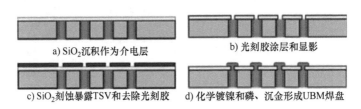

a) SiO_2 沉积作为介电层　　　　b) 光刻胶涂层和显影

c) SiO_2 刻蚀暴露 TSV 和去除光刻胶　　d) 化学镀镍和磷、沉金形成 UBM 焊盘

图 9.6　TSV 顶部进行 ENIG 电镀的工艺流程

9.2.5 顶层芯片铜柱和锡焊帽的制造

图 9.7 展示了在芯片顶部制造铜柱和锡焊帽的工艺流程。首先在晶圆上沉积一层 SiO_2 钝化层和一层铝膜金属化层，如图 9.7a 所示，然后旋涂一层 2μm 厚的光刻胶并进行显影，如图 9.7b 所示，之后刻蚀铝膜形成金属焊盘并去除光刻胶，如图 9.7c 所示，图 9.7d～f 显示沉积另一个钝化层并进行显影，然后溅射黏附/种子层的钛/铜，如图 9.7g 所示。为了在表面形成电镀掩模，需要涂上一层 10μm 厚的光刻胶并进行显影，如图 9.7h 所示，然后依次电镀铜/锡，如图 9.7i 所示，图 9.7j 显示去除光刻胶，且钛/铜黏附/种子层被回蚀回去。完成电镀后需要进行回流，重塑焊料微凸点，实现均匀的凸点高度。

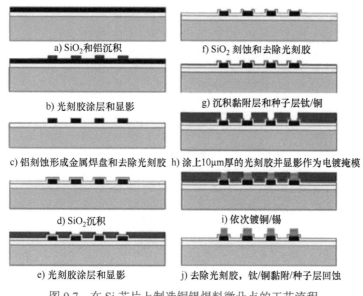

图 9.7 在 Si 芯片上制造铜锡焊料微凸点的工艺流程

9.2.6 TSV 的 DRIE

在 DRIE 过程中，许多变量会影响 TSV 的高深宽比，如 SF_6/C_4F_8 的流速、刻蚀/钝化活性时间、电极功率和自动压力阀的方向等[8, 9]。研究发现，在这些参数中刻蚀/钝化循环时间的比值（R_t）是最关键的参数。为了评估 R_t 的影响，图 9.8 设计了一个具有方形通孔的试验件。需要指出的是，由于与其他结构共享掩模，设计的方形通孔宽度为 10μm，节距为 25μm。图 9.9a、b 分别显示了 $R_t = 1.8$ 时刻蚀的 TSV 的横截面，以及上端的放大图像。侧壁宽度小于 7μm（原设计值为 15μm），最窄的宽度（w）只有约 4μm。在这种刻蚀条件下的底部切口约为 2.5μm，并且在 TSV 的上端出现了一个较大的弓形。图 9.9c、d 分别显

示了 $R_t = 1$ 时刻蚀的 TSV 的横截面和上端的放大图像。整体侧壁是相当平直的，最窄的宽度为 14μm。底部切口小于 0.3μm，并且没有出现明显的弓形。为了研究 R_t 对刻蚀后 TSV 侧壁的影响，本研究进行了一系列的实验。图 9.10 说明了 R_t 对 R_w 的影响（R_w 为最窄侧壁宽度和预期侧壁宽度之间的比率）。在这些实验中，其他刻蚀参数保持不变，见表 9.1。从图 9.10 可知，较大的 R_t（较长的刻蚀循环时间或较短的钝化循环时间）会导致侧壁变薄；对于特定的 R_t，侧壁在 TSV 的底部变得更宽。结果就是 TSV 会变成带状形状。

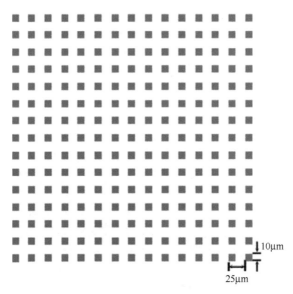

图 9.8　用于评估 TSV 刻蚀工艺的试验件布局

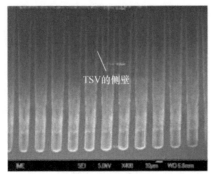

a) $R_t=1.8$ 时刻蚀的TSV的横截面，侧壁变得比7μm更窄(设计值为15μm)

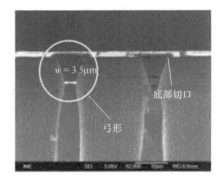

b) $R_t=1.8$ 时TSV的顶部形状

图 9.9　刻蚀循环和钝化循环时间的比值 R_t 对 TSV 底部切口和形状的影响

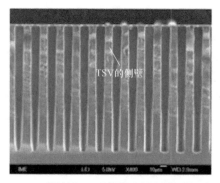
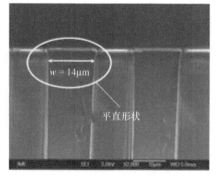

c) $R_t=1$时蚀刻的TSV的横截面,侧壁是平直的   d) $R_t=1$时TSV的顶部形状

图 9.9　刻蚀循环和钝化循环时间的比值 R_t 对 TSV 底部切口和形状的影响（续）

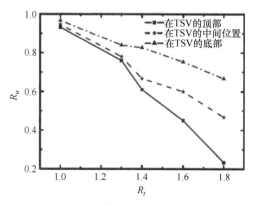

图 9.10　TSV 不同位置的 R_w 和 R_t 的关系

表 9.1　TSV 的 DRIE 工艺参数

参数		值
平板功率 /W		12
功率提升速率 /（W/min）		0.1
刻蚀循环中的气体流量 /sccm	SF_6	130
	O_2	10
钝化循环中的气体流量 /sccm	C_4F_8	100

上述实验结果可以解释如下：在硅刻蚀循环中，硅被各向同性地刻蚀，如图 9.11a 所示；而在钝化循环中，侧壁被一薄层聚合物保护，如图 9.11b 所示。在接下来的刻蚀步骤中，首先通孔底部的聚合物会被刻蚀掉，然后底部的硅可以被刻蚀下来，如图 9.11c 所示。刻蚀和钝化循环交替进行，在刻蚀特征的侧壁上形成扇形缺口。如果 R_t 适当优化，可以实现如图 9.11d 所示的平直侧壁。然而，如果 R_t 太大，侧壁上的聚合物不够厚，则无法在刻蚀周期内保护侧壁。因

此，扇形缺口变得更大，如图 9.11e 所示。此外，由于硅刻蚀各向同性，在 SiO_2 硬掩膜下面的区域出现了底部缺口，并随着刻蚀时间的增加而增加。特别是对于高深宽比的通孔刻蚀，到达通孔入口的离子通量比到达通孔底部的离子通量要大。因此，通孔入口处的硅更容易被各向同性地刻蚀。结果是随着刻蚀过程的进行，通孔入口处的弓形变得更大，如图 9.11f 所示。为了减小弓形，应该减小 R_t，这意味着减小刻蚀周期或增加钝化周期，特别是对于高深宽比通孔来说，为了在硅刻蚀周期中保护侧壁，应确保侧壁上的钝化层足够厚。

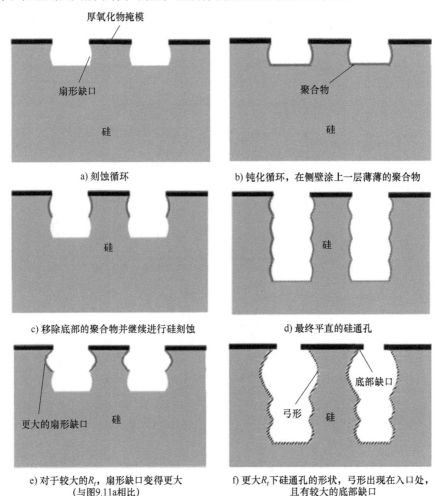

图 9.11 TSV DRIE 工艺图示

9.2.7 侧壁的钝化

在 TSV 形成之后，厚度为 2μm 的 PETEOS 在晶圆的正面沉积下来，形成一个绝缘层。图 9.12 展示了钝化的结果。可以发现，在 TSV 的顶部和中间的侧壁，阶梯覆盖分别为 0.6μm 和 0.2μm。在 TSV 的底部，PETEOS 的厚度为 40nm。

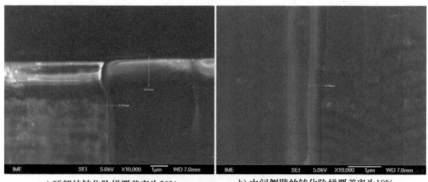

a) 顶部的钝化阶梯覆盖率为30%　　b) 中间侧壁的钝化阶梯覆盖率为10%

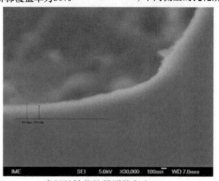

c) 底部的钝化阶梯覆盖率为 2%

图 9.12　钝化结果

9.2.8　自下而上的电镀

图 9.13 展示了铜电镀结果的顶视图。电镀后，晶圆顶部出现了铜的覆盖层，如图 9.13a～c 所示。在去除覆盖层后，有一些空隙留在 TSV 的顶部，如图 9.13c 所示。产生空隙的原因是 TSV 中铜的填充率不同，部分 TSV 顶部表面被相邻的铜覆层所覆盖，如图 9.13c 所示。因此，这些 TSV 不能被完全填充。

为了完全消除空隙，晶圆被进一步减薄 30～50μm。如图 9.14 所示，所有的 TSV 都被铜完全填充，没有任何空隙。在镀铜后进一步减薄硅片将导致更高的成本。因此，需要优化铜填充工艺，以实现更均匀的铜镀。

图 9.15 显示了减薄 30μm 后（最终厚度为 170μm）的 TSV 的横截面。如图 9.15a、b 所示，所有的 TSV 都被铜填充，而在 TSV 的顶部仍存在一些微小的空隙。在进一步减薄晶圆时，这些微小的空隙可以被完全去除。如图 9.15c 所示，当使用溅射的钛层作为附着层时，电镀的铜与硅通孔侧壁没有分离。为了获得更好的附着力和扩散屏障，还可以使用 TiW、Ta 或 TaN 这些材料。

| 异构集成技术

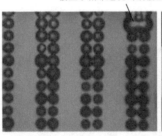 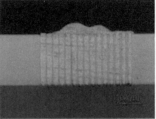

a) CMP前　　　　　　　　b) 镀铜后、CMP前的TSV截面图(没有空隙)

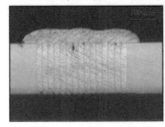 

c) 镀铜后、CMP前的TSV截面图，　　d) CMP后，TSV顶部有少量空隙
TSV顶部有空隙

图 9.13　铜电镀结果的顶视图

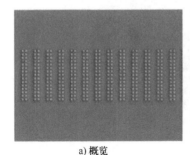

a) 概览　　　　　　　　　　b) 放大的图像

图 9.14　晶圆再减薄 30μm 后，TSV 中的所有空隙都被去除

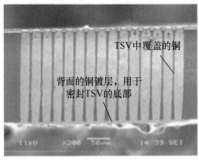 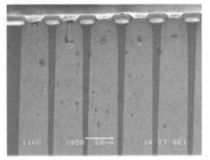

a) 概览　　　　　　　　　　b) TSV上端放大图（由于抛光和
清洁过程，横截面上有一些污点）

图 9.15　镀铜和减薄至 170μm 后的 TSV 横截面，显示了 TSV 上端一些微小的空隙

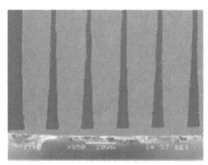

c) TSV 底面放大图

图 9.15　镀铜和减薄至 170μm 后的 TSV 横截面，显示了 TSV 上端一些微小的空隙（续）

从图 9.15a 可以看出，为了密封 TSV 的底部，在晶圆的背面有大约 4μm 厚的铜镀层。这一层可以在镀铜后用 CMP 方法去除。

在铜 CMP 工艺之后，即图 9.5k 工艺步骤后，使用光学 FSM Echoprobe TM 413 和红外干涉仪测量晶圆的翘曲。测量得到的平均晶圆翘曲度小于 100μm。由于 TSV 的总面积小于整个晶圆面积的 1%，由 TSV 引起的晶圆翘曲很低。

为了评估 TSV 的电气性能，测量了 TSV 的直流电阻。测量的平均 TSV 电阻为 63mΩ。同时初步测量了 TSV 和硅基板之间的绝缘性能。TSV 的电容和泄漏电流分别由精密 LCR 表和安捷伦 4156C 精密半导体参数分析仪测量。在 1MHz 时，平均电容约为 2.4pF，而在 10V 偏置电压下，平均泄漏电流低于 1×10^{-10}A。

9.2.9　ENIG 电镀结果

图 9.16a 显示了在 TSV 顶部的 ENIG 焊盘俯视图。在电镀之后，由于钝化层上径向镍的生长，ENIG 焊盘的直径比 TSV 的直径大 4μm 左右[10]。图 9.16b 显示了在 TSV 顶部的 ENIG 焊盘的横截面，图 9.16c 显示了放大的视图。

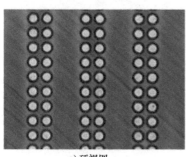
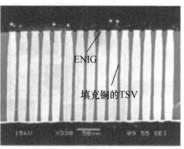

a) 顶视图　　　　　　　　　　　b) 横截面图

图 9.16　在 TSV 顶部的 ENIG UBM 层的扫描电子显微镜图像

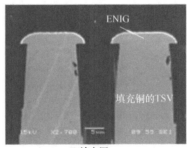

c) 放大图

图 9.16 在 TSV 顶部的 ENIG UBM 层的扫描电子显微镜图像（续）

9.2.10 铜锡合金焊凸点的制造结果

电镀后，进行回流焊以重塑微凸点，同时获得均匀的凸点高度。回流温度为 265℃。在回流过程中，使用美国 Indium 公司的 WS3543 助焊剂，在回流之后，用去离子（Ic）水清洗晶圆以去除助焊剂的残留物。图 9.17 显示了制造的铜柱+锡焊帽和铝焊盘的 SEM（扫描电子显微镜）图像，图 9.18 显示了焊凸点的 FIB（聚焦离子束）图像。图 9.18 还表明，回流后形成的 IMC 的厚度约为 1.5μm，并占据了近 50% 的焊料体积。然而，剩余的锡仍然足够用于组装两个硅芯片。

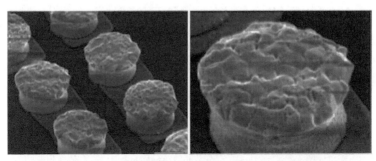

a) 内存芯片上电镀CuSn之后

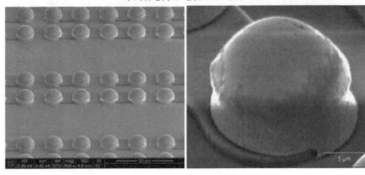

b) 内存芯片上CuSn回流焊之后

图 9.17 CuSn 焊料微凸点的 SEM 图像

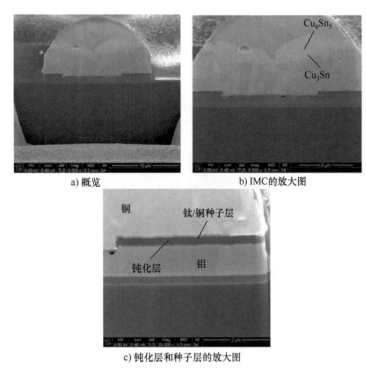

图 9.18 顶层芯片回流后的铜柱 + 锡焊帽的 FIB 图像

9.2.11 组装结果

在制作完成顶部芯片上的 CuSn 焊料微凸点和底部芯片上 TSV 顶部的 ENIG UBM 焊盘后，使用 FC15 型倒装芯片键合机将顶部芯片和底部芯片连接在一起。连接条件为：压力 20MPa，底基板温度 300℃，臂温度 350℃，连接时间 60s。组装过程无助焊剂参与，因为在芯片连接之后，顶部芯片和底部芯片之间的间隙只有 10μm 左右。因此，如果使用助焊剂，很难清洗干净，并且在进行底部填充之后，间隙内会形成很多空洞。在芯片连接完成之后，再使用细填料尺寸的材料作为下填料，如长濑（Nagase）ChemteX 公司的 T693/R3434iHx-2 或纳美仕（Namics）公司的 U8443-14 型下填料，用于填充两个硅芯片之间的间隙[2]。图 9.19 显示了在选定的键合条件下连接在一起的顶部芯片和底部芯片的横截面。可见连接是成功的。

此外还需进行剪切测试，以评估组装的连接强度。这里使用市售的剪切测试仪（DAGE-SERIES-4000-T，Dage Precision Industries Ltd.）进行测试。通过剪切探头施加到组件上的力在平行于测试装置表面的方向上进行剪切。该测试的剪切速度为 100μm/s，剪切探头的高度为 400μm。在这种键合条件下，接头的平均剪切强度为 12.1MPa。在之前已经研究并介绍过了微焊点的温度循环测试[2]（-40℃↔125℃，保持15min，以15℃/min的速率变温）。11个样品中的1个样品在进行到300次循环时失效，而所有其他样品都通过了1000次循环直到测试停止。另外，使用制造好的TSV转接板进行芯片键合的可靠性还有待进一步研究。机械模型表明，在顶部芯片和底部芯片之间使用底部填充，微焊点处的普通应力大大降低，并且基板核心厚度对微焊点处的应力影响很小[11]。

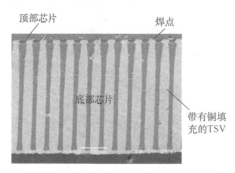

图9.19 顶部芯片和底部芯片连接的横截面

9.2.12 总结和建议

本章介绍了采用TSV结构进行的芯片之间异构集成技术的发展。一些重要的结果和建议总结如下[2, 12]：

1）采用TSV的内存芯片和CPU芯片面对面互连实现了最高的带宽和电气性能。

2）内存芯片和CPU芯片的键合可以通过晶圆对晶圆的键合来实现，从而能获得最大的生产量。可惜的是，大多数内存芯片都是长方形的，而CPU芯片往往是正方形的。另外，它们的尺寸差异也很大。因此这些因素对晶圆间的键合提出了重大挑战。内存芯片到CPU晶圆的键合或许能成为一种替代方案。然而这种方案会长时间占用键合机。另一种方法则是通过内存芯片到CPU芯片的键合，但是这种方法只能获得最小的生产量。

3）目前已成功研发了节距25μm、焊盘直径15μm的焊料微凸点阵列。这种超细节距和高密度的凸点阵列可用于高密度互连的异构集成。

4）目前已经能采用DRIE方法在底部芯片（晶圆）中制造深宽比大于10的TSV。在DRIE工艺流程中发现，当刻蚀循环时间比钝化循环时间长80%时，底部切口尺寸会大于2μm，并呈现出明显的弓形。

5）在镀铜之后，TSV顶部会存在一些空隙，可以通过进一步减薄晶圆来去除这些空隙。铜填充工艺仍有待优化，以便在TSV中形成更为均匀的铜镀层。

6）在优化的键合条件下，Cu/Sn焊料微凸点和TSV顶部的ENIG焊盘之间可以实现良好的连接，剪切强度为12.1MPa。

9.3 无 TSV 的芯片到芯片异构集成

9.3.1 试验件与制造方法

图 9.20 所示结构是一种低成本的封装，适用于高性能、高密度、低功耗的应用场景，并且也很可能满足高带宽应用。该结构中的散热器是可选的。芯片与芯片的叠放是面对面的（最短距离），信号、电源、地信号等可以很容易地通过焊点进入下一层。这种异构集成是最具成本效益的封装结构（存储器作为子芯片，逻辑电路作为母芯片）。需要强调的是，子母芯片中都没有使用 TSV。

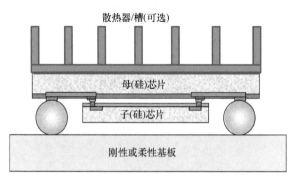

图 9.20 不使用 TSV 的芯片级异构集成

试验件是一个堆叠的硅芯片模块，如图 9.21 所示[13]。该模块由晶圆制造的顶部子芯片和底部母芯片组成。在子芯片上，涂有 AuSn 焊料的铜柱制成的凸点。在母芯片上，子芯片的凸点采用沉镍浸金的 UBM，而下一级互连（如基板）则采用 AuSn 焊凸点。对于本测试中使用的两个芯片，焊盘以交替模式互连，以便在将它们焊接在一起时提供菊花链连接。子芯片上有 20 个焊盘排列成两排，如图 9.21 所示。硅堆叠模块关键元件的材料和尺寸见表 9.2。该封装模块最后连接到刚性或柔性基板（见图 9.20）。

AuSn 焊料特别适合光电和医疗领域设备的封装。大多数医疗和光电设备必须在无助焊剂工艺中进行焊接，可以使用 AuSn 焊料，如图 9.22 所示。此外，80%Au 和 20%Sn（重量占比）的焊料因具有以下优点而被广泛使用，即高强度、高耐腐蚀性、高抗疲劳性。因此，Au20Sn 焊料系统通常被选作无助焊剂覆晶封装的互连材料。除了 AuSn 焊料，SnAg 焊料也是一种选择。

AuSn 焊料可以通过多种方法制造，如电子束蒸发、电子沉积以及浆料和焊料预成型。在这些方法中，蒸发焊料可以更准确地控制尺寸和位置。因此，在 UBM 图形化之后，沉积一系列交替的 Au 和 Sn 层以形成 AuSn 层状结构。使用电子束蒸发工艺进行 AuSn 沉积的一些优点包括：沉积后形成的氧化物量将减少；可以精确控制焊料的厚度和位置；Au 沉积速率很高，并且可以在 200mm 晶圆上实现厚度均匀的 AuSn 层。

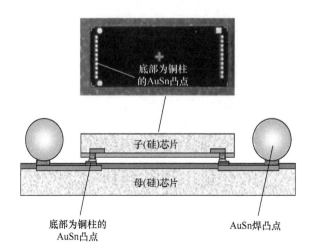

图 9.21　堆叠硅模块（子芯片和母芯片）和子芯片的细节图

表 9.2　芯片对芯片集成模块关键元件的材料和几何尺寸

模块类型	堆叠硅模块	
测试裸片	子芯片	母芯片
芯片尺寸 /mm	3.405 × 1.34 × 0.06	4.793 × 1.34 × 0.13
焊盘开孔	FC 焊盘 30μm	FC 焊盘 30μm
焊盘节距 /μm	100	100
凸点类型和高度	铜柱上 Au80Sn20，23.5μm	Au80Sn20，125μm
UBM	铝	化学镀 NiAu

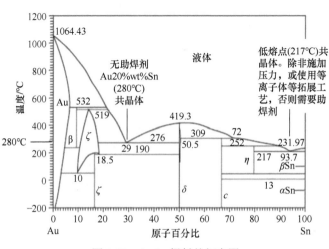

图 9.22　AuSn 焊料的相态图

9.3.2 试验件的制造

制造试验件采用了直径为 200mm 的 p 型硅（100）晶圆，其上覆有 5000Å 的 SiO_2 绝缘薄膜。图 9.23 展示了在子芯片上制作 AuSn 焊凸点的流程示意图。第一步是在硅晶圆上沉积一层 AlCu 层，然后用菊花链图形化做金属焊盘并刻蚀 AlCu 层。其次，沉积 5000ÅSiO_2/5000ÅSiN 的钝化层，然后使用干法刻蚀对钝化层进行图形化以打开金属焊盘。溅射钛/铜作为导电层，用于后续的微型铜柱电镀工艺。然后使用热辊层压机层压 20μm 厚的干膜光刻胶，使用 EVG 640 接触掩模对准器进行超紫外线 UV 曝光显影，继而光刻以形成 UBM 焊盘。这些图形化通孔使用 RENA 200mm 晶圆电镀工具填充高度为 18～20μm 的铜，该工具是一种喷泉型（杯型）工具，其中使用基于 Sperolyte $CuSO_4$ 的溶液（Atotech Pte，Ltd.）作为电解质溶液。然后使用电子束蒸发器交替沉积 2200Å

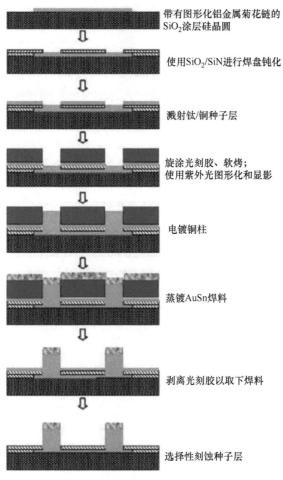

图 9.23　子芯片晶圆上 AuSn 焊料凸点的制造工艺

异构集成技术

厚度的金层和 2000Å 厚度的锡层，总共沉积 16 层，以在微型铜柱上实现 3.5μm 厚的 AuSn 凸点。这些 AuSn 层沉积在整个晶圆表面上，但干膜层上的 AuSn 可以在干膜剥离过程中使用剥离工艺去除，并且由于这些物质与铜的黏附性，AuSn 仅保留在铜柱上层。

干膜剥离后，使用湿法刻蚀将钛/铜层逐层地刻蚀掉，完成 AuSn 焊凸点制作，如图 9.23 所示。这层 AuSn 在组装时的回流工艺中将会融化并形成均匀的 Au20Sn 焊点。图 9.24 显示了 AuSn 焊凸点，3.5μm 厚的 AuSn 通过蒸镀工艺沉积在 20μm 厚的铜柱上。AuSn 焊锡凸点沿子芯片两侧以 100μm 的节距分布，并且非常均匀。

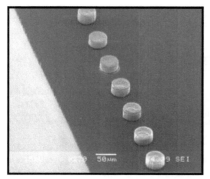

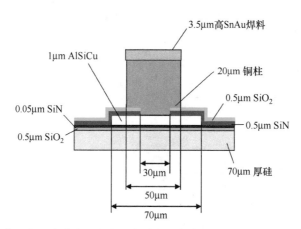

图 9.24　使用电子束蒸发工艺在子芯片晶圆上制造 AuSn 焊凸点的尺寸和均匀性

图 9.25 显示了母芯片上的凸点，即 AuSn 焊凸点（本节将不讨论）和化学镀 NiAu 的 UBM。由于 AlCu 金属焊盘已经基于菊花链和其钝化开口进行了图形化，母芯片的制作工艺与子芯片相同。之后将化学镀 NiAu 层沉积为 UBM，如图 9.25 所示。

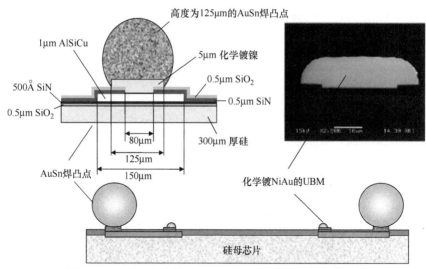

图 9.25　母芯片上化学镀 NiAu 的 UBM 的尺寸和细节特写

9.3.3　芯片到晶圆的组装方法

AuSn 焊凸点的子芯片在母芯片上的组装是采用芯片对晶圆（C2W）键合方法；首先将子芯片晶圆切割成多个单一芯片，然后挑选良好的子芯片将它们固定到性状良好的母芯片晶圆上，如图 9.26 所示。

蒸镀形成的AuSn凸块

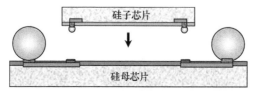

通过倒装芯片键合机将子芯片与母芯片对齐

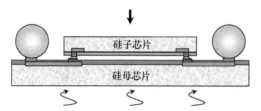

在键合期间加热使AuSn之间互连

图 9.26　芯片到芯片模块的整体组装流程

键合过程使用 Karl Suss 的倒装芯片键合机 FC150 完成,如图 9.27 所示。组装流程通过非回流焊的凸点进行,并在键合期间使用氮气吹扫。在键合工艺前,对子芯片和母芯片(晶圆上的)进行氩溅射清洗。AuSn 倒装芯片键合的最大挑战是在化学镀 NiAu 的 UBM 键合焊盘上实现良好的 AuSn 润湿,使键合后凸点和 UBM 之间挤压的 AuSn 焊料最少。除了凸点高度、平面度外,这里还研究了组装过程中对齐的精度、键合力、键合温度和键合时间对接触电阻、AuSn 润湿和 AuSn 焊点形状等的影响。

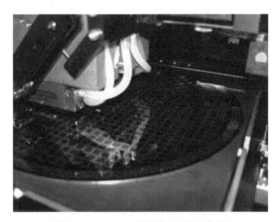

图 9.27　FC150 型倒装芯片键合机

9.3.4　凸点高度平面度

在倒装芯片连接工艺前表征金凸点高度非常重要,因为不均匀的凸点高度会导致接触不良甚至开路。图 9.28 显示了在可接受范围内的凸点高度测量值,芯片级允许最大值为 2.5μm,晶圆级最大为 1.5μm。

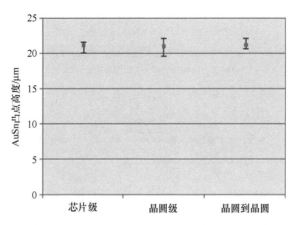

图 9.28　在不同晶圆上 AuSn 凸点高度的测量值

9.3.5 对齐精度

在传统倒装芯片的焊接工艺中,焊料凸点浸入助焊剂中并附着在基板焊盘上,然后再进行回流。在烘箱回流过程中,芯片上的焊料凸点能够与基板上的焊盘自对准并形成良好的互连。然而,使用热压键合的方法时,焊料无法在组装过程中自对准。因此,键合对准精度变得至关重要,尤其是在使用带有焊料凸点的大芯片时,需要具有良好的对准和键合精度的精密设备才能在基板焊盘上形成具有稳定接触电阻和良好润湿性的 AuSn 互连。

互连的接触电阻的大小可以用不同的对准精度来表征,如图 9.29 所示,在倒装芯片键合过程中施加了 0.8kg 的键合力。如图 9.30 所示,与 100% 对齐的 AuSn 凸点与化学镀 NiAu 的焊盘相比,未对准的 AuSn 凸点与化学镀 NiAu 的焊盘之间的接触电阻和 AuSn 润湿情况均存在明显差异。这种结合方式导致焊盘上的 AuSn 润湿性差,同时电阻也会更大。造成这种情况的原因之一是在键合后,AuSn 焊凸点与化学镀 NiAu 的 UBM 没有自行对准。如果错位偏移量很大,则 AuSn 焊料会无法润湿化学镀 NiAu 的 UBM。另一方面,对于 100% 对齐的键合,在化学镀 NiAu 的 UBM 上能够观察到较为稳定的接触电阻阻值和良好的 AuSn 润湿性。测量得出:两个菊花链上的接触电阻比未对准键合测量的电阻更低且更一致。经过精确对准的互连横截面如图 9.31 所示。

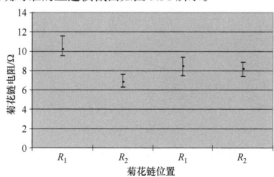

图 9.29 芯片到晶圆键合对准精度与接触电阻测量的关系

a) 子芯片未对齐的情况　　　　b) 母芯片未对齐的情况

图 9.30 AuSn 没能润湿的情况

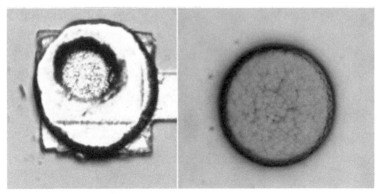

c) 子芯片100%对齐的情况　　　　d) 母芯片100%对齐的情况

图 9.30　AuSn 没能润湿的情况（续）

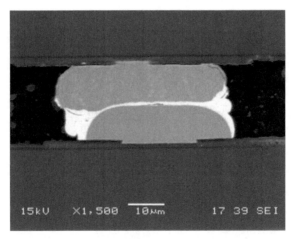

图 9.31　精确对准的 AuSn 焊点的横截面

9.3.6　芯片到晶圆的实验设计（DoE）

1. 实验设计中的三个变量

在解决了组装过程中的不同问题后，根据三个变量进行了试验件的实际组装。表 9.3 评估了三个不同的键合力、两个键合温度和时间的组合，以确定实现一致的测量电阻、良好的 AuSn 润湿和 AuSn 连接所需的键合力和键合温度。实验中选择的三个键合力分布在 0.4 ~ 1kg 之间，见表 9.4。图 9.32 表征了装配良率，并通过菊花链网络对接触电阻进行了测量。

第 9 章 芯片到芯片堆叠的异构集成

表 9.3 C2W DoE 键合评估考虑的三个变量

序号	键合力 /kg	键合温度 /℃	键合时间 /s
A	0.4	290	15
B	0.4	315	15
C	0.8	290	15
D	0.8	315	15
E	0.1	290	15
F	0.1	315	15

表 9.4 不同键合力、温度和时间组合下的 C2W 键合

变量	取值	相应测量对象
键合力 /kg	0.4，0.8，1	接触电阻
键合温度 /℃	290，315	AuSn 焊点形状
键合时间 /s	15	AuSn 润湿情况

图 9.32 通过母芯片上的菊花链测量电阻

2. 实验结果

键合力对互连电阻和 AuSn 接触点的形成有显著影响。图 9.33 显示在 0.8kg 和 1kg 键合力下可以实现接触电阻的稳定。然而 1kg 的键合力是不可行的，因为这会导致铝焊盘 UBM 界面处的芯片钝化且开裂，如图 9.34 所示。此外，高键合力容易挤出凸点和键合焊盘之间的 AuSn 共晶焊料，并导致键合焊盘上的 AuSn 润湿性不佳。图 9.35 显示了在 1kg 键合力下的 AuSn 组装的横截面图。研究结果还表明，接触电阻对键合力的敏感性比对键合温度和键合时间这两个变量的敏感性更强。

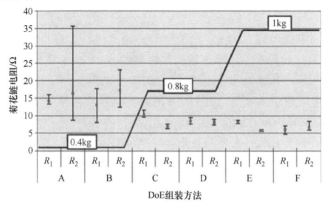

图 9.33 实验中各种组装方法的接触电阻测量值

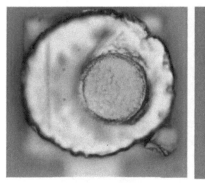 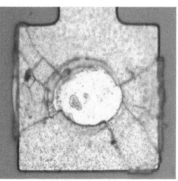

a) 子芯片　　　　　　　　　　b) 母芯片

图 9.34　在高键合力（1kg）作用下铝焊盘 UBM 界面处的芯片钝化裂纹

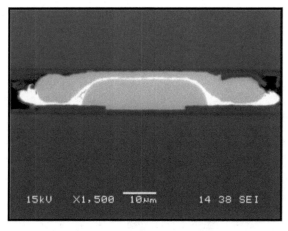

图 9.35　高键合力（1kg）作用下的 AuSn 组装的横截面

9.3.7　可靠性试验与结果

本节对一组试验件（表 9.3 中序号 D，无下填料）进行 -40～125℃的热循环（TC）测试，分析 AuSn 互连的热疲劳特性[12]，在可靠性试验前通过在 100 次热循环测试中每 25 次循环测量一次接触电阻，评定失败的标准是菊花链的电阻值连续开路（∞）。如图 9.36 所示，本组样品通过了热循环测试，在 100 次循环中没有出现连续开路的现象。对于 0.8kg 的键合力，观察到接触电阻的最小变化小于 10%，表明与化学镀 NiAu UBM 焊盘直接接触的同质 AuSn 共晶焊料的互连方案是稳健的并且能够承受 TC 条件。因此在 0.8kg 的键合力下，序号 D（键合温度 315℃，键合时间 15s）可以实现接触电阻的稳定性和良好的可靠性。

第 9 章 芯片到芯片堆叠的异构集成

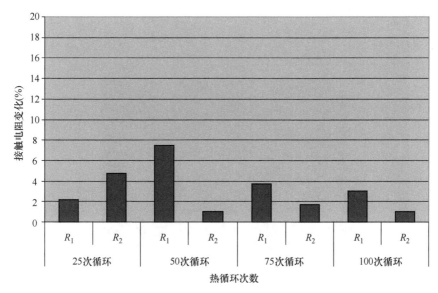

图 9.36 温度从 −40℃ 到 125℃ 变化时接触电阻的变化比例

9.3.8 3D IC 封装与 SnAg 互连

图 9.37 展示了在相同的子芯片和母芯片上使用不同的焊接方式。如图 9.38 所示,在子芯片上使用 Sn3wt%Ag 和铜柱代替 AuSn 焊料。母芯片上的芯片级封装(CSP)焊球使用 Sn37wt%Pb。该部分的重点在于子芯片和母芯片之间的 SnAg 焊点的可靠性[13]。在组装之前需要将带有 40μm 高的 SnAg 和铜柱微凸点的子芯片晶圆减薄至 70μm,然后将带有 200μm 高的 CSP SnPb 和 UBM 微凸点

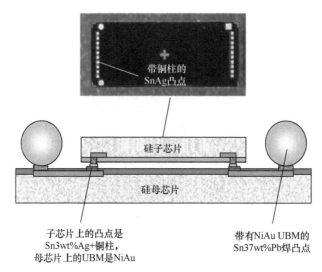

图 9.37 采用 SnAg 无铅焊料的堆叠硅模块(子芯片和母芯片)

的母芯片晶圆减薄至 300μm，如图 9.39 所示。图 9.40 展示了使用下填料组装的横截面。注意：焊点是自然回流，并且由于熔融焊料的表面张力导致焊点的形状非常圆滑，与 AuSn 焊点的 C2W 键合方式有很大不同。

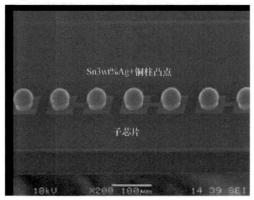

图 9.38　子芯片上的无铅 Sn3wt%Ag+ 铜柱（约 40μm）凸点

带有40μm高SnAg+铜柱微凸点的子芯片晶圆减薄至70μm

带有200μm高CSP SnPb+UBM 微凸点的母芯片晶圆减薄至300μm

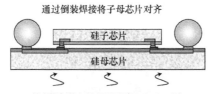

图 9.39　使用 SnAg 焊料的整体芯片到芯片模块组装过程

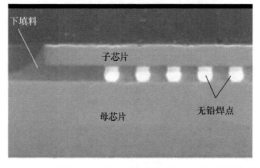

图 9.40　通过 SnAg 焊点组装在母芯片上的子芯片

第9章 芯片到芯片堆叠的异构集成

组装后可以对样品进行无偏置的加速应力测试（uHAST）：130℃，85%相对湿度（RH），96h；MST L3：30℃，60%RH，192h，260℃ 3次回流；MST L1：85℃，85%RH，168h，260℃ 3次回流；HTS：125℃，1000h；TC：125～-55℃，15min保温，15℃/min升温，1000次循环。1000次循环后失效样品的典型图像如图9.41所示。裂纹在金属间化合物（IMC；Cu_6Sn_5）和块状焊料之间的界面角附近形成，然后通过焊点传播。

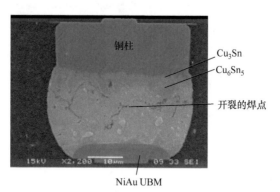

图9.41 高温储存测试、1级和3级湿敏测试以及1000次热循环测试后的不合格样品

9.3.9 总结和建议

一些重要的总结和建议汇总如下[13, 14]：

1）子芯片和母芯片的面对面互连使用焊凸点代替了TSV，降低了成本，适用于移动和便携的产品，如医疗设备等。

2）为了确保可以在测试芯片上施加均匀的压力并形成良好的AuSn互连，需要实现芯片级均匀的凸点共面性（±3.5μm）。结果表明，使用高键合力（1kg）组装样品是不可行的，这个大小的键合力会导致铝焊盘-UBM界面处的芯片钝化开裂。

3）此外高键合力会挤出凸点和焊盘之间的AuSn共晶焊料，导致键合焊盘上的AuSn润湿不良。

4）热循环测试的结果表明，在0.8kg的键合力下，无须下填料即可实现稳定的接触电阻（±9Ω）和良好的可靠性（超过100次热循环）。

5）这项工作的额外价值在于为使用AuSn无助焊剂焊料凸点技术建立倒装芯片组装的行业提供参考。

参考文献

[1] Polka, L. A., H. Kalyanam, G. Hu, and S. Krishnamoorthy, "Package Technology to Address the Memory Bandwidth Challenge for Tera-scale Computing", *Intel Technology Journal*, Vol. 11, No. 3, 2007, pp. 197–206.

[2] Yu, A. B., A. Kumar, S. W. Ho, W. Y. Hnin, J. H. Lau, C. H. Khong, P. S. Lim, X. W. Zhang, D. Q. Yu, N. Su, B. R. Chew, M. C. Jong, T. C. Tan, V. Kripesh, C. Lee, J. P. Huang, J. Chiang, S. Chen, C.-H. Chiu, C.-Y. Chan, C.-H. Chang, C.-M. Huang, and C.-H. Hsiao, "Development of fine pitch solder microbumps for 3-D chip stacking", *Proceedings of 10th Electronics Packaging Technology Conference*, Singapore, December 2008, pp. 387–392.

[3] Yu, D. Q., H. Oppermann, J. Kleff, and M. Hutter, "Stability of AuSn eutectic solder cap on Au socket during reflow", *Journal of Materials Science: Materials in Electronics*, Vol. 20, No. 1, pp. 55–59, 2009.

[4] Douglas, M. A., "Trench etch process for a single wafer RIE dry etch reactor". U.S. Patent 4 855 017 and 4 784 720 and F. Laermer and A. Schilp, "Method of anisotropically etching silicon". U. S. Patent 5 501 893.

[5] Charbonnier, J., S. Cheramy, D. Henry, A. Astier, J. Brun, N. Sillon, A. Jouve, S. Fowler, M. Privet, R. Puligadda, J. Burggraf, and S. Pargfrieder, "Integration of a temporary carrier in a TSV process flow", *Proceedings of 59th Electronic Components and Technology Conference*, San Diego, CA, May 2009, pp. 865–871.

[6] Premachandran, C. S., N. Rangnathan, S. Mohanraj, C. S. Choong, and K. I. Mahadevan, "A vertical wafer level packaging using through hole filled via interconnects by lift off polymer method for MEMS and 3-D stacking applications", *Proceedings of 55th Electronic Components and Technology Conference*, Lake Buena Vista, FL, May–Jun. 2005, pp. 1094–1099.

[7] Pang, X., F., T. T. Chua, H. Y. Li, E. B. Liao, W. S. Lee, and F. X. Che, "Characterization and management of wafer stress for various pattern densities in 3-D integration technology," *Proceedings of 60th Electronic Components Technology Conference*, Las Vegas, NV, June 2010, pp. 1866–1869.

[8] Ayon, A. A., R. Bratt, C. C. Lin, H. H. Sawin, and M. A. Schmidt, "Characterization of a time multiplexed inductively coupled plasma etcher", *Journal of the Electrochemical Society*, Vol. 146, No. 1, 1999, pp. 339–349.

[9] Chen, K., A. A. Ayon, X. Zhang, and S. M. Spearing, "Effect of process parameters on the surface morphology and mechanical performance of silicon structures after deep reactive ion etching (DRIE)", *Journal of Microelectromechanical Systems*, Vol. 11, No. 3, June 2002, pp. 264–275.

[10] Iwasaki, T., M. Watanabe, S. Baba, M. Kimura, Y. Hatanaka, S. Idaka, and Y. Yokoyama, "Development of 30 micron pitch bump interconnections for COC-FCBGA", *Proceedings of 56th Electronic Components Technology Conference*, San Diego, CA, 2006, pp. 1216–1222.

[11] Khong, C., A. Yu, A. Kumar, X. Zhang, V. Kripesh, T. T. Chun, J. H. Lau, and D.-L. Kwong, "Sub-modeling technique for thermomechanical simulation of solder microbumps assembly in 3-D chip stacking", *Proceedings of 11th Electronics Packaging Technology Conference*, Singapore, December 2009, pp. 591–595.

[12] Yu, A. B., J. H. Lau, S. Ho, A. Kumar, W. Hnin, W. Lee, M. Jong, et al., "Fabrication of High Aspect Ratio TSV and Assembly with Fine-Pitch Low-Cost Solder Microbump for Si Interposer Technology with High-Density Interconnects", *IEEE Transactions on CPMT*, Vol. 1, No. 9, September 2011, pp. 1336–1344.

[13] Lim, S., V. Rao, H. Yin, W. Ching, V. Kripesh, C. Lee, J. H. Lau, J. Milla, and A. Fenner, "Process Development and Reliability of Microbumps", *IEEE/ECTC Proceedings*, December 2008, pp. 367–372. Also, *IEEE Transactions on Components and Packaging Technology*, Vol. 33, No. 4, 2010, pp. 747–753.

[14] Vempati, S., S. Nandar, C. Khong, Y. Lim, V. Kripesh, J. H. Lau, B. P. Liew, K. Y. Au, S. Tamary, A. Fenner, R. Erich, and J. Milla, "Development of 3D Silicon Die Stacked Package Using Flip-Chip Technology with Micro Bump Interconnects", *IEEE/ECTC Proceedings*, San Diego, CA, 2009, pp. 980–987.

第 10 章

CIS、LED、MEMS 和 VCSEL 的异构集成

10.1 引言

本章将介绍 CIS（CMOS 图像传感器）、LED（发光二极管）、MEMS（微机电系统）、VCSEL（垂直腔表面发射激光器）和 PD（光电二极管）探测器的异构集成，重点介绍这些异构集成的一些示例。

10.2 CIS 异构集成

CIS 的基本功能是将光（光子）转换成电信号（电子）。CIS 在便携、移动、可穿戴和自动化产品中拥有巨大的市场，也是 IoT（物联网）中的关键元件。如智能手机和平板计算机中的摄像头采用了 CIS，汽车的机械视觉也采用了 CIS。一般来说，一个 CIS 包括微透镜阵列、晶体管与金属布线和 PD[1-7]。本节将介绍 3D CIS 和集成电路（IC）的异构集成，重点介绍 3D CIS 和 IC 的堆叠[3,4]，以及 CIS 和处理器 IC 的 3D 混合集成[5]。

10.2.1 前照式 CIS 和背照式 CIS

CIS 有两种不同的类型，即 FI（前照式）-CIS 和 BI（背照式）-CIS。对于 FI-CIS，微透镜矩阵在前面，晶体管和金属布线层在中间，PD 探测器在距离晶圆前表面较深的底部（背面），如图 10.1 所示。硅基板上的晶体管和金属布线层会反射部分光线，因此 PD 只能接收剩余的入射光。

BI-CIS 包含相同的单元，但是在制造过程中通过翻转硅片使晶体管和金属布线层位于 PD 层的背面再对硅片反面进行减薄，因此光可以如图 10.2a 所示不经过晶体管和金属布线层直接射入 PD 层。与 FI-CIS 相比，BI-CIS 能够将入射光子被俘获的概率从大约 60% 提升到 90% 以上[7]。然而，BI-CIS 会导致诸如串扰等问题，会引起噪声、暗电流和相邻像素块之间的混色从而造成图像退

异构集成技术

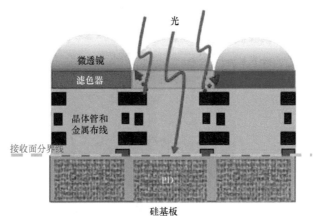

图 10.1 FI-CIS，部分光被晶体管和金属布线层阻挡（反射）

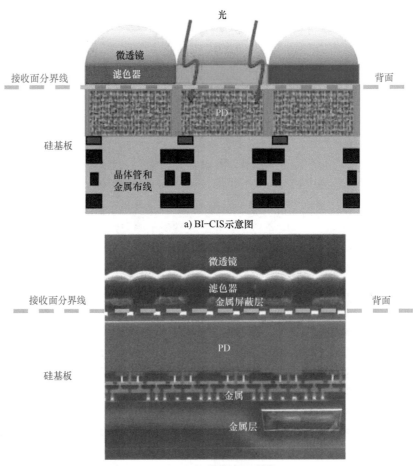

a) BI-CIS示意图

b) BI-CIS横截面SEM图像

图 10.2 BI-CIS

化。为了克服这个问题,索尼公司开发出一种独特的 PD 结构[2]。在被照射侧上方,在 PD 间形成减少像素阵列之间串扰的金属光罩(如图 10.2b 所示)来实现良好的分色。

10.2.2 3D CIS 和 IC 混合集成

如图 10.2 所示 BI-CIS[1, 2],其像素部分和逻辑电路部分整合成一个由硅基板支撑的单颗芯片,如图 10.3a 所示。索尼公司将 CIS 像素芯片和逻辑电路芯片(即用逻辑电路芯片代替硅基板)分开并以 3D 的方式将它们进行堆叠,如图 10.3b 所示。

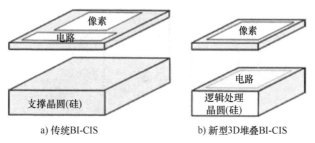

图 10.3 传统 BI-CIS 与新型 3D 堆叠 BI-CIS 结构对比

1. 结构

图 10.3a 展示了传统的 BI-CIS[1, 2] 结构,而图 10.3b 展示了新型 3D 堆叠 BI-CIS[3] 结构。可以看出新 BI-CIS 结构包含两块芯片,即 CIS 像素芯片和逻辑电路芯片,它们通过如图 10.4 所示芯片边缘的 TSV(硅通孔)实现垂直互连。这种新设计的优势是相同尺寸的 CIS 像素芯片上能够放置更多的像素(或者采用更小的芯片尺寸能实现相同数量的像素),以及 CIS 像素芯片和逻辑电路芯片可以采用不同的工艺技术分别制造。结果是 CIS 芯片的尺寸减小了 30%,而逻辑电路芯片的规模从 500k 门增加为 2400k 门[3]。

TSV 的数量达数千个,其作用包含信号、电源和接地。在像素阵列区域没有 TSV 存在。列 TSV 位于像素芯片的比较器和逻辑电路芯片的计数器之间。行 TSV 位于 CIS 芯片的行驱动器和逻辑电路芯片的行解码器之间,如图 10.4 所示。这种 TSV 排列能够减小噪声的影响,并使得 CIS 芯片的制造更容易。如为了减小噪声的影响,比较器被布置在 CIS 像素芯片上,并通过索尼公司的成熟工艺技术进行制造,而不是布置在逻辑电路芯片上。

2. CIS 像素晶圆和逻辑 IC 晶圆的制造

CIS 像素芯片采用索尼公司传统的 1P4M BI-CIS(90nm)工艺技术。逻辑电路芯片采用成熟的 1P7M 逻辑电路制造工艺技术。像素芯片和逻辑芯片的尺寸基本相同。索尼公司并没有公开其 TSV 和封装技术。在此猜想是将 CIS 晶圆上的 CIS 芯片硅绝缘层和逻辑晶圆上的逻辑芯片硅绝缘层进行键合(非常类似

基于 SiO_2 和 SiO_2 共价键合的晶圆对晶圆键合）。在晶圆键合后制造 TSV 并填充铜。图 10.5 和图 10.6 显示了 3D CIS 像素芯片和逻辑 IC 芯片集成的 SEM（扫描电子显微镜）横截面图。可以看到，上部为 BI-CIS 芯片，底部为逻辑芯片，BI-CIS 晶圆和逻辑晶圆采用绝缘层对绝缘层（晶圆对晶圆）键合（见图 10.5），CIS 芯片和逻辑芯片通过 TSV 互连（见图 10.6）。

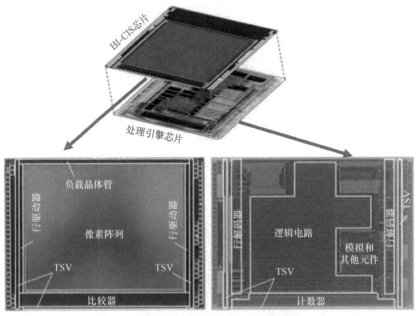

图 10.4 3D CIS 像素芯片和逻辑 IC 集成

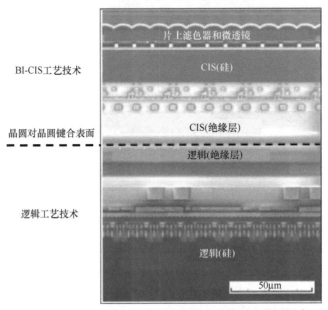

图 10.5 CIS（绝缘层）晶圆和逻辑（绝缘层）晶圆键合

第 10 章　CIS、LED、MEMS 和 VCSEL 的异构集成

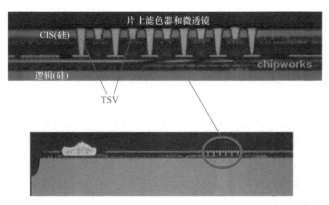

图 10.6　连接 CIS 像素芯片和逻辑电路芯片的 TSV

3. 混合键合

值得一提的是还有一种晶圆对晶圆键合技术，即混合键合，它能够同时实现晶圆两边的金属焊盘和绝缘层的键合。索尼公司最先在 HVM[4] 中采用铜 - 铜直接混合键合。索尼公司为 2016 年发布的三星 Galaxy S7 制造了 IMX260 背照式 CMOS 图像传感器（BI-CIS）。电测试结果显示其强健的铜 - 铜直接混合键合技术实现了显著的连通性能和可靠性。图像传感器的性能也非常出色。IMX260 BI-CIS 的截面如图 10.7 所示。可以看出，与参考文献 [3] 中的索尼 ISX014 堆叠式摄像机传感器不同，TSV 被排除而 BI-CIS 芯片和处理器芯片间的互连通过铜 - 铜直接键合实现。信号从封装基板通过引线键合传输至处理器芯片的边缘。

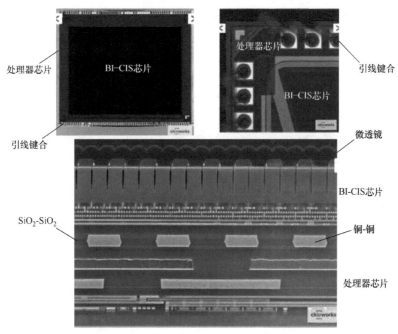

图 10.7　索尼公司铜 - 铜混合键合的 BI-CIS 截面图

铜-铜直接混合键合封装工艺过程首先要进行表面清洗，去除金属氧化物和晶圆表面 SiO_2 或 SiN 活化（通过湿法清洗和等离子体活化），以获得高的键合强度。然后，在常温和典型的洁净室环境中通过光学对准使得晶圆接触。第一次热处理（100~150℃）用于增强晶圆表面 SiO_2 或者 SiN 表面的结合力同时降低由于硅、铜和 SiO_2 或者 SiN 之间热膨胀系数不匹配导致的界面应力。然后，通过施加更高的温度和压力（300℃、25kN、10^{-3}Torr、氮气环境）30min，使铜在界面发生扩散并在界面处进行晶粒生长。键合后在 300℃、氮气气氛和大气环境压力下退火 60min。这种工艺步骤能够在铜和 SiN 或者 SiO_2 间同时形成无缝连接，如图 10.7 所示。

10.2.3 3D IC 和 CIS 异构集成

1. 结构

图 10.8 为参考文献 [5] 中展示的 3D CIS 和 IC 的异构集成结构。其中包括 CIS、协处理器 IC 和玻璃载板。CIS 的输入/输出端口数是 80，协处理器 IC 的端口数是 164。CIS 和协处理器 IC 的尺寸不同。CIS 的尺寸是 5mm×4.4mm，IC 的尺寸是 3.4mm×3.5mm。CIS 和 IC 分别如图 10.9a、b 所示，它们采取如图 10.8 所示的正面对反面的方式键合。CIS 和 IC 通过带 SnAg 焊帽的铜柱互连。TSV 位于 CIS 中，它通过 CIS 和协处理器 IC 上带焊凸点的 RDL 实现与基板的连接。

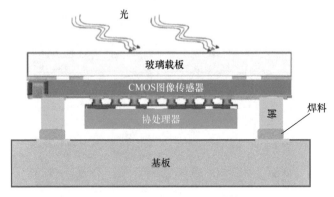

图 10.8 3D CIS 和 IC 的异构集成

2. 协处理器晶圆制造工艺流程

图 10.10 显示了协处理器晶圆的制造工艺流程。在 BEOL（后端工艺）工序后，通过 PVD（物理气相沉积）溅射阻挡/种子层（钛/铜）。接下来，通过一种 BOT（轨迹键合）光刻和 ECD（电化学沉积）来制造 RDL，再通过化学气相沉积（CVD）SiN（500nm）和低应力 SiO_2（600nm）作为钝化层。铜柱的直径为 20μm，铜柱（高 12μm）通过在钛（100nm）阻挡层和铜（400nm）种子层上电镀制造。然后，电镀镍（2μm）和 SnAg（2μm）。最后 260℃ 时回流。图 10.11

第 10 章　CIS、LED、MEMS 和 VCSEL 的异构集成

显示了 RDL 和其上的微铜柱的 SEM 图片。

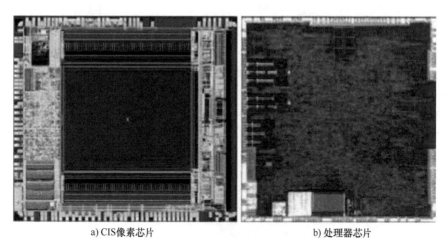

a) CIS 像素芯片　　　　　　　　b) 处理器芯片

图 10.9　CIS 和 IC

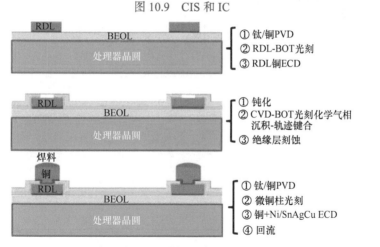

图 10.10　处理器芯片制造工艺流程

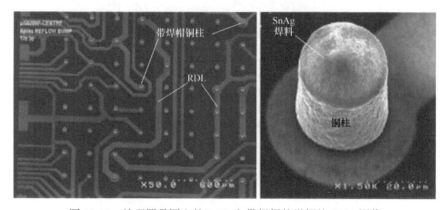

图 10.11　处理器晶圆上的 RDL 和带焊帽的微铜柱 SEM 图像

255

3. CIS 晶圆的制造工艺流程

CIS 晶圆的制造工艺流程如图 10.12 所示。在 BEOL 工序后，将 CIS 晶圆的正面和玻璃载板晶圆进行键合。玻璃晶圆既作为机械（支撑）载板也是封装结构的光学部件。键合采用 Shin-Etsu MicroSi 公司生产的 7μm 厚的 SiNRTM 黏结胶，键合区域的图形在像素阵列四周呈条带状，如图 10.13 所示。接下来将 CIS 晶圆的从背面减薄至 90μm。在 CIS 晶圆的背面制造 TSV（直径 60μm）和 RDL。然后电镀 25μm 直径的微铜柱（高 12μm）和焊帽（2μm 厚的镍和 2μm 厚的 SnAg），在 260℃时回流。图 10.14 显示了带有焊帽（Ni/SnAg）的微铜柱和 RLD 的 SEM 图（这些带焊帽铜柱用于将来和顶部芯片上对应的铜柱连接）。

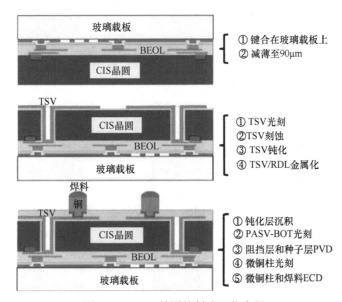

图 10.12 CIS 晶圆的制造工艺流程

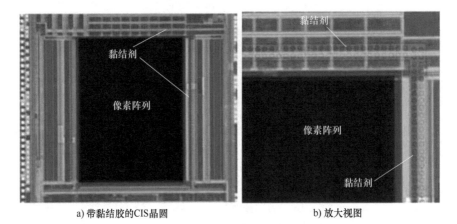

a) 带黏结胶的CIS晶圆　　　　　　　b) 放大视图

图 10.13 采用黏结胶键合

第 10 章　CIS、LED、MEMS 和 VCSEL 的异构集成

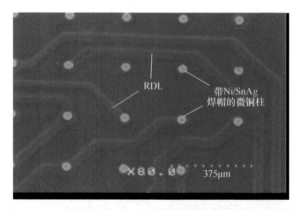

图 10.14　CIS 晶圆上 RDL 和带焊帽微铜柱的 SEM 图像

4. 最终组装

最终的组装方式至少有两种。一种是将协处理器 IC 芯片先堆叠 CIS 晶圆上（芯片到晶圆封装或者 C2W），然后 CIS 晶圆上进行植焊球。还有一种是先在 CIS 晶圆上植焊球，接着将 IC 芯片和 CIS 晶圆（C2W）键合（见表 10.1）。图 10.15 显示了组装工艺（先植焊球）中的 3D 原型：已组装的协处理器 IC 和未处理的表面。

表 10.1　CIS 和处理器 IC 组装工艺中先堆叠和先植焊球的对比

先堆叠	先植焊球
协处理器晶圆减薄	CIS 助焊剂涂覆
协处理器晶圆划片	CIS 植焊球
协处理器芯片点胶	CIS 晶圆回流
协处理器芯片和 CIS 晶圆键合	协处理器晶圆减薄
回流和清洗	协处理器晶圆划片
CIS 芯片助焊剂涂覆	协处理器晶圆助焊剂点涂
CIS 晶圆植焊球	协处理器芯片和 CIS 晶圆键合
CIS 晶圆回流	CIS 晶圆回流
底部填充胶注入	底部填充胶注入
CIS 划片	CIS 晶圆划片

图 10.15　组装工序中的 3D 原型图（先植焊球）：已组装的协处理器 IC 和未处理的表面

10.2.4　总结和建议

一些重要成果和建议总结如下：

1）介绍了若干现实的 3D CIS 和 IC 异构集成案例。

2）希望在不久的将来更多本领域的案例得到发布。

3）对于晶圆对晶圆混合键合 W2W（同时进行 Cu-Cu 键合和 SiO_2 对 SiO_2 键合），图 10.7 方案更适合 3D CIS 和 IC 异构集成，因为处理器芯片和 CIS 芯片的尺寸相差不大。

10.3　LED 异构集成

由于白光 LED（发光二极管）的快速发展和环保压力，每颗 LED 封装发出的光量在过去的十年里增大了超过 30 倍。本小节将讨论如参考文献 [8-12] 所述的 3D LED 和 IC 的异构集成封装，包括单颗或者多颗 LED 与某种有源 IC（集成电路）芯片，如 ASIC（专用 IC）、LED 驱动器、处理器、存储器、RF（射频）传感器和电源管理器，通过 3D 方式或者采用无源转接板的 2.5D 方式进行集成；介绍并讨论了这些封装的组装工艺，具体讨论了 2.5D IC 和 LED 集成，如采用带空腔的硅基板进行荧光粉涂覆和铜填充的 TSV 进行互连的 LED 封装，以及带空腔和 TSV 的硅基板上的 LED。

10.3.1　采用带空腔和铜填充 TSV 的硅基板 LED 封装

1. 结构

图 10.16 显示了带有用于荧光粉涂覆空腔和用铜填充 TSV 进行互连的硅基

板的截面图[9, 10]。硅基板厚度大约400μm，正反面有3μm厚的低温氧化层。空腔的尺寸为1.3mm×1.3mm×0.22mm。TSV的直径为100μm并填充铜。TSV铜柱顶部突出硅片30μm并镀锡。

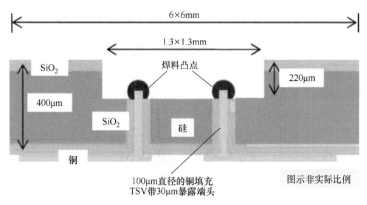

图 10.16　带 TSV 和空腔的硅基板

2. 关键工艺步骤

LED 封装的关键组装工艺流程如图 10.17 所示。首先，实施深反应离子刻蚀（DRIE）在硅片背面刻蚀出盲孔，在正面刻蚀出腔体，如图 10.17a 所示。接下来进行绝缘层、阻挡层和种子层沉积以及通孔电镀铜填充，如图 10.17b 所示。然后用 KOH（无机化合物氢氧化钾分子式）和 BOE（缓冲的氧化物刻蚀剂，缓冲的氢氟酸溶液）进一步刻蚀腔体，使得正面的 TSV 填充铜柱顶部露出，如

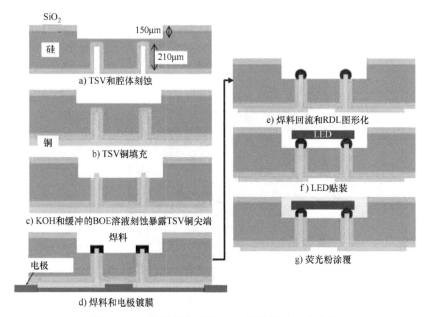

图 10.17　采用硅基板的 LED 封装制造工艺流程

图 10.17c 所示。在 TSV 铜柱顶部镀锡，接着再在背部电镀电极，如图 10.17d 所示。锡膏回流并在硅片背面进行 RDL 图形化和电镀，如图 10.17e 所示。接下来拾取、放置倒装芯片（1mm×1mm×0.07mm）蓝光 LED 器件并回流形成焊点，如图 10.17f 所示。最后，再涂覆荧光粉胶填充腔体，如图 10.17g 所示。

图 10.18a 显示了铜填充 TSV 尖端的图像。图 10.18b 显示了铜填充 TSV 尖端电镀焊料的图像（焊料回流前），图 10.18c 显示了铜填充 TSV 尖端电镀焊料的图像（焊料回流后）。可以看出，电镀和回流工序实施效果非常好。图 10.19a 和图 10.19b 分别显示了 LED 封装的顶部视图和横截面视图。图中能够清晰地看到硅基板、RDL、腔体和带焊凸点的 TSV 顶部（LED 芯片贴装之前）。

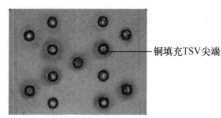

a) 铜填充TSV尖端

 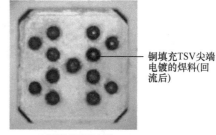

b) 铜填充TSV尖端电镀的焊料(回流前)　　c) 铜填充TSV尖端电镀的焊料(回流后)

图 10.18　铜填充 TSV 尖端的图像

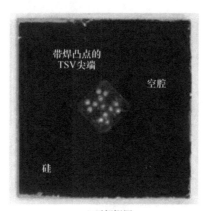

a) 顶部视图

图 10.19　带有空腔和尖端有焊凸点的 TSV 硅基板顶部视图和截面视图

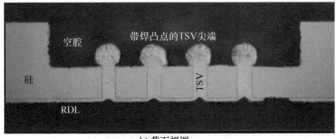

b) 截面视图

图 10.19 带有空腔和尖端有焊凸点的 TSV 硅基板顶部视图和截面视图（续）

图 10.20a 通过示意图显示了刮印干燥的 YAG:Ce（铈掺杂钇铝石榴石）黄色荧光粉来封装 LED 芯片并填充空腔。首先，将少量的（1μL）紫外固化胶滴入空腔，并用紫外灯预固化 10s，紫外灯照射功率密度为 75mW/cm^2。将荧光粉涂在硅基板上，并用橡皮刮刀使其填充印满空腔。最后，将紫外固化胶固化 6min 使其固定荧光粉并完成 LED 芯片的封装。图 10.20b、c 分别显示了荧光粉印制前和印制后的空腔图像。

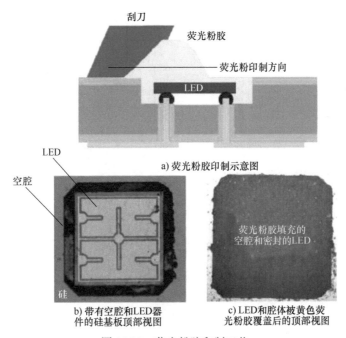

图 10.20 荧光粉胶印制工艺

图 10.21a 和图 10.21b 分别显示了最终 LED 封装的顶部视图和横截面视图。可以看出，LED 芯片嵌入空腔并被黄色荧光粉覆盖，盖帽在空腔内成型并形成一个平面的轮廓。

a) LED顶部视图

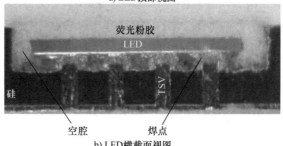

b) LED横截面视图

图 10.21　带有空腔硅基板和荧光粉胶覆盖的 LED 顶部视图和横截面视图

图10.22a 和图10.22b 分别显示了没有荧光粉和有荧光粉的 LED 点亮的图片。可以看出，黄色荧光粉将蓝光 LED 转换成白光 LED。

a) 未涂覆荧光粉

b) 涂覆荧光粉(白光)

图 10.22　点亮的未涂覆 / 涂覆荧光粉的蓝光 LED

10.3.2　基于 TSV 的 LED 晶圆级封装

1. 结构

图 10.23 显示了一种将 LED 器件封装在硅基板上空腔中的晶圆级封装示意图[11, 12]。可以看出，硅基板正面有空腔用来容纳 LED 器件，通过 TSV 实

第 10 章　CIS、LED、MEMS 和 VCSEL 的异构集成

现 LED 和芯片背面 RDL 的互连。硅基板上方为带有黄色荧光粉的玻璃盖板。图 10.24 为参考文献 [11，12] 试验件使用的 1W 的 LED 芯片。其中，LED 芯片的尺寸为 885μm×885μm，正极焊盘和负极焊盘的尺寸分别为 100μm×200μm 和 100μm×100μm。

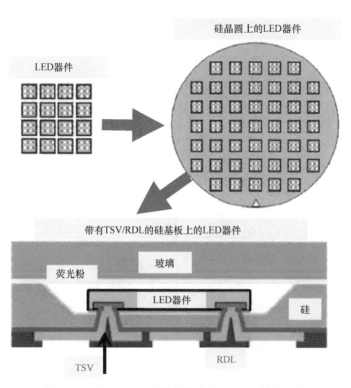

图 10.23　用于 LED 封装的带空腔和 TSV 的硅基板

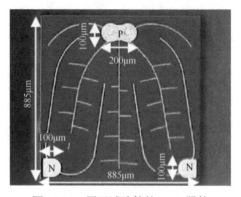

图 10.24　用于试验件的 LED 器件

2. 关键工艺步骤

图 10.25 显示了前面提出的 LED 封装的组装工艺流程。首先，通过 DRID 在硅片正面刻蚀出空腔，如图 10.25a 所示。接下来溅射反射金属层，如图 10.25b 所示。然后用聚合物胶将 LED 芯片粘贴在硅基板空腔的底部，如图 10.25c 所示。带荧光粉的光学玻璃晶圆和带 LED 的硅基板间进行键合（晶圆对晶圆），如图 10.25d 所示。接下来将硅基板背面研磨减薄，如图 10.25e 所示。通过光刻胶、掩模版、光刻、图形化和 DRIE 制造 TSV，如图 10.25f 所示。通过旋涂和激光打孔在硅基板背面制造掩蔽层和互连窗口，如图 10.25g 所示。然后溅射种子层、光刻并电镀 RDL，如图 10.25h 所示。最终，通过光刻工艺制造阻焊层，如图 10.25i 所示。各个工艺步骤的图片如图 10.26 所示。

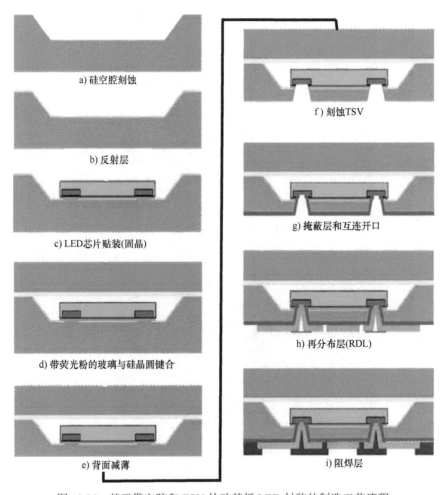

图 10.25 基于带空腔和 TSV 的硅基板 LED 封装的制造工艺流程

第 10 章 CIS、LED、MEMS 和 VCSEL 的异构集成

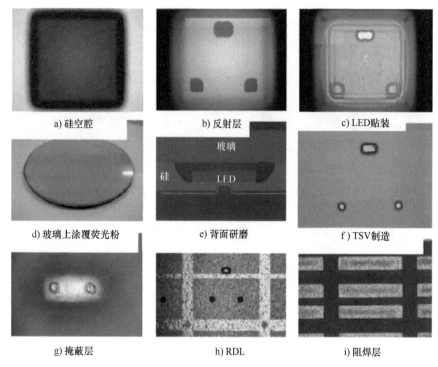

图 10.26　LED 封装工艺流程图片

封装好的晶圆通过划片切割成单个 LED 封装（2000μm × 2000μm）。图 10.27 为不带荧光粉的 LED 封装的横截面图。可以看出，LED 芯片粘贴在硅基板空腔的底部，玻璃键合在硅基板的正面，LED 芯片上的焊盘和 RDL 通过 TSV 连接，接触区域长度约为 20μm，掩蔽层将硅基板和 RDL 隔离开，能够阻止阴极和阳极之间的短路。

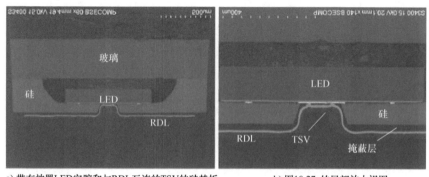

图 10.27　不带荧光粉的 LED 封装结构的截面 SEM 图

265

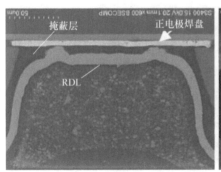
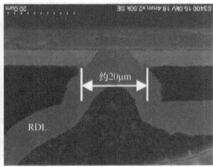

c) 图10.27b的局部放大视图　　　　　　d) 图10.27c的局部放大视图

图 10.27　不带荧光粉的 LED 封装结构的截面 SEM 图（续）

3. 性能

图 10.28a 显示了在一个 MCPCB（金属芯印制电路板）上的带荧光粉 LED 封装。MCPCB 的背面和一个散热器通过 TIM（热界面材料）相连，如图 10.28b 和图 10.28c 所示。带荧光粉的 LED 点亮照片如图 10.28d 所示。可以看出 LED 发出了混杂着白色和蓝色的光，意味着荧光粉的含量 / 涂覆工艺需要进行改进。

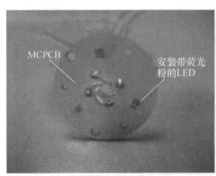

a) 贴装在MCPCB上　　　　　　　　b) 背面涂覆TIM材料

c) 安装散热器　　　　　　　　　　d) 通电

图 10.28　MCPCB 上带荧光粉的 LED 封装

图 10.29 显示了不同工作电流下 Labsphere（蓝菲光学）积分球测试的结果。带荧光粉 LED 封装模块的工作电流为 50mA、200 mA 和 350mA。可以看出，光功率随着输入电流的增大而增加，工作电流为 350mA 时 LED 的光功率为 189.98mW，大约是原始 LED 芯片的 85%（这表明通过晶圆级封装实现了可靠的电互连），LED 的光效随着输入电流的增大而降低（这表明 LED 芯片的结温随着输入电流增大而升高）。因此，这里展示的 LED 封装系统的热管理需要进行优化。

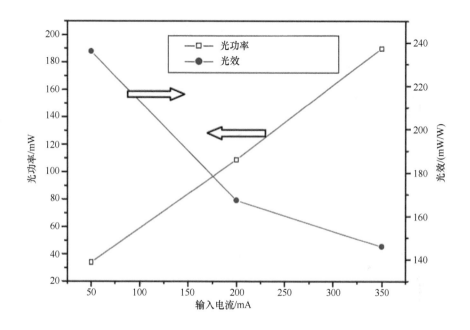

图 10.29　光效和光功率相对试验件输入电流的关系

10.3.3　总结和建议

一些重要成果和建议总结如下[9-12]：

1）介绍和讨论了一些 2.5D IC 和 LED 集成案例。基本上，集成结构中包含带容纳 LED 芯片的空腔的硅基板（硅转接板）和用于将 LED 芯片与下一级进行互连的 TSV。

2）大量的工作仍然有待开展。然而，用硅转接板支撑 LED 芯片封装的可行性得到验证。

3）下一步将通过带 TSV 的有源芯片取代硅基板（硅转接板），实现一种真正的 IC 和 LED 的 3D 集成。

10.4 MEMS 异构集成

MEMS 即微机电系统，其意为采用微加工技术将机械部件如传感器、执行器和电子器件集成在一块普通的硅基板上[13-21]。传统电子 IC（集成电路）如 CMOS 制造所采用的一些基本技术和工艺步骤，如刻蚀、图形化、掺杂和互连，通过选择性地刻蚀掉硅片的一部分或者增加新的结构层来制造 MEMS 器件，同样可用于制造（微加工）微机械元件。然而，MEMS 器件和 IC 器件之间有一个本质的区别，即 MEMS 器件必须物理上可移动。MEMS 器件采用电子器件去驱动其机械元件（部件）！

大部分人认为智能手机之后的下一个大事件就是 IoT（物联网）。IoT 有时也被称为 M2M（机器到机器）革命，MEMS 是其中一组重要的机器，将在 IoT 繁荣的扩张中将发挥关键的作用。如 MEMS 传感器使得设备可以收集和数字化现实世界的数据，并将其分享到互联网上。IoT 为 MEMS 市场带来了一个重要的新增长机会。

本小节将对基于 TSV 的 RF MEMS 器件晶圆级封装、基于 FBAR 振荡器的晶圆级封装及基于低温键合的 3D IC MEMS 异构集成进行介绍。

10.4.1 基于 TSV 的 RF MEMS 器件晶圆级封装

1. 结构

图 10.30 为一个 RF MEMS 器件[15, 16]采用晶圆级封装的横截面图和鸟瞰图，包括 MEMS 晶圆（RF MEMS 器件、带有 RDL 的高阻硅（HR-Si）晶圆和用于密封和焊盘键合的 AuSn 焊料）和盖板晶圆（盖板晶圆中有空腔、TSV、RDL 和焊凸点）。它是一个 2.5D MEMS 和 IC 集成。参考文献[15, 16]的主要目的是研究在和带 TSV 盖板晶圆封装时 RF-MEMS 晶圆的插入损耗。如图 10.30 所示，文中设计和制造了两种不同的 CPW（共面波导）结构（1mm CPW 和 2mm CPW）。

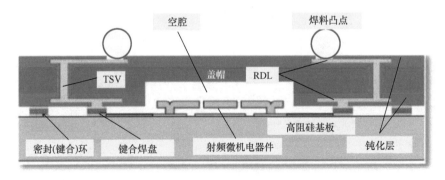

图 10.30 MEMS 晶圆级封装示意图（一个典型的带 TSV 的 RF-MEMS 封装设计布局，不同长度的 CPW 结构）

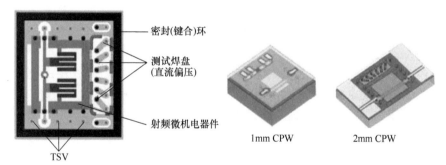

图 10.30 MEMS 晶圆级封装示意图（一个典型的带 TSV 的 RF-MEMS 封装设计布局，不同长度的 CPW 结构）（续）

共面波导线采用铝（厚 1.5μm）制成。一共有三种可以考虑的钝化方案：3kÅ 多晶硅、2μm 低应力氧化层和 1kÅSiN。密封环和键合焊盘上的 UBM（凸点下金属化层）为 0.1μm 钛 /0.2μm 铂 /2μm 金 /2.15μm 锡 /500Å 金。

2. 关键工艺步骤

盖帽晶圆的制造工艺流程如图 10.31 所示。首先，通过 DRIE 制备 TSV

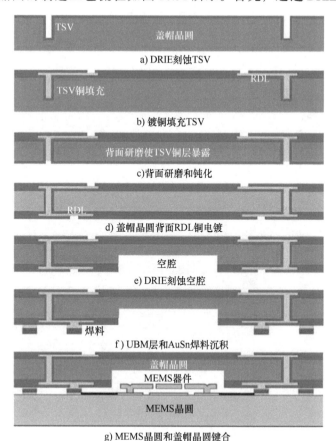

图 10.31 带 TSV 的盖帽晶圆和 W2W 键合制造工艺流程示意图

异构集成技术

（60μm 直径、200μm 深），如图 10.31a 所示。接下来通过电镀进行 TSV 填充和 RDL 制备，如图 10.31b 所示。然后对盖帽晶圆背部研磨减薄至 200μm，使得 TSV 的顶部暴露出来并进行钝化，如图 10.31c 所示。在盖帽晶圆背面电镀 RDL 铜层，如图 10.31d 所示。通过 DRIE 将空腔刻蚀至 100μm 深，如图 10.31e 所示。UBM 和 AuSn 焊点金属沉积，如图 10.31f 所示。薄晶圆的处理是必要的。在背部研磨暴露出 TSV 之前，盖帽晶圆的正面和一个支撑晶圆（载体）用 BSI 的 HT10.10 黏结胶临时键合在一起。焊点沉积结束后，晶圆通过热滑移法进行解键合（分开）。

图 10.31g 显示了 MEMS 晶圆和盖帽晶圆键合（W2W）的示意图，键合条件如图 10.32 所示。可以看出，AuSn 焊料的焊接温度为 280℃，键合时间为 30min，键合压力为 10kN。图 10.33 显示了键合后晶圆（盖帽晶圆在顶部，

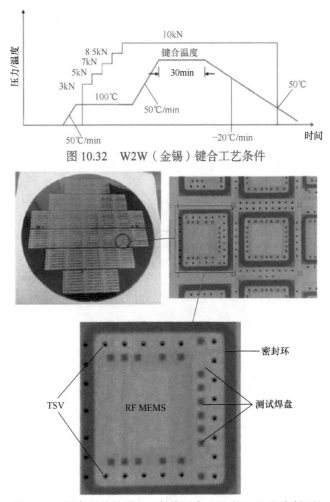

图 10.32　W2W（金锡）键合工艺条件

图 10.33　键合后的晶圆（X 射线图中显示出 TSV 和密封环）

MEMS 晶圆在底部）的光学图像和 X 光图像，可以看到没有出现什么异常。图 10.34 显示了键合封装的 X 光和 SEM 图像，可以看出密封环中没有空洞和分层。图 10.35 显示了键合后的 MEMS 封装图像。可以看出 TSV 中没有空洞，TSV 结构的绝缘层、阻挡层和种子层不存在分层。

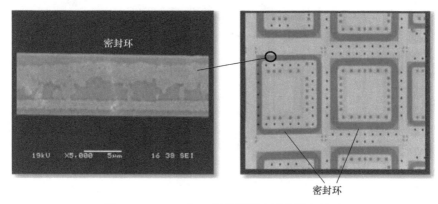

图 10.34　SEM 和 X 射线图像中显示的密封环

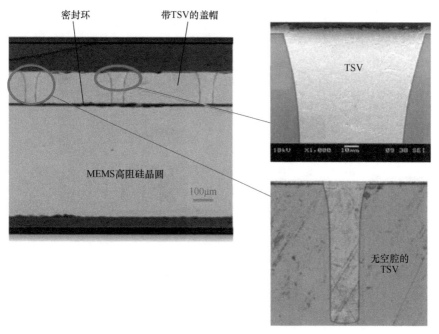

图 10.35　盖帽晶圆和 MEMS 晶圆键合后的截面图，图中可以看到 AuSn 密封环和无空腔的 TSV

3. 性能

在开始 19GHz 谐振频率前，测量 1mm CPW 线在不同频率下封装的插入损耗，如图 10.36 所示。插入损耗在 5~20GHz 频率范围内为 0.12~0.2dB，1dB 截止频率为22.7GHz。2mm CPW 线的插入损耗在5～25GHz 频率范围内为0.14～0.3dB，1dB 截止频率为27.4GHz，如图10.37 所示。由于在小于20GHz 频率范围内可接受的封装损耗小于0.2dB，1dB 截止频率大于20GHz。因此，1mm CPW 的设计是可行的。

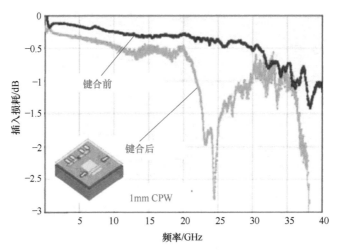

图 10.36　1mm CPW MEMS 封装设计的插入损耗

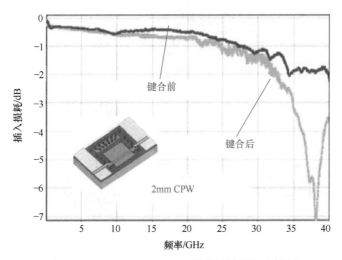

图 10.37　2mm CPW MEMS 封装设计的插入损耗

10.4.2　基于 FBAR 振荡器的晶圆级封装

1. 结构

图 10.38 展示了基于 FBAR（薄膜体声波谐振器）的振荡器[17, 18]的晶圆级封装。图 10.38a 展示了封装结构的盖帽晶圆（上半部分）。可以看出其中有一块

IC 元件（330μm×285μm）、6 个用于外部互连的 TSV、2 个内部互连点（ICP）和一个用于 FBAR 的小空腔。图 10.38b 显示了封装结构的 FBAR 晶圆（下半部分）。可以看出其中有一个 FBAR、6 个外部焊盘（对应前面的 6 个 TSV）、2 个内部互连点（从盖帽上的电路到 FBAR）和一个空腔（用于 IC 元件）。这是一个真实的 3D MEMS 和 IC 异构集成。FBAR 晶圆和盖帽晶圆都采用高阻硅以减少串扰和电容损耗。

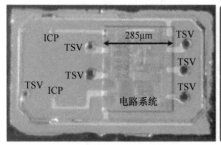 

a) 用于 FBAR 封装的带有 IC 器件、TSV、内部互连和空腔的盖帽晶圆

b) 带有 FBAR 焊盘、内部互连和容纳 IC 元件空腔的 FBAR 晶圆

图 10.38　FBAR 气密性封装照片

2. 关键工艺步骤

图 10.39 为不含 IC 元件的 FBAR 封装制造工艺流程。对于 FBAR 晶圆，首先确定并制造空腔并在空腔中填充牺牲层氧化物，如图 10.39a 所示。制造 FBAR/ZDR（零漂移谐振器），如图 10.39b 所示。在焊盘和密封环上沉积 Au 并图形化，接着去除牺牲层氧化物，如图 10.39c 所示。对于盖板晶圆，采用标准的微加工工艺制造空腔，采用 DRIE 技术刻蚀 TSV。在内部焊盘、外部焊盘、密封环和 TSV 的侧壁电镀金。最后，通过金 - 金扩散键合将盖板晶圆和 FBAR 晶圆键合，如图 10.39d 所示。

图 10.40 为 FBAR 封装中带有 IC 元件的盖帽晶圆制造工艺流程。IC 元件采用标准双极工艺制造，除了中间层电介质被保留在场中直到元件制造完成，如图 10.40a 所示。接下来移除场电介质直到硅表面，如图 10.40b 所示。采用标准的微加工工艺制造空腔和 DRIE 技术刻蚀 TSV。接下来在焊盘和 TSV 侧壁镀金，再将晶圆翻转过来进行键合，如图 10.40c 所示。TSV 并没有进行填充。

图 10.41a 为带 IC 元件和 TSV 的盖帽晶圆与 FBAR 晶圆进行金 - 金扩散键合的示意图。键合后封装结构的截面 SEM 图像如图 10.41b 所示。可见所有的关键部件，如 FBAR、IC 元件、TSV、盖帽、密封环和焊盘位置都正确。详细描述和机械与电性能，可参阅参考文献 [17，18]。

| 异构集成技术

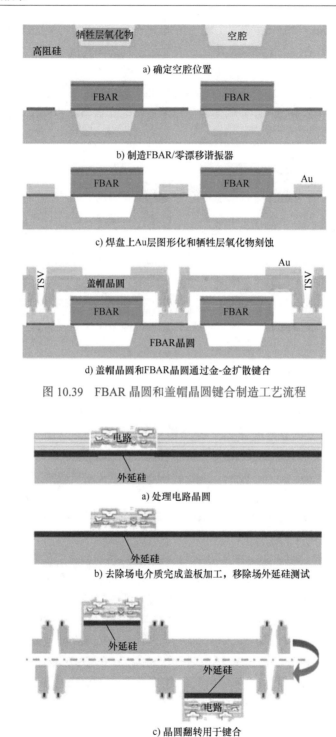

图 10.39　FBAR 晶圆和盖帽晶圆键合制造工艺流程

图 10.40　FBAR 盖帽晶圆制造工艺流程

第 10 章　CIS、LED、MEMS 和 VCSEL 的异构集成

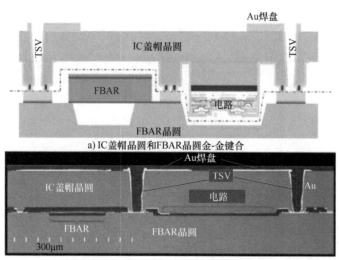

a) IC盖帽晶圆和FBAR晶圆金-金键合

b) 带IC的盖帽和FBAR MEMS键合的封装结构截面SEM图

图 10.41　带 IC 元件和 TSV 的盖帽晶圆与 FBAR 晶圆进行金 - 金键合

10.4.3　基于焊料的 3D MEMS 封装低温键合

大部分现有的芯片对晶圆（C2W）和晶圆对晶圆（W2W）键合方法采用的工艺温度高于 300℃。然而，在键合的过程中，MEMS 器件已经释放（有一些悬空结构，如薄膜、梁和悬臂梁等），需要采用低温键合降低由于键合结构间的热膨胀系数不匹配对微结构造成的损伤（弯曲更小）。室温下的熔融键合（如铜对铜）能够满足要求，但其键合表面必须非常平整和干净（1 级洁净室），并不适合大规模生产。另一方面，通过合理地设计焊盘和 UBM（凸点下金属化层）以及选择焊料材料，实现 180℃和更低的键合温度是可行的[19-21]。在键合中和键合后，焊料内部和 UBM 间相互反应，形成的熔点远高于焊料的 IMC（金属间化合物）。

这一特性非常适合 3D MEMS 和 IC 异构集成。例如，当一个 MEMS 器件和一块 ASIC 晶圆通过低熔点焊料键合后，所有的键合（焊料互连）区域形成具有很高再熔温度的 IMC。当盖板晶圆和 ASIC 晶圆（已经与 MEMS 器件键合）时，ASIC 和 MEMS 器件间的互连层不会产生回流。此外，当整个 3D MEMS 和 IC 异构集成模块采用 SMT（表面贴装技术）无铅焊接（260℃）技术贴装到 PCB 上后，ASIC 和 MEMS、盖板和 ASIC 之间的焊点互连结构不会回流。本节展示了将低温焊料用于无功能 3D MEMS 和 ASIC 异构集成封装（如图 10.42 所示）的可行性。

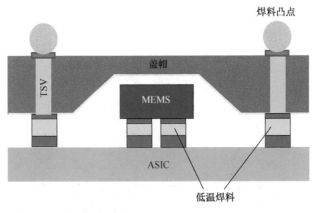

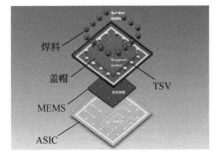
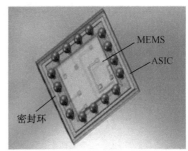

图 10.42 3D MEMS 和 ASIC 异构集成实验模型

1. 不同芯片尺寸的 3D IC 和 MEMS 集成

图 10.43 显示了无功能 MEMS、ASIC 和盖板的尺寸和材料。从图中可知，

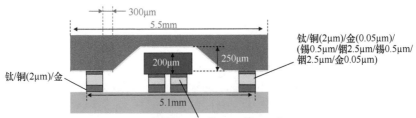

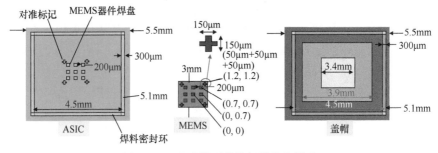

图 10.43 实验模型的几何形状和尺寸

MEMS 的尺寸为 3mm×3mm×200μm，有 9 个焊盘（200μm×200μm），如图 10.43 和图 10.44 所示；ASIC 尺寸为 5.5mm×5.5mm，ASIC 上密封环的宽度为 300μm，沿着 ASIC 每条边方向的长度为 5.1mm，如图 10.43、图 10.45 和图 10.46 所示；键合焊盘（对于 MEMS 器件）和密封环都用低温焊料涂覆。

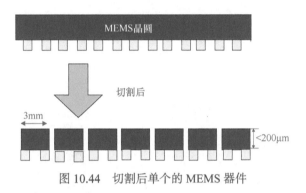

图 10.44 切割后单个的 MEMS 器件

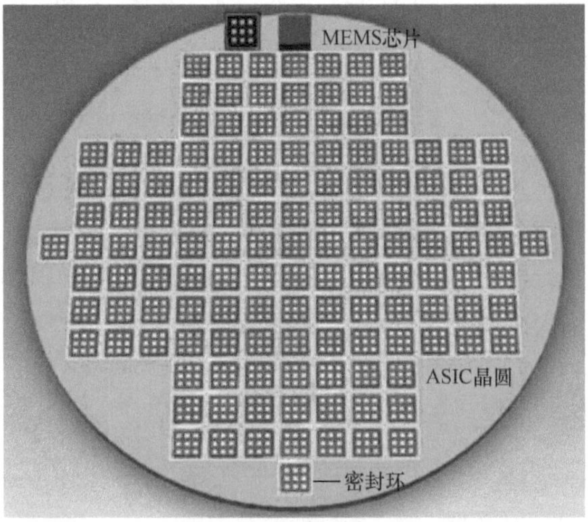

图 10.45 显示 MEMS 器件位置的 ASIC 晶圆示意图

与本示例展示的情况一样，通常 MEMS 和 ASIC 的芯片尺寸在 3D MEMS 和 IC 异构集成中并不一样。因此，将这两种器件进行晶圆对晶圆键合可能并不是一个好的主意，互连通常采用芯片对芯片或者芯片对晶圆键合的方式实现。本小节讨论了一个仅使用（Ti/Cu/Au）UBM 的 MEMS 芯片通过低温焊料（AuInSn）和 ASIC 晶圆键合的案例（见图 10.43）。

2. 盖帽晶圆上的空腔和 TSV

图 10.47 和图 10.48 展示了一个采用 KOH（氢氧化钾）湿法工艺刻蚀空腔的直径 200mm 盖帽晶圆及其典型截面图。可以看出，KOH 刻蚀产生的壁十分

平滑。盖帽晶圆的TSV采用DRIE（深反应离子刻蚀）工艺加工。空腔和TSV也可以通过激光进行加工，如图10.49所示。尽管其加工质量不如KOH刻蚀好，然而对于大部分应用场景，激光加工可以适用。激光加工得到的TSV表面状况非常粗糙，并可能会影响电性能。因此，通常条件下，盖帽晶圆上的TSV采用DRIE工艺加工。

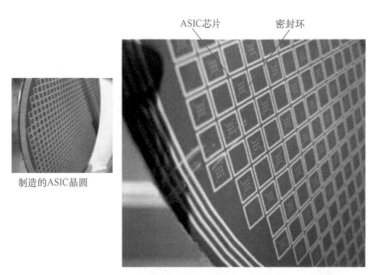

图10.46　标注了MEMS器件和密封环位置的ASIC晶圆

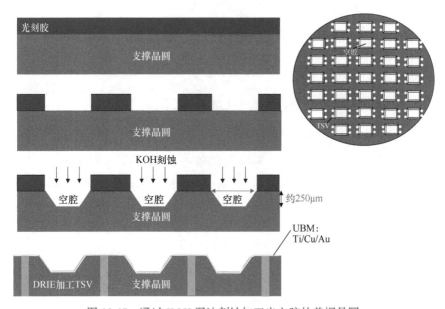

图10.47　通过KOH湿法刻蚀加工出空腔的盖帽晶圆

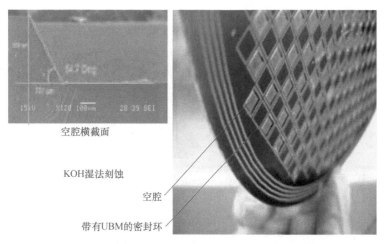

图 10.48　盖帽晶圆湿法（通过 KOH）工艺加工结果

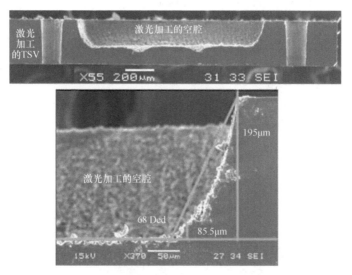

图 10.49　通过激光加工的空腔和 TSV 的盖帽晶圆

3. MEMS 芯片和 ASIC 晶圆（C2W）键合

图 10.50 展示了无功能的 3D ASIC 和 MEMS 异构集成的组装工艺。可以看出，ASIC 晶圆上加工了密封环和焊盘，密封环和焊盘上覆盖有 UBM 和焊料。MEMS 芯片在 FC150 型倒装焊机中被拾取并放置在 ASIC 晶圆上，如图 10.50 和图 10.51 所示。键合参数为：键合压力 6MPa，键合工具（MEMS）温度 200℃，ASIC 晶圆温度 90℃，键合时间 40s。以上操作是在封闭的氮气环境下进行。图 10.52 为键合的横截面图。图中显示没有对准错位存在。

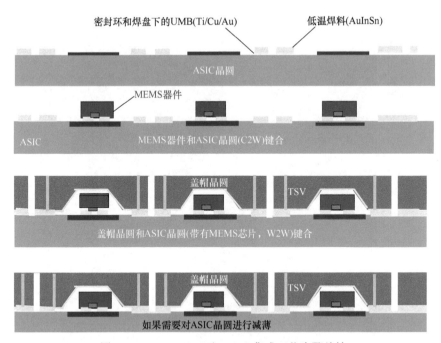

图 10.50　3D MEMS 和 ASIC 集成工艺步骤总结

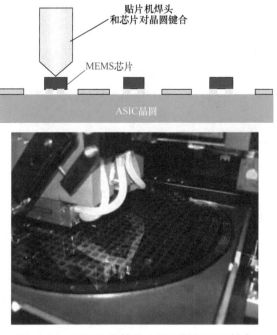

图 10.51　MEMS 器件和 ASIC 晶圆 C2W 键合

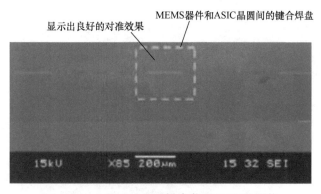

图 10.52 MEMS/ASIC 键合截面的 SEM 图像

4. 带有 MEMS 芯片的 ASIC 晶圆和盖帽晶圆（W2W）键合

如前所述，200mm 直径盖帽晶圆上的腔体通过 KOH 刻蚀工艺加工。盖帽晶圆上的密封环表面只有 UBM。图 10.53 为盖帽晶圆和带有 MEMS 器件的 ASIC 晶圆键合的原理图，采用 EVG 的键合设备，键合压力为 9kN，温度为 200℃，键合时间为 20min。图 10.54 显示了带有 MEMS 器件的 ASIC 晶圆和盖帽晶圆进行晶圆对晶圆键合的截面 SEM 图和 X 光照片。可以看出，MEMS 芯片和 ASIC 芯片的键合焊盘对准良好，焊料密封环中没有空腔。图 10.55 显示了剪切强度测试结果，所有的芯片都符合规范要求。图 10.55 也显示了在焊凸点制造和划片之后一个独立的封装结构的图像。

图 10.53 盖帽晶圆和带有 MEMS 器件的 ASIC 晶圆的 W2W 键合

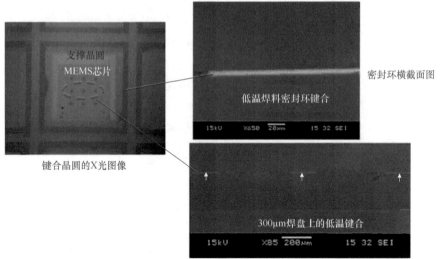

图 10.54 盖帽晶圆和 ASIC 间的密封环在 X 射线和 SEM 下显示的图像

键合压力	温度	时间	剪切强度 (盖帽-ASIC)	剪切强度 (MEMS-ASIC)
9kN	200℃	20m	71.6MPa	9.2MPa

带有植入焊球的完整封装

图 10.55 用于性能表征测试的完整 3D MEMS 和 ASIC 集成封装结构

10.4.4 总结和建议

一些重要成果和建议总结如下[13-21]：

1）IC 芯片、MEMS 器件和 TSV 转接板的异构集成应用越来越多。

2）大部分研究论文盖帽中有 TSV 结构但没有任何器件（2.5D MEMS 和 IC 集成）。

3）一些论文中采用 TSV 转接板来支撑 MEMS 器件（2.5D MEMS 和 IC 集成）。

4）有部分论文介绍了关于 FBAR MEMS 器件和带有 TSV 和 IC 元件的盖帽的键合，以及采用低温焊料实现 3D MEMS 和 ASIC 异构集成的键合。

10.5 VCSEL 和 PD 的异构集成

本节将介绍和描述嵌入 PCB 或者有机层压基板的低成本（带有裸芯片）和高性能（光、电、热和机械）光电系统及其设计[22-31]。这些系统包括一块 PCB（或基板）、嵌入式光聚合物波导、嵌入式 VCSEL（垂直腔表面发射激光器）、嵌入式驱动芯片、嵌入式串行器、嵌入式 PD 探测器、嵌入式 TIA（跨阻放大器）、嵌入式反串行器、嵌入式均热板和一个散热器。下面首先介绍一个简单系统的设计和制造。

10.5.1 嵌入式板级光互连

1. 结构

图 10.56 显示了一个采用传统 PCB 制造工艺制造的嵌入波导的单通道 OECB（光电路板）。OECB 由 4 层电路层和 1 层嵌入在一个 60μm 厚的基板下面的光学层组成。加工 2 个 100μm 直径的光学孔引导光束从 VCSEL 到 45°反射镜耦合器上。同样地，从光波导中出来的光束通过 45°反射镜耦合器转向并经过光学孔被光电探测器接收。

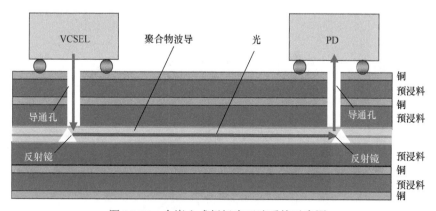

图 10.56　全嵌入式板级光互连系统示意图

10cm 长的嵌入式聚合物波导由 70μm 粗的线芯和 15μm 厚的顶部与底部覆层组成。在波导的两个角上用 90°金刚石切割刀加工出 2 个 45°的反射镜耦合器。这些反射镜将从 VCSEL 发出的垂直路径上的光束转换成沿着平面方向并进入波导。

2. 聚合物波导

可多种制造工艺制造柔性的聚合物波导，如光刻[32, 33]、RIE（反应离子刻蚀）、热压印和软注塑[34]。其中软注塑法成本低，并且适合于制造微纳米结构[32-36]。软注塑工艺适合由单步工艺复制出 3D 结构，也适合批量生产或者少量的样品原型制造。

采用紫外光固化的含氟丙烯酸酯材料来制造波导的过程分为四个步骤，如图 10.57 所示。首先，在支撑基板的覆铜聚酰亚胺薄膜上表面旋涂一层底披覆层材料（WIR30-RI）。接着将 PDMS（聚二甲基硅氧烷）模具放置在覆层材料的顶部。然后填充芯材料，并在氮气环境中紫外光固化。将软模具从聚合物基板上移除。最后，在线芯上再旋涂一层覆层材料作为顶披覆层并通过紫外光固化。

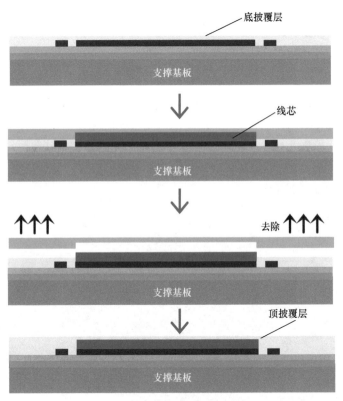

图 10.57　聚合物波导制造步骤

3. 制造 45° 微反射镜

为了在波导和有源光学器件之间建立 3D 光互连，通常采用集成微反射镜来辅助光线偏转，从而使光线能传入/传出波导。因此，微反射镜是 3D 光互连的一个重要的基本组成部分[37]。到目前为止，微反射镜的制造工艺方法包括激光烧蚀[24, 38]、灰度光刻[39]、切割[40]、X 射线光刻[41]、反应离子刻蚀[42]和注塑[43]。本小节将讨论切割法。

在完成波导制造之后，接下来的工序是用 90° 金刚石切割刀采用机械切割的方法制造 45° 微反射镜。在制造微反射镜时需要注意一些关键设计问题，包括反射镜的定位、倾斜角度，反射镜的角度和反射镜表面的粗糙度，因为这些问题会影响光耦合效率。因此，选择了一组不同的切割参数用于制造 45° 微反射镜，见表 10.2。表 10.2 提供了主轴转速和进给速率的数据，这些参数影响微反射镜的表面粗糙度和角度。制造的结果将决定哪一组切割参数能够达到预期的要求。

表 10.2 微反射镜切割的切割参数

样品	主轴转速/(r/min)	进给速率/(mm/s)
X1	35000	1
X2	35000	5
X3	45000	1
X4	45000	5

将用基板支撑的光波导安装在切割机上，然后使用对准标记进行对准，如图 10.58 所示。在切割线的位置切割微反射镜，制造 45° 的微反射镜。图 10.59 为带有微反射镜的光波导的侧视图。

在完成 45° 微反射镜制造后，接下来的工序是确保微反射镜的角度在规定范围内。为了得到微反射镜的角度数据，对光波导的截面进行分析，并用 SEM 测定。利用 SEM 采集到图像后，利用 3D 图形软件测量微反射镜的角度，如图 10.60 所示。测量角度的步骤如下：

1）画一条与波导表面平面平行的基准线。

2）沿着反射镜的边缘画出镜面线。画出 2 条镜面线，相交于一点。

3）通过两条镜面线的交点画一条和基准线垂直的分界线，形成一个 90° 的角。

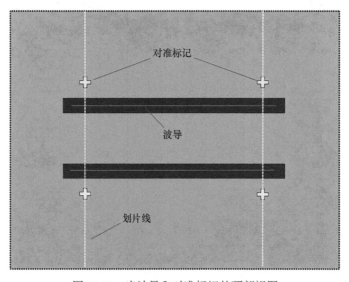

图 10.58 光波导和对准标记的顶部视图

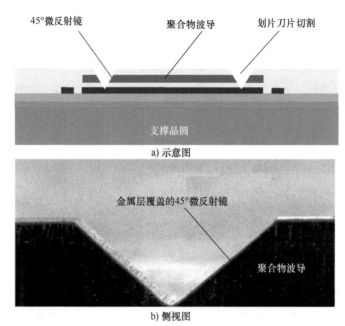

图 10.59　45°微反射镜示意图与侧视图

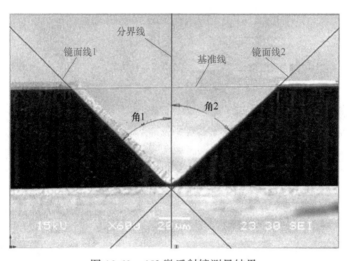

图 10.60　45°微反射镜测量结果

4）使用 3D 图形软件的测量工具，测量从镜面线 1 到分界线间的夹角，即角 1，如图 10.60 所示。对于角 2，测量镜面线 2 到分界线间的夹角。

最终，将角度测量结果（角 1 和角 2）根据之前确定的切割参数集制成表格。

没有完美的 45°微反射镜，制造缺陷将会影响耦合效率。由于这种缺陷结构，反射镜面积变小，耦合光源也无法准确地偏转到波导线芯中并耦合到接收器，从而增加了光的损耗。因此，对波导的制造工艺进行分析。为了确定这种缺陷结构是由于波导的固化条件引起的，进行了对比实验。两种波导分别采用不同的固化条件，一种是完全固化，一种是部分（半）固化。接着切割波导制造微反射镜。部分固化的波导有缺陷结构，如图 10.61a 所示，而完全固化的波导呈现无缺陷的完美的 45° 微反射镜形状，如图 10.61b 所示。因此，波导必须完全固化，波导的最佳固化条件为在氮化环境下，紫外光强度为 15~18mW/cm^2，固化时间为 1.5h。

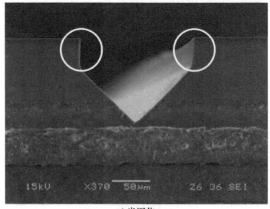

a) 半固化

b) 完全固化

图 10.61　45° 微反射镜侧面视图

产生缺陷的原因是旋转速率和进给速率的配比提供了足够的去除速率，从而将 V 形沟槽尖端的波导材料去除。然而，由于顶部表面去除量的增加，速率的配比不再满足去除速率。因此，在切割过程中，切割刀片将波导材料往旁边推而不是将波导材料移除。于是，当切割载荷移除后，波导材料恢复产生如图 10.61a 显示的缺陷。结果表明，波导材料在切割过程中发生了弹性变形。这一现象对于完全固化的波导材料不会发生，如图 10.61b 所示。因此，可以认为这是由于未完全固化的波导材料仍然处于软的弹性状态。

在 SEM 下发现另外一个问题。从波导的俯视图中，如图 10.62a 所示，可以观察到反射镜有轻微的倾斜。反射镜的位置和倾斜角度会影响光耦合效率，结果导致光损耗增加。为了修正波导线芯定位过程中的位移误差，采用精密对准键合机减小波导倾斜。如图 10.62b 所示为波导位置的测量。

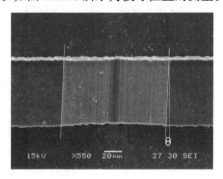

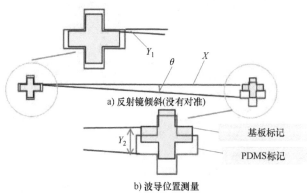

图 10.62　反射镜倾斜与波导位置测量

图 10.58（波导基板俯视图）中，使用一组对准标记来对准 PDMS 软模具，PDMS 软模具用来制造波导的线芯。在 PDMS 模具上，有一组和波导基板上相似的对准标记。这两组标记相互叠加，形成一个完美的没有误差的对准场景。但误差总是会产生，可通过采用对准系统来最大限度地减小错位。对于正常放置的 PDMS 模具，会存在错位导致波导倾斜。表 10.3 显示了人工手动对准放置 PDMS 模具的不同样本的对准偏差。倾斜角度导致了镜面角的倾斜。

表 10.3　沿 Y 方向对准偏差的结果（对准系统）

样品	$Y_1/\mu m$	$Y_2/\mu m$	$\theta/(°)$
1	0	3	0.008
2	1	3	0.006
3	1	4	0.008
4	0	1	0.003
平均			0.006

当切割机沿着直线运动时，参照对准标记与波导材料相垂直，这个角度是由于线芯的位置导致反射镜倾斜形成的，所以线芯的位置至关重要，因此，使用高精度的对准设备放置线芯可以减少倾斜。由于反射镜的角度倾斜，耦合光不会完全耦合到线芯。由于角度误差，将会损失少量的耦合光。

使用对准系统能够极大地改善对准问题。表 10.3 列出了对准系统的几组测量结果。根据表 10.3 中的结果，反射镜在对准系统下的倾斜平均角度为 0.006°，从而最大限度地减小倾斜角度并改善镜面角度，降低光耦合损失；人工对准的反射镜倾斜角度为 0.36°，见表 10.4。

表 10.4　沿 Y 方向对准偏差的结果（手动放置）

样品	$Y_1/\mu m$	$Y_2/\mu m$	$\theta/(°)$
1	10	123	0.32
2	15	44	0.30
3	8	54	0.45
平均			0.36

不同的主轴转速和进给速率影响镜面角度和镜面的粗糙度。使用表 10.2 给出的一组切割参数进行切割后，选择一个最优的切割参数。表 10.5 列出了前文提到的采用 SEM 和 3D 图形软件测量的不同切割参数下微反射镜角度数据。根据表 10.5，获得能够满足要求角度的切割参数为主轴转速 35000r/min、进给速率 1mm/s。该切割参数下加工的微反射镜，角 1 为 45.23°、角 2 为 45.85°，都在 45°±1° 范围内。当切割参数为主轴转速 45000r/min、进给速率 5mm/s 时，微反射镜角度超出了要求范围，影响光耦合效率。镜面角度会比 45° 偏大或者偏小太多，导致耦合光不能精确地偏转到波导的线芯中，从而影响光损耗。

表 10.5　微反射镜的角度测量结果

样品	主轴转速/(r/min)	进给速率/(mm/s)	角 1/(°)	角 2/(°)
X1	35000	1	45.23	45.85
X2	35000	5	46.79	45.73
X3	45000	1	39.93	46.58
X4	45000	5	43.98	42.39

接下来测量微反射镜表面粗糙度。表 10.6 显示，与其他几组结果相比，粗糙度最佳结果为切割参数的主轴转速为 35000r/min、进给速率为 1mm/s 时获得的。在微反射镜表面选取 10 个不同区域进行测量并得到微反射镜表面粗糙度的平均值。表 10.7 给出了反射镜表面 10 个不同区域的粗糙度测量结果。10 个不同区域粗糙度平均值为 316nm，标准差为 0.006nm。其余的切割参数，即使加工获得的粗糙度和 316nm 接近，但是无法满足对角度的要求。因此，主轴转速

35000r/min、进给速率为 1mm/s 是首选的切割参数。

表 10.6 微反射镜粗糙度测量结果

样品	主轴转速 /(r/min)	进给速率 /(mm/s)	粗糙度 /nm
X1	35000	1	316
X2	35000	5	355
X3	45000	1	340
X4	45000	5	389

表 10.7 主轴转速区域序号 35000r/min 和进给速率 1mm/s 时的粗糙度测量结果

区域序号	粗糙度 /μm	区域序号	粗糙度 /μm
1	0.313	7	0.317
2	0.307	8	0.322
3	0.312	9	0.325
4	0.309	10	0.324
5	0.312	平均值	0.316
6	0.315	标准差	0.006

4. OECB 组装工艺

图 10.63 显示了聚酰亚胺薄膜上的微反射镜和聚合物波导的光学图像。组装 OECB 需要解决两个关键挑战，即波导和预浸料之间的层压黏合；波导反射镜、PCB 垂直光通道和光器件在封装时层与层之间的对准精度。考虑到这两点，设计和开发了[28]如图 10.64 和图 10.65 所示的制造工艺。

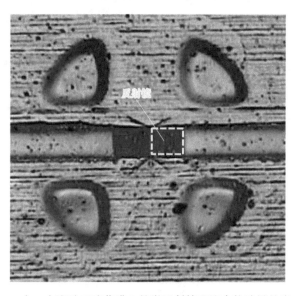

图 10.63 在一个聚酰亚胺薄膜上的微反射镜和聚合物波导的光学图像

第 10 章 CIS、LED、MEMS 和 VCSEL 的异构集成

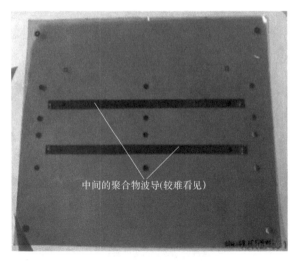

图 10.63　在一个聚酰亚胺薄膜上的微反射镜和聚合物波导的光学图像（续）

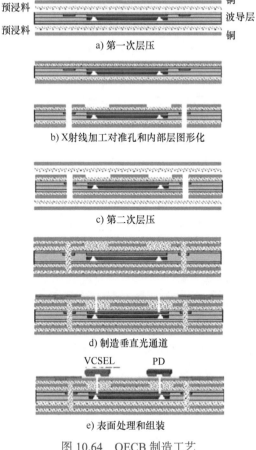

a) 第一次层压

b) X射线加工对准孔和内部层图形化

c) 第二次层压

d) 制造垂直光通道

e) 表面处理和组装

图 10.64　OECB 制造工艺

| 异构集成技术

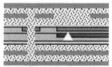
a) 层压的OEPC

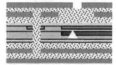
b) 紫外激光刻蚀铜层形成通道加工窗口

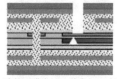
c) 贴装干膜和CO_2激光加工盲孔

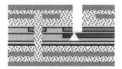
d) 刻蚀底部铜层和剥离干膜

图 10.65 垂直光通道制造工艺

在层压前，对波导层表面进行表面预处理，如等离子处理、微刻蚀或者浮石抛光等。这些预处理工艺可以增加表面粗糙度。因此，波导层和预浸料之间的黏附力也会相应增加。然后，将波导层置于两片带有铜箔的预浸料层中间进行堆叠，进行层压工艺处理，形成铜 - 预浸料 - 波导 - 预浸料 - 铜的三明治结构，上下两层带有铜箔的预浸料层分别称为 L2/L3，如图 10.64 所示。层压工艺条件为温度 200℃、时间 100min。层压后，采用 X 射线检测对准标记并钻出基准孔用于 L2/L3 的图形化，L1/L4 层的层压工艺也采用相同的方法。同时，用 X 射线钻出基准孔，继续钻孔形成通孔并镀铜。然后形成垂直的光学通道使得反射镜暴露出来。最后，进行外层图形化、表面处理，一个 4+1 层的 OECB 制造完成。

5. 垂直光通道制造工艺

垂直光通道的加工是整个工艺流程中最重要的一步。这里对一种特殊的工艺进行了设计和讨论。讨论的 3 个关键点为如何实现从顶层与位于 PCB 内部层的微反射镜的对准、如何控制通道侧壁的粗糙度，以及如何控制通道的形状。特殊的制造工艺如图 10.65 所示。

首先，用紫外激光在所需的垂直光通道位置加工出一个保形铜掩膜，将干膜贴在表面。紫外激光可以直接精确地定位并钻孔。采用 CO_2 激光器烧蚀预浸料并形成一个盲孔，通过调整激光参数可以优化壁面的粗糙度。接着用闪蚀工艺去除底层的铜，剥离干膜，形成垂直的光通道。最后，采用传统的 HDI（高密度互连）工艺并完成 OECB 的制造。

6. 最终组装

OECB 制造完成后，采用 SMT（表面贴装技术）完成最后的组装。焊膏为 SAC305，最高温度为 240℃。VCSEL 和 PD 的焊盘尺寸只有 80μm，因此需要采用显微镜进行焊接。所选的 VCSEL 能够发射波长为 850nm 的激光光束，频率可以达到 5Gbit/s，发射器开口直径为 13μm，光耗散角约为 17°。组装完成的样品如图 10.66 所示，其尺寸为 12cm×5cm。测试装置的原理如图 10.67 所示，部分测试结果见参考文献 [26, 28, 44]。

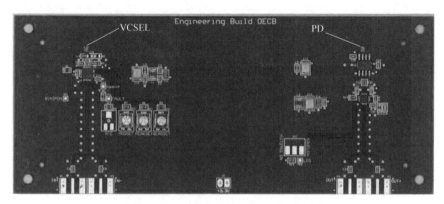

图 10.66　组装完成的 OECB 的顶视图

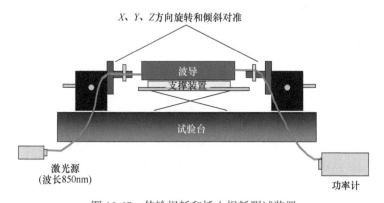

图 10.67　传输损耗和插入损耗测试装置

10.5.2　嵌入 OECB 的 3D 异构集成

1. 结构

针对超薄、超轻、高性能和低成本的应用场景，参考文献 [29-31] 提出了一种集成了光、电 IC 的 3D 混合光电系统，如图 10.68 所示。整个系统由采用

异构集成技术

FR4（环氧树脂）材料（或有机基板）制成的刚性 PCB、嵌入式光学聚合物波导、嵌入式 VCSEL、嵌入式 LD（激光驱动器）、嵌入式串行器、嵌入式 PD 探测器、嵌入式 TIA、嵌入式反串行器和包含或者不包含散热器的嵌入式均热板组成。VCSEL、LD 和串行器裸芯片堆叠在一起，然后贴装到 PCB 中嵌入式光学聚合物波导的一端。值得注意的是，由于 VCSEL 和光学聚合物波导的反射镜非常接近，因此光学透镜是可选的。类似地，PD 探测器、TIA 和反串行器的裸芯片也被堆叠起来然后贴装在 PCB 中嵌入式光学聚合物波导的另一端。这两个 3D 堆叠的芯片组都采用特殊的底部填充胶封装，如透明的聚合物。

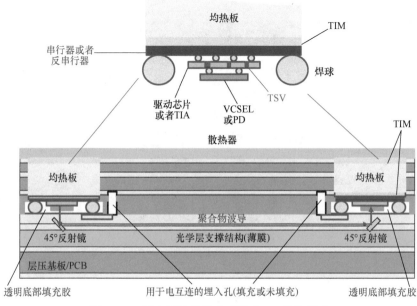

图 10.68　光电互连的嵌入式混合 3D 异构集成

2. 组装工艺

如图 10.68 所示，VCSEL 裸芯片（光从芯片背部发出）通过倒装焊接的方式与带 TSV 结构的 VCSEL 驱动芯片表面的材料实现互连，VCSEL 驱动芯片通过焊凸点倒装在串行器芯片上。较大的焊凸点通过焊接材料焊接在还处于整晶圆形式的串行器芯片上。

在将 3D 混合 IC 芯片组划片后，将芯片组安装在 PCB（或者基板）上聚合物波导的上方。可能需要专门的密封材料如透明的聚合物来保护芯片组。如果需要，可以通过 TIM（热界面材料）将均热板或者某些导热材料黏结到串行器芯片背面。同样地，如果需要，可以用 TIM 将散热器黏结到均热板的顶部。

相似地，光电探测器芯片通过焊凸点倒装在 TIA 芯片上，然后 TIA 芯片通过焊凸点倒装在反串行器芯片上。芯片组的热管理技术和 VCSEL 芯片组所用到的热管理技术相同。本案例中，在 VCSEL 芯片和 VCSEL 驱动芯片、VCSEL 驱动芯片和串行器芯片、PD 芯片和 TIA 芯片、TIA 芯片和反串行器芯片之间的空间可能需要填充底部填充胶。

3. 案例

如图 10.68 所示，对一种基于板级应用的低成本带电和光 IC 的 3D 混合光电系统进行设计和分析[29-31]。整个系统包括一块由 FR4 制成的刚性 PCB、嵌入式光学聚合物波导、一个 10Gbit/s 的 VCSEL（0.31mm × 0.4mm × 0.2mm，2.2V，33mW，16W/cm^2）、一个 10.7Gbit/s 的 LD（2mm × 2mm × 0.3mm，3.3V，0.35W，6.35W/cm^2）、一个 10Gbit/s 16:1 串行器和多路转换器（4.5mm × 4.5mm × 0.8mm，3.3V，2.5W，2.5 ~ 6.7W/cm^2）、一个 10Gbit/s PD 探测器（0.31mm × 0.4mm × 0.23m）、一个 10.7G bit/s 跨阻抗放大器或者 TIA（2mm × 2mm × 0.3mm）、一个 1∶16 反串行器或解复用器（4.5mm × 4.5mm × 0.8mm，3.3V，2.5W，2.5 ~ 6.7W/cm^2）和散热器组成。本案例给出了所提出的 3D 光电系统的光学、热学和结构设计的原理及其指标，并对分析结果进行讨论。

4. 光学设计原理、仿真和结果

除了均热板/散热器外，图 10.68 显示 3D 系统最顶层包含一个被称为串行器的实现由并联到串行的电转换器。串行器将 16 位 622Mbit/s 的并行信号转换成输入到 3D 系统下一层（VCSLE LD）的 10Gbit/s 串行信号。输入串行器和 VCSEL LD 的信号采用晶圆级 RDL 沉积技术重新分配路线。VCSEL 本身通过焊料焊接在 LD 上，LD 将输入信号转换成 VCSEL 的电驱动信号。

图 10.68 所显示的接收器示意图和发射器示意图相似，除了其内部 IC 是接收元件。3D 堆叠的发射机和接收机可用于板级光互连。发射器输出光信号在聚合物波导一端进行耦合，然后从聚合物波导的另一端耦合输出到接收元件，如图 10.68 所示。为了降低成本，本案例中光学耦合设计采用直接耦合（无透镜）的方式。光信号从 VCSEL 通过一端的反射镜直接耦合到波导中，然后通过另一端的反射镜直接耦合到光电二极管。对于直接耦合，由于光信号的发散性质，为了达到所需要的耦合效率必须保持尽可能短的光传输距离是至关重要的。

基于一个给定的结构，光传输的最小距离是由光电组件的组装设计决定的。对于引线键合组件，最小距离或者高度由键合的线弧高决定，该值通常为 200μm。另一方面，对于倒装的背照式光电组件，在不损坏光电元件有源区域的情况下，最小高度可以尽可能小（常规的焊凸点约为 100μm，微凸点约为 25μm[29, 30]）。

光波导的设计如图 10.69 所示。波导嵌入 FR4 PCB 中，形成完全的平面设计。波导的截面设计如图 10.69 所示，线芯和披覆层的折射率分别为 1.5622 和 1.5544。在波导的两端都有一个面外反射器或者反射镜，能够以 90°的角度反射光信号。

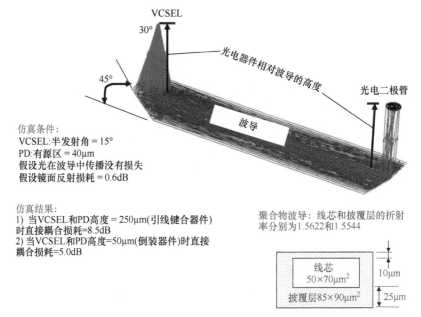

图 10.69　显示光信号直接耦合的仿真模型

光学仿真使用商业软件 ASAP，仿真模拟使用布局如图 10.69 所示。仿真中 VCSEL 发出的光束发散角为 30°，光电二极管的有源区域长度为 40μm。波导两端反射镜的损耗均设定为 0.6dB，光在波导中的损耗在仿真模型中被忽略。使用上述参数，当高度（VCSEL 和光电二极管相对波导的距离）为 250μm 和 50μm 时，耦合损耗分别为 8.5dB 和 5dB。图 10.70 显示了其他高度下的耦合损耗。同预想的一样，高度越高，耦合损耗越大。

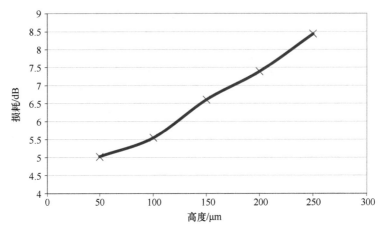

图 10.70　耦合损耗与 VCSEL 高度之间的关系曲线

典型的 VCSEL 发出的激光功率为 3dBm，GaAs（砷化镓）光电二极管在

850μm 下的响应率为 0.6A/W。假设波导损耗为 0.1dB/cm，那么 10cm 场波导的光损耗为 1dB。同时假设 10Gbit/s 接收器要求的最小输入光功率为 13dBm，那么最大可接收的光耦合损耗约为 15dB。因此，对于 10Gbit/s OECB 设计来说，直接耦合方案是可行的。

5. 热设计原理、分析和结果

选择光子封装结构不仅需要认识到关键的成本问题和光学性能，还需要关注其热学性能和机械可靠性方面的问题。由于光电系统的结构是通过采用低成本的裸芯片（VCSEL、PIN、LD、TIA、串行器和反串行器）堆叠形成，如图 10.68 所示，所以热管理是一个关键的问题。系统由高功耗的 10Gbit/s 激光器（2.2V，33mW，16W/cm^2）、10.7Gbit/s LD 和 TIA 芯片（3.3V，0.35W，6~35W/cm^2）、10Gbit/s 1∶16 串行器和 10Gbit/s 16∶1 反串行器（3.3V，2.5W，2.5~6.7W/cm^2）组成，因此，如何在芯片烧毁之前把热量从芯片中提取出来是一个挑战。特别是 VCSEL 工作需要一个稳定的热环境，偏离期望的工作温度会导致发出光的波长发生改变，从而降低 VCSEL 的性能。在本研究中，10Gbit/s VCSEL 激光器工作的目标温度为 85℃。图 10.71 展示了用于对一个有机基板上的基于 3D 混合 IC 异构集成的光电互连进行详细分析的模型。它由一端的 VCSEL、LD、串行器和另一端的 PIN、TIA 和反串行器组成（各元件的尺寸在前面已经提到）。因为发射器和接收器结构相似，这里对结构的一半进行建模。可以看到模型里有 3 层铜层、4 层 FR4 树脂层、1 层聚合物波导、1 层光学层（波导

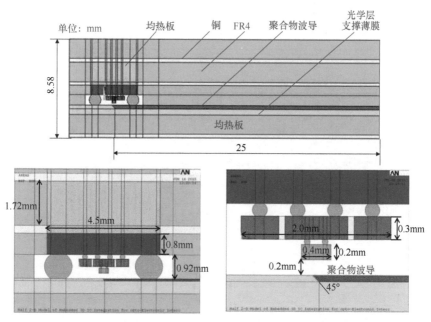

图 10.71 采用有机基板的嵌入式 3D 异构集成光电互连结构的有限元模型

层）支撑薄膜和 1 个铜均热板。铜层厚度为 0.3mm；FR4 树脂层厚度为 1.72mm；聚合物波导层厚度为 0.085mm；光学层支撑薄膜厚度为 0.5mm。聚合物波导的长度为 50mm，整个 OECB 的厚度为 8.37mm。结构中包含的材料的物性参数见表 10.8。

表 10.8 用于 3D 混合集成模拟材料的物性参数

组件	PCB	散热片	多路转换器	激光驱动器	VCSEL	铜层	焊料	聚合物波导
材料	FR4	Cu	Si	Si	GaAs	Cu	63Sn-37Pb	聚合物波导
热导率/[W/(m·K)]	0.8	390	150	150	68	390	50.9	1.5
TCE(ppm/K)	15	17	2.7	2.7	5.6	17	24.5	251
模量/MPa	21000	110000	131000	131000	85000	11000	25000	50
泊松比	0.18	0.34	0.28	0.28	0.3	0.34	0.35	0.3

热仿真模型的边界条件为在 VCSEL 芯片上施加 1W 的热量，环境温度为 25℃，如图 10.72 所示。在工作时，大多数 VCSEL 将会以每秒数百万（或者更多）个周期的频率打开和关断。由于 VCSEL 结构的热时间常数明显地大于其工作时通电/断电的周期，因此 VCSEL 结构可以认为在一个接近均匀的工作温度保持平衡。保守假设 VCSEL 具有连续的 1W 热量产生，产生的热量在整个 VCSEL 结构中均匀分布。此外，由于这是一个线性传热分析，可以对模型施加单位热量（1W）进行计算。在实际应用中，大多数 VCSEL 的功耗远小于 1W。封装结构和空气间存在对流传热。在封装结构的所有外表面施加热对流边界条

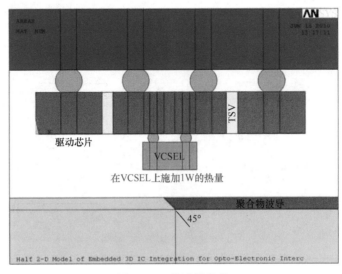

图 10.72 热边界条件

件,其中环境温度为 25℃,对流换热系数为 0.3W/(m²·K)。对于嵌入的结构部分,由于埋入 PCB 中,VCSEL 芯片和空气之间没有热传递。

图 10.73 显示了半个 OECB 模型和发射器所在关键区域的温度分布。可以看到,最高温度区域位于 VCSEL 处,其幅值为 64℃,低于允许的规范(85℃)。因此,VCSEL 能够按照预想的性能工作。

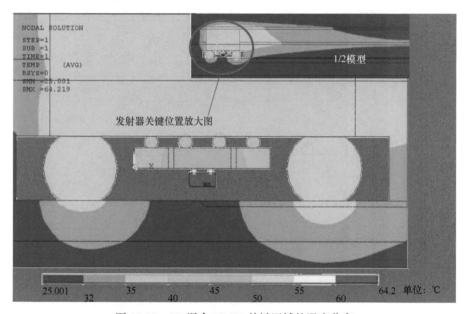

图 10.73 3D 混合 IC SiP 关键区域的温度分布

6. 结构设计原理、分析和结果

如前面提到的一样,选择光子封装结构不仅需要认识到关键的成本问题和光学性能,还需要关注其机械可靠性方面的问题。如图 10.68 所示,由于光电封装结构采用不同材料性能制造的低成本裸芯片进行堆叠,如 CTE(热膨胀系数),在环境压力条件下,如 $-25℃ \leftrightarrow 140℃$ 温度循环,每个循环 1h,结构的整体或者局部存在热膨胀不匹配所引起的 VCSEL 相对聚合物波导反射镜的结构偏转,以及在 OECB 焊点上产生应力和应变是关键问题。因此,需要关注的主要问题是 VCSEL 和聚合物波导上反射镜间的失配。本研究中允许的失配量为 2μm。

焊球的一个作用是增加 3D 堆叠系统的隔离高度,从而缓解作用在焊点上的应力和应变。然而,考虑到无透镜的 VCSEL 激光器的光学性能,隔离高度应尽可能小。因此,有必要对焊球的高度进行设计使得结构具有足够的柔性(对于焊点的可靠性)并仍然保持在目标光学性能范围内。在本研究中,每个温度循环的应变和应力目标分别为 2% 和 45MPa。仿真模型和之前的热仿真模型完全相同,不同之处是中心线处施加了对称边界条件。热循环边界条件如图 10.74

所示。由温度曲线可以看出，将整个OECB置于−25℃下持续15min，然后用15min向上升温至140℃，接着降温至−25℃。升温和降温时间均为15min。参考温度为25℃，即无热应力状态下的温度。因为本小节的重点是热膨胀效应，所以仿真中没有考虑残余应力。

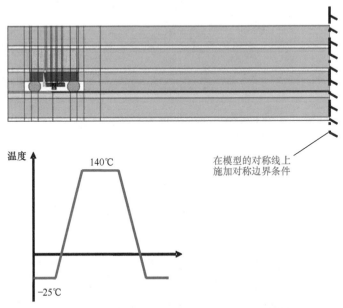

图10.74 用于力学分析的几何和温度载荷边界条件

图10.75显示了由于温度循环和聚合物波导上反射镜和VCSEL间的失配造成的OECB中发射器的关键区域未变形的形状（图10.75a）和变形后的形状（图10.75b）。图10.76显示了不同聚合物波导长度（25μm、50μm和100μm）的失配模拟结果。从结果可知，聚合物波导长度越长，失配越大。但是，失配都在允许的范围（2μm）内。

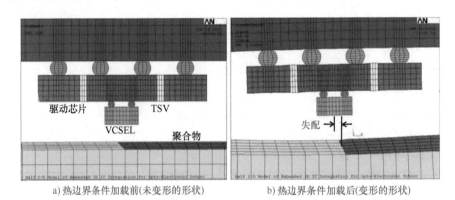

图10.75 发射器未变形和变形的形状（放大100倍）

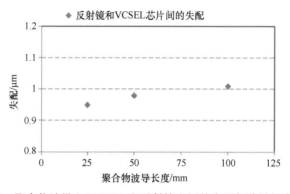

图 10.76 聚合物波导上 VCSEL 和反射镜之间的失配与波导长度的关系

图 10.77 显示了作用在焊点上的等效应变（上部）和应力（下部）分布。同样地，它们的最大值（0.14% 和 35Mpa）在允许的规范范围内，说明焊点在大多数工作情况下是可靠的。

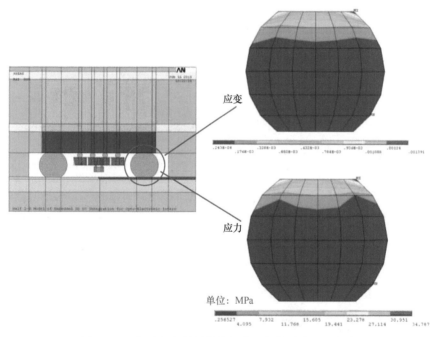

图 10.77 作用在发射器关键焊点上的等效应变和应力

10.5.3 总结和建议

本小节提出了一种通用的低成本和高（光学、电学、热学和机械）性能的 3D 混合 IC 光电异构集成系统，该系统嵌入在具有光学聚合物波导的 PCB 或者有机层压基板中。此外，还提供了一个特殊的设计，并通过仿真进行演示。一些重要成果和建议总结如下[29, 30]：

1）在 OECB 上的 VCSEL 和 PD 异构集成正在进入批量生产，尤其是在背板应用中。

2）光学分析表明，目前的光聚合物波导截面设计能够满足光耦合可接受的损耗（15dB）。因此，直接耦合方案（无透镜）适用于 10Gbit/s 的光学板级互连。

3）热分析表明，基于当前的设计，当向 10Gbit/s VCSEL 施加 1W 的热量时，VCSEL 的最高温度只有 65℃，低于规范的要求（85℃）。

4）结构分析表明，基于当前的设计，将嵌入式 3D 混合 IC 异构集成置于环境温度循环载荷中（-25℃ ↔140℃），VCSEL 和聚合物波导反射镜间的最大失配（1μm）在规范允许的范围内（2μm）。波导的长度越长，失配越大。

5）结构分析表明，基于当前的设计，整个 3D 混合 IC SiP（系统级封装）置于环境温度循环载荷（-25℃ ↔140℃）中，在关键的焊点区域的最大等效应变（0.14%）和应力（35MPa）低于规范允许的最大值（2% 和 45MPa）。

参考文献

[1] Iwabuchi, S., Y. Maruyama, Y. Ohgishi, M. Muramatsu, N. Karasawa, and T. Hirayama, "A Back-Illuminated High-Sensitivity Small-Pixel Color CMOS Image Sensor with Flexible Layout of Metal Wiring", *Proceedings of IEEE/ISSCC*, San Francisco, CA, February 2006, pp. 16.8.1–16.8.8.

[2] Wakabayashi, H., K. Yamaguchi, M. Okano, S. Kuramochi, O. Kumagai, S. Sakane, M. Ito, et al., "A 1/2.3-inch 10.3 Mpixel 50 frame/s Back-Illuminated CMOS Image Sensor", *Proceedings of IEEE/ISSCC*, San Francisco, CA, February 2010, pp. 411–412.

[3] Sukegawa, S., T. Umebayashi, T. Nakajima, H. Kawanobe, K. Koseki, I. Hirota, T. Haruta, et al., "A 1/4-inch 8 Mpixel Back-Illuminated Stacked CMOS Image Sensor", *Proceedings of IEEE/ISSCC*, San Francisco, CA, February 2013, pp. 484–486.

[4] Kagawa, Y., N. Fujii, K. Aoyagi, Y. Kobayashi, S. Nishi, N. Todaka, et al., "Novel Stacked CMOS Image Sensor with Advanced Cu2Cu Hybrid Bonding", *Proceedings of IEEE/IEDM*, December 2016, pp. 8.4.1–8.4.4.

[5] Coudrain, P., D. Henry, A. Berthelot, J. Charbonnier, S. Verrun, R. Franiatte, N. Bouzaida, et al., "3D Integration of CMOS Image Sensor with Coprocessor Using TSV Last and Micro-Bumps Technologies", *Proceedings of IEEE/ECTC*, Las Vegas, NV, May 2013, pp. 674–682.

[6] Rhodes, H., D. Tai, Y. Qian, D. Mao, V. Venezia, W. Zheng, Z. Xiong, et al., "The Mass Production of BSI CMOS Image Sensors", *Proceedings of International Image Sensor Workshop*, 2009, pp. 27–32.

[7] Tetsuo Nomoto1, Swain, P. K., and D. Cheskis, "Back-Illuminated Image Sensors come to the Forefront," *Photonics Spectra*, http://www.photonics.com/Article.aspx?AID=34685.

[8] Lau, J. H., R. Lee, M. Yuen, and P. Chan, "3D LED and IC Wafer Level Packaging", *Journal of Microelectronics International*, Vol. 27, No. 2, 2010, pp. 98–105.

[9] Zhang, R., R. Lee, D. Xiao, and H. Chen, "LED Packaging using Silicon Substrate with Cavities for Phosphor Printing and Copper-filled TSVs for 3D Interconnection", *Proceeding of IEEE/ECTC*, May 2011, pp. 1616–1621.

[10] Zhang, R., and R. Lee, "Moldless Encapsulation for LED Wafer Level Packaging using Integrated DRIE Trenches", *Journal of Microelectronics Reliability*, Vol. 52, 2012, pp. 922–932.

[11] Chen, D., L. Zhang, Y. Xie, K. Tan, and C. Lai, "A Study of Novel Wafer Level LED Package Based on TSV Technology", *IEEE Proceedings of ICEPT*, August 2012, pp. 52–55.

[12] Xie, Y., D. Chen, L. Zhand, K. Tan, and C. Lai, "A Novel Wafer Level Packaging for White

Light LED", *IEEE Proceedings of ICEPT*, August 2013, pp. 1170–1174.
[13] Lau, J. H., "Design and Process of 3D MEMS Packaging", *IMAPS Transactions, Journal of Microelectronics and Electronic Packaging*, First Quarter Issue, Vol. 7, 2010, pp. 10–15.
[14] Lau, J. H., C. K. Lee, C. S. Premachandran, and Yu Aibin, *Advanced MEMS Packaging*, McGraw-Hill, New York, NY, 2010.
[15] Sekhar, V., J. Toh, J. Cheng, J. Sharma, S. Fernando, and B. Chen, "Wafer Level Packaging of RF MEMS Devices Using TSV Interposer Technology", *Proceedings of IEEE/EPTC*, Singapore, December 2012, pp. 239–243.
[16] Chen, B., V. Sekhar, C. Jin, Y. Lim, J. Toh, S. Fernando, and J. Sharma, "Low-Loss Broadband Package Platform With Surface Passivation and TSV for Wafer-Level Packaging of RF-MEMS Devices", *IEEE Transactions on CPMT*, Vol. 3, No. 9, September 2013, pp. 1443–1452.
[17] Pang, W., R. Ruby, R. Parker, P. W. Fisher, M. A. Unkrich, and J. D. Larson, III, "A Temperature-Stable Film Bulk Acoustic Wave Oscillator", *IEEE Electron Device Letters*, Vol. 29, No. 4, April 2008, pp. 315–318.
[18] Small, M., R. Ruby, S. Ortiz, R. Parker, F. Zhang, J. Shi, and B. Otis, "Wafer-Scale Packaging For FBAR Based Oscillators", *Proceedings of IEEE International Joint Conference of FCS*, 2011, pp. 1–4.
[19] Premachandran, C. S., J. H. Lau, X. Ling, A. Khairyanto, K. Chen, and M. Pa, "A Novel, Wafer-level Stacking Method for Low-chip Yield and Non-uniform, Chip-size Wafers for MEMS and 3D SiP Applications", *Proceedings of IEEE/ECTC*, May 2008, pp. 314–318.
[20] Chen, K., C. Premachandran, K. Choi, C. Ong, X. Ling, A. Ratmin, J. H. Lau, et al., "C2W Low Temperature Bonding Method for MEMS Applications", *Proceedings of IEEE/EPTC*, December 2008, pp. 1–7.
[21] Lee, C., A. Yu, L. Yan, H. Wang, J. Han, Q. Zhang, and J. H. Lau, "Characterization of Intermediate In/Ag Layers of Low Temperature Fluxless Solder based Wafer Bonding for MEMS Packaging", *Journal of Sensors Actuators A: Physical*, Vol. 154, 2009, pp. 85–91.
[22] Lau, J. H., Y. Lim, T. Lim, G. Tang, K. Houe, X. Zhang, P. Ramana, et al., "Design and Analysis of 3D Stacked Optoelectronics on Optical Printed Circuit Boards", *Proceedings of SPIE, Photonics Packaging, Integration, and Interconnects VIII*, Vol. 6899, San Jose, CA, January 19–24, 2008, pp. 07.1–07.20.
[23] Ramana, P., H. Kuruveettil, B. Lee, K. Suzuki, T. Shioda, C. Tan, J. H. Lau, et al., "Bi-directional Optical Communication at 10 Gb/s on FR4 PCB using Reflow Solderable SMT Transceiver", *IEEE/ECTC Proceedings*, May 2008, pp. 244–249.
[24] Lim, T. G., B. Lee, T. Shioda, H. Kuruveettil, J. Li, K. Suzuki, J. H. Lau, et al., "Demonstration of high frequency data link on FR4 PCB using optical waveguides", *IEEE Transactions of Advanced Packaging*, Vol. 32, May 2009, pp. 509–516.
[25] Lim, L, C. Teo, H. Yee, C. Tan, O. Chai, Y. Jie, J. H. Lau, et al., "Optimization and Characterization of Flexible Polymeric Optical Waveguide Fabrication Process for Fully Embedded Board-level Optical Interconnects", *IEEE/EPTC Proceedings*, December 2008, pp. 1114–1120.
[26] Chai, J., G. Yap, T. Lim, C. Tan, Y. Khoo, C. Teo, J. H. Lau, et al., "Electrical Interconnect Design Optimization for Fully Embedded Board-level Optical Interconnects", *IEEE/EPTC Proceedings*, December 2008, pp. 1126–1130.
[27] Teo, C., W. Liang, H. Yee, L. Lim, C. Tan, J. Chai, J. H. Lau, et al., "Fabrication and Optimization of the 45° Micro-mirrors for 3-D Optical Interconnections", *IEEE/EPTC Proceedings*, December 2008, pp. 1121–1125.
[28] Chang, C., J. Chang, J. H. Lau, A. Chang, T. Tang, S. Chiang, M. Lee, et al., "Fabrication of Fully Embedded Board-level Optical Interconnects and Optoelectronic Printed Circuit Boards", *IEEE/EPTC Proceedings*, December 2009, pp. 973–976.
[29] Lau, J. H., M. S. Zhang, and S. W. R. Lee, "Embedded 3D Hybrid IC Integration System-in-Package (SiP) for Opto-Electronic Interconnects in Organic Substrates", *ASME Paper IMECE2010-40974*.
[30] Lau, J. H., M. S. Zhang, and S. W. R. Lee, "Embedded 3D Hybrid IC Integration System-in-Package (SiP) for Opto-Electronic Interconnects in Organic Substrates," *ASME Transactions, Journal of Electronic Packaging*, Vol. 133, September 2011, pp. 1–7.
[31] Lau, J. H., S. W. Lee, M. Yuen, J. Wu, C. Lo, H. Fan, and H. Chen, "Apparatus Having an Embedded 3D Hybrid Integration for Optoelectronic Interconnects in Organic Substrate". US

Patent No: 9,057,853, Date of Patent: June 16, 2015.

[32] Matsubara, T., K. Oda, K. Watanabe, K. Tanaka, M. Maetani, Y. Nishimura, S. Tanahashi, "Three Dimensional Optical Interconnect on Organic Circuit Board", *IEEE Electronic Components and Technology Conference*, May 2006, pp. 789–794.

[33] Immonen, M., M. Karppinen, J. Kivilahti, "Fabrication and Characterization of Polymer Optical Waveguuides with Integrated Micromirrors for 3-D Board Level Optical Interconnects", *IEEE Transactions on Electronics Packaging Manufacturing*, Vol. 28, No. 4, 2005, pp. 304–311.

[34] Choi, C., L. Lin, Y. Liu, J. Choi, L. Wang, D. Hass, J. Magera, et al., "Flexible Optical Waveguide Film Fabrications and Optoelectronic Devices Integration for Fully Embedded Board-Level Optical Interconnects", *IEEE Journal of Lightwave Technology*, Vol. 22, No. 9, 2004, pp. 2168–2176.

[35] Wang, L., X. L. Wang, J. Choi, D. Haas, J. Magera, and R. T. Chen, "Low-loss Thermally Stable Waveguide with 45° Micromirrors Fabricated by Soft Molding for Fully Embedded Board-level Optical Interconnects", *Proceedings of SPIE*, Vol. 5731, 2005, pp. 87–93.

[36] Xia, Y., and G. M. Whitesides, "Soft-lithography", *Annual Review Of Materials Science*, Vol. 28, 1998, pp. 84–153.

[37] Park, J., E.-D. Sim, and Y.-S. Beak, "Improvement of Fabrication Yield and Loss Uniformity of Waveguide Mirror", *IEEE Photonics Technology Letters*, Vol. 17, No. 4, April 2005, pp. 807–809.

[38] Kim, J. T., B. C. Kim, M. Jeong, and M. Lee, "Fabrication of a Micro-Optical Coupling Structure by using Laser Ablation", *Journal of Materials Processing Technology*, Vol. 146, 2004, pp. 163–166.

[39] Garner, S., S.-S. Lee, V. Chuyanov, A. Chen, A. Yacoubian, W. Steier, and L. Dalton, "Three-Dimensional Integrated Optics using Polymers", *IEEE Journals Of Quantum Electronics*, Vol. 35, No. 8, August 1999, pp. 1146–1155.

[40] Kim, J.-S., and J.-J. Kim, "Stacked Polymeric Multimode Waveguide Arrays for Two-Dimensional Optical Interconnects," *IEEE Journal of Lightwave Technology*, Vol. 22, No. 3, March 2004, pp. 840–844.

[41] Kim, J.-S., and J.-J. Kim, "Fabrication of Multimode Polymeric Waveguides and Micromirrors using Deep X-ray Lithography", *IEEE Photonics Technology Letters*, Vol. 16, No. 3, March 2004, pp. 798–800.

[42] Kim, J.-H., and R. T. Chen, "A Collimation Mirror in Polymeric Planar Waveguide Formed by Reactive Ion Etching", *IEEE Photonics Technology Letters*, Vol. 15, No. 3, March 2003, pp. 422–424.

[43] Lehmacher, S., and A. Neyer, "Integration of Polymer Optical Waveguides into Printed Circuit Boards", *Electronics Letters*, Vol. 36, No. 12, June 2000, pp. 1052–1053.

[44] Chandrappan, J., H. Kuruveettil, T. Wei, C. Liang, P. Ramana, K. Suzuki, T. Shioda, et al., "Performance Characterization Methods for Optoelectronic Circuit Boards", *IEEE Transactions on CPMT*, Vol. 1, No. 3, March 3, 2011, pp. 318–326.

第 11 章

异构集成的发展趋势

11.1 引言

半导体行业已经确定了五个主要增长引擎（应用），即移动终端、高性能计算（HPC）、汽车（特别是自动驾驶汽车）、物联网（IoT）和大数据（尤其是云计算）。而诸如 5G、人工智能（AI）及机器学习（ML）等系统技术驱动力正在推动这五种半导体应用快速增长。

封装技术人员正使用各种封装方法，如引线键合、倒装芯片、积层基板、PoP（层叠封装）、WLCSP（晶圆级芯片封装）、FOW/PLP（扇出晶圆/面板级封装）、TSV（硅通孔）、2.5D/3D IC 集成、HBM（高带宽存储器）、多芯片组件/系统级封装（SiP）/异构集成、小芯片和 EMIB（嵌入式多裸芯互连桥）来容纳（封装）半导体器件以满足这五类主要应用。由于 5G、AI、ML 的驱动，半导体集成密度增加，焊盘节距减小，芯片尺寸增大，给封装技术人员带来了许多挑战（机遇）。

11.2 异构集成的发展趋势

异构集成是封装技术之一，它使用多种封装技术将不同材料与功能的芯片/元件，以及来自不同的无晶圆厂、代工厂、晶圆尺寸和特征尺寸的设计，集成到系统或子系统中，而不是将大部分功能集成到单个芯片中，并追求更精细的特征尺寸，如片上系统（SoC）。接下来的几年将会看到更多更高层次的异构集成，至于其上市速度，则取决于其性能、外形因子、功耗、信号完整性和/或成本。这些不同的芯片/元件间的通信由 RDL（再布线层）实现[1,2]。

11.2.1 有机基板上的异构集成

一般对于大批量生产（HVM）来说，70% 的异构集成 RDL 应该在有机基板上，并且金属线宽和间距≥10μm（这些异构集成实际上大部分是 SiP）；不超过 5% 的异构集成 RDL 同样在有机基板上，但金属线宽和间距均<10μm。

大多数有机基板上的异构集成是使用 SMT（表面贴装技术）设备组装。板上芯片引线键合、板上焊凸点倒装芯片与其他 SMT 元件批量回流焊、嵌入式 IPD（集成无源器件）、无芯基板、引线上凸点、嵌入式线路基板和类似 PCB 基板的应用将增加。有机基板可以是刚性的或柔性的，其范围从普通 PCB 到高密度 PCB。

积层有机封装基板已被广泛使用，其制程主要是 SAP（半加成工艺）。目前，两个核心层以及具有 10μm 金属线宽和间距 RDL 的 12 个积层（6-2-6）已实现量产。思科（Cisco）公司[3]提出的有机转接板正在为进入量产寻找出路。新光（Shinko）公司提出的 i-THOP 基板在积层顶部具有薄膜层（2μm 金属线宽和间距）[4, 5]，目前正在努力提升良率以降低损失。

11.2.2 非有机基板上的异构集成

一般对于量产来说，25% 的异构集成 RDL 是在其他（非有机）基板上完成的，如硅基板（无源 TSV 转接板或有源 TSV 转接板或两者）、硅基板（桥）、扇出 RDL 基板和陶瓷基板（本书没有讨论）。RDL 的金属线宽和间距通常非常小，可低至亚微米量级。

1. 硅基板上的异构集成（TSV 转接板）

硅基板（无源 TSV 转接板或有源 TSV 转接板）上的异构集成 RDL 制作工艺是通过 PECVD（等离子增强化学气相沉积）和镶嵌铜+CMP（化学机械抛光）完成。目前，可实现至少 4 层 RDL，最小节距 0.4μm，金属线宽和节距可低至亚微米量级。这就是所谓的 2.5D IC 集成技术[6, 7]。在本书所有讨论的封装技术中，这种封装技术价格最高。然而，这种封装技术可以应用于超精细间距、超高密度、超高 I/O、超大芯片，可以满足非常高性能的应用。

2. 硅基板（桥）上的异构集成

硅基板（桥）上的异构集成 RDL 制作工艺取决于金属线宽和间距要求。如果线宽和间距≥2μm，则采用聚合物和 ECD（电化学沉积）+刻蚀制作工艺就足够。反之，如果线宽和间距<2μm，则需要采用 PECVD 和镶嵌铜+CMP 制作工艺。硅基板（桥）上异构集成的目标是为了消除 TSV 转接板。RDL 金属线宽和间距可以非常小。

3. RDL 扇出（芯片先置）基板上的异构集成

扇出（芯片先置）基板上的异构集成 RDL 制作工艺是在嵌入芯片的 EMC（环氧模塑料）重组晶圆上进行聚合物和 ECD+刻蚀。这些 RDL 将取代（消除）微凸点、芯片到晶片的键合、清洗、下填料和固化，以及 TSV 转接板。但是，金属线宽和间距不能下降到 < 2μm，并且肯定不会达到亚微米量级。

4. RDL 扇出（芯片后置）基板上的异构集成

扇出（芯片后置或 RDL 先置）基板上的异构集成 RDL 制作工艺是在临时裸露的玻璃（晶圆或面板）载体上进行聚合物和 ECD+ 刻蚀。这些步骤之后是晶圆凸点、助熔、芯片到晶圆或芯片到面板黏合、清洁、下填料、EMC 压缩成型等。这些 RDL 将取代（消除）TSV 转接板。然而，金属线宽和间距不能 < 2μm。相对于芯片先置扇出技术，芯片后置扇出技术的优点是不存在芯片位移问题，由此可以获得更高的封装组装良率，并且不必扔掉已知良好的芯片（KGD）；缺点是成本较高，以及更多的工艺步骤导致有更多的机会使得封装组装良率损失。

11.2.3 各种异构集成的应用

如何选择不同的异构集成？这取决于应用。最重要的指标（选择标准）是异构集成 RDL 的金属线宽和间距。如面向 HPC 应用，由 AI 驱动的半导体技术，异构集成 RDL 的金属线宽和间距必须是超精细的（< 2μm 或低至亚微米）。在这种情况下，需要硅基板，如 TSV 转接板。采用 PECVD 和镶嵌铜 +CMP 制作的桥互连硅基板也能完成这项工作。

另一方面，面向移动应用，由 5G 驱动的半导体技术，异构集成 RDL 的金属线宽和间距在 10μm 量级。在这种情况下，采用积层有机基板互连射频（RF）芯片和调制解调器芯片就足够，当然采用扇出基板会更好。

11.2.4 各种异构集成的应用范围

图 11.1 展示了异构集成在各种基板上的应用范围（尺寸和引脚数）。可以看出，硅基板（TSV 转接板）上异构集成具有最大数量的引脚数（> 100000），同时也具有最大基板尺寸（1200mm^2）；硅基板（桥）上异构集成具有 4000 引脚数及基板尺寸 ≤ 1200mm^2；扇出（芯片先置）基板上异构集成具有 2500 引脚数及基板尺寸 ≤ 625mm^2；扇出（芯片后置）基板上异构集成具有 5000 引脚数及基板尺寸 ≤ 1400mm^2；以及有机基板上异构集成具有 6000 引脚数及基板尺寸高达 3000mm^2。

11.2.5 总结和建议

一些重要的结果和建议总结如下。
1）RDL 是异构集成最重要的组成部分。
2）各种异构集成的选择取决于应用，最重要的选择标准是 RDL 的金属线宽和间距。

3）75% 的异构集成在有机基板上进行（其实大部分是 SiP）。25% 的异构集成在其他基板上进行，如硅（TSV 转接板）、硅（桥）、扇出 RDL 和陶瓷。

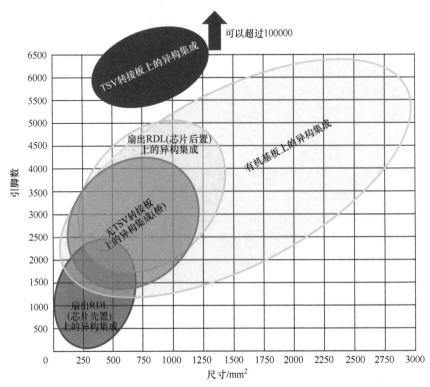

图 11.1 异构集成 RDL 在有机基板、硅基板（TSV 转接板）、硅基板（桥）、扇出 RDL 基板（芯片先置）和扇出 RDL 基板（芯片后置）上的基板尺寸和引脚数

4）为了促进/普及异构集成，标准是必要的。美国国防部高级研究计划局（DARPA）的微电子常用异构集成及知识产权产品复用策略（CHIPS）正朝着正确的方向前进。

5）复杂的异构集成系统迫切需要 EDA（电子设计自动化）工具进行自动系统分区和设计。如通过使用 EDA 工具，UCSB 和 AMD[8] 提出了未来非常高性能的系统，如图 11.2 所示。该系统包括中央处理器单元（CPU）小芯片和几个 GPU（图像处理器单元）小芯片及无源 TSV 转接板和/或具有 RDL 有源 TSV 转接板上的 HBM（高带宽存储器）。

第 11 章 异构集成的发展趋势

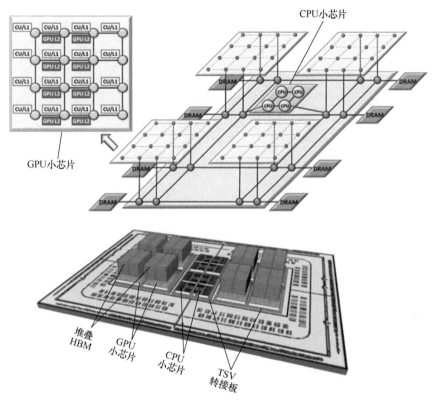

图 11.2 UCSB/AMD GPU 小芯片、CPU 小芯片及 TSV 转接板上的 HBM

参考文献

[1] Lau, J. H., P. Tzeng, C. Lee, C. Zhan, M. Li, J. Cline, et al., "Redistribution Layers (RDLs) for 2.5D/3D IC Integration", *IMAPS Transactions, Journal of Microelectronic Packaging*, Vol. 11, No. 1, First Quarter 2014, pp. 16–24.

[2] Lau, J. H., "Redistribution-Layers for Heterogeneous Integrations", *Chip Scale Review*, January/February 2019, pp. 20–25.

[3] Li, L., P. Chia, P. Ton, M. Nagar, S. Patil, J. Xue, J. DeLaCruz, M. Voicu, J. Hellings, B. Isaacson, M. Coor, and R. Havens, "3D SiP with organic interposer for ASIC and memory integration", *Proceedings of IEEE/ECTC*, May 2016, pp. 1445–1450.

[4] Shimizu, N., Kaneda, W., Arisaka, H., Koizumi, N., Sunohara, S., Rokugawa, A., and T. Koyama, "Development of Organic Multi Chip Package for High Performance Application", *IMAPS International Symposium on Microelectronics*, Orlando, FL, September 30–October 3, 2013, pp. 414–419.

[5] Oi, K., Otake, S., Shimizu, N., Watanabe, S., Kunimoto, Y., Kurihara, T., Koyama, T., Tanaka, M., Aryasomayajula, and Z. Kutlu, "Development of New 2.5D Package With Novel Integrated Organic Interposer Substrate With Ultra-Fine Wiring and High Density Bumps", *IEEE 64th Electronic and Components Technology Conference*, Orlando, FL, May 27–30, 2014, pp. 348–353.

[6] Lau, J. H., *3D IC Integration and Packaging*, McGraw-Hill, New York, 2016.

[7] Lau, J. H., *Through-Silicon Via (TSV) for 3D Integration*, McGraw-Hill, New York, 2013.

[8] Stow, D., Y. Xie, T. Siddiqua, and G. H. Loh, "Cost-effective design of scalable high-performance systems using active and passive interposers", *Proceedings of IEEE/ACM International Conference on Computer-Aided Design*, November 2017, pp. 728–735.